Gold Placer Deposits In NorthEast Alaska

By: Richard Lampright

ABOUT THE AUTHOR

The author has been a recreational prospector since 1960 when he first picked up a gold pan to try his luck at a roadside gold-panning concession in the Black Hills of South Dakota. He found 14 flakes of gold that afternoon and put them in a specimen bottle for safe-keeping. Several days later, on a hike in the Badlands, the bottle slipped from his pocket never to be seen again. Although the loss was a huge disappointment, he was hooked forever.

In 1969 he joined the Air Force and was stationed in California. The very next weekend he was prospecting the "Motherlode". During the following 34 years, the Air Force gave him the opportunity to prospect for gold in California, Colorado, Georgia, Australia, Philippines, Nevada, Utah, Arizona and Alaska.

He arrived at Elmendorf AFB, Alaska in the Fall of 1992 and spent his spare time during the winters in the local US Geological Survey Library researching Alaska gold placer locations. Initially, his research concentrated on the area around Anchorage. In 1995, with the support and encouragement of Steve Herschbach, part owner of Alaska Mining and Diving Supply in Anchorage, he published his first book titled "Gold Placer Deposits Near Anchorage Alaska". Buy the time he retired from the Air Force in 2004, he had accumulated a vast amount of information on gold placer deposits which has since been compiled in an additional 12 books covering the entire state.

He now spends the winters in Colorado and places south, but returns to Alaska during the summer months to prospect and continue with his research.

Hard Cover and Digital Books can be purchased directly from me at **https://www.IronFireGold.com**. See last page in this book for more information.

Published by:
IRON FIRE PUBLICATIONS
Huntsville, Utah 84317
Phone: 303-258-0363
Email: lampright@mac.com

Web Site: **https://www.IronFireGold.com**

ISBN 9780964536647
Key Title: Gold Placer Deposits In NorthEast Alaska

REPRINTED IN U.S.A.
Iron Fire Publications.
Nederland, Colorado 80466

DISCLAIMER

This document is compiled from United States and Alaska State government sources which the author believes to be accurate. Nevertheless, no warranties, either expressed or implied are intended to be created by this publication. Questions or issues including but not limited to mining regulations, permit regulations, land status and land access should be directed to the appropriate agencies or governing bodies contained herein. **If you are challenged by a person or party claiming to be the owner(s) or caretaker(s) of lands described within this document, treat them as such and respect their request(s). Chances are they are right!**

It is the responsibility of the reader to obtain all necessary federal, state and local permits prior to operating any mining equipment on any of the locations contained herein. The author assumes no risk or liability for the issuance, revocation, suspension or termination of any federal, state or local permit. It is the readers responsibility to comply with all federal, state and local laws.

Readers must understand that mining, in general, and mining locations can be dangerous with many unforeseen hazards. The user of the information contained herein accepts full responsibility for any liability or injury incurred through the use of the information contained within this document.

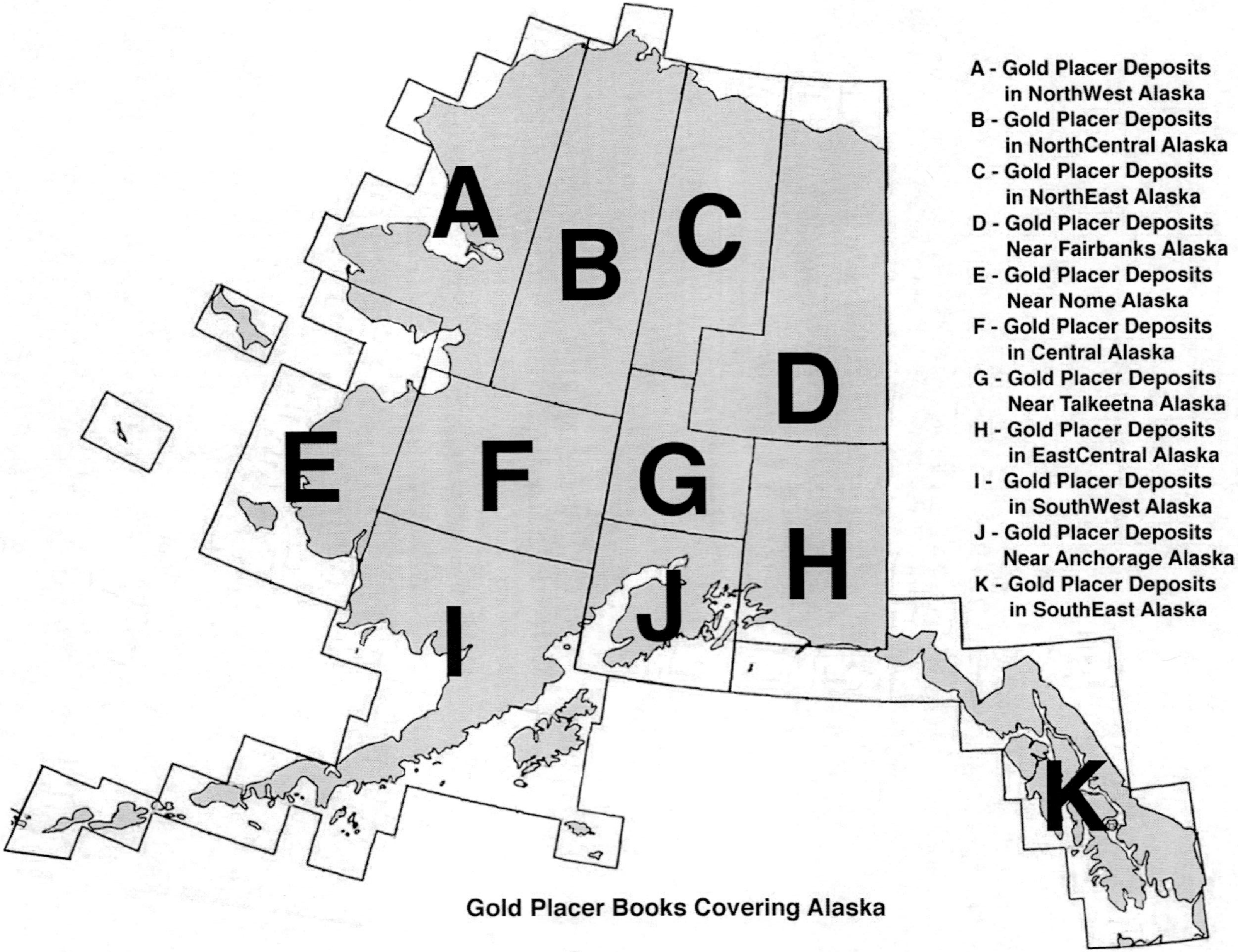

Gold Placer Books Covering Alaska

PREFACE

There is no occupation that man is so anxious to enter with so little knowledge as gold mining.

Recreational prospecting used to be a low risk venture. A resurgence of mining activity and an increase in agencies attempting to regulate this activity has made it difficult for the recreational miners. Many excellent places are still available to the novice. However, to determine whether they are open to mineral entry or have valid claims filed on them requires a significant amount of research. The individual not willing to invest a little time in research must rely on pure luck.

A prospector in Alaska may operate a suction dredge with up to a 4" suction nozzle without a permit unless the dredge is operated in a stream that supports salmon. In this case, the operator must obtain a permit from the Department of Fish and Game. A larger suction dredge may require a permit from one or more of the following agencies: Environmental Protection Agency, U. S. Forest Service, U. S. Army Corp of Engineers, Dept of Fish and Game, or the Alaska Division of Mining. This list of agencies is not meant to be all-inclusive. Permit requirements vary from year to year.

The permitting process generally requires a legal description, including a map or drawing, of the area being applied for. Without this information the applicant may be denied a permit. In addition to the information above, the EPA also requires the name of the mining district the operation will be located in.

The process of searching through the Bureau of Land Management (BLM) computer database for the lack of a valid claim on a perspective stream is difficult, time consuming and frustrating, without the having the appropriate information at hand. The BLM computer can also provide a prospective miner with the claim owner's name, address, and telephone number that may be useful if attempting to lease or purchase a claim in a particular area.

For every prospector, blinded by the dream of striking it rich and armed with little or no knowledge, there is at least one unscrupulous individual waiting to take advantage. Each year worthless mining claims are bought and sold in Alaska; many to people in the lower 48 states. The material contained in this book will not cure blindness but it can provide enough knowledge to increase the chance of realizing the dream while reducing the risk of foolish investments of both time and money.

There are 24 million acres of land, containing valuable minerals, located in Alaska that are open to mineral entry and another 218 million acres which are closed to mineral entry, but not necessarily closed to prospecting (Bottage, Agency Session Presentation, October 30, 1986, p. 7). If you are fortunate enough to locate a piece of ground, worth the trouble to file or lease a claim on, the information contained herein will make the process much easier, especially, if the claim is located on federal land.

Gold Placer Deposits In NorthEast Alaska contains information on 185 gold bearing streams located in the in Beaver, Bettles, Chandalar, Tanana and Wiseman Quadrangles.

The Dalton Highway

The Dalton Highway is 414 miles of gravel that starts from the Elliot Highway and runs to Deadhorse near Prudhoe Bay. The highway was named after James William Dalton an arctic oil engineer. It was built to get supplies from the Yukon River to the Trans-Alaskan Oil Pipeline. Construction began in April 1974. The highway was restricted to non-public use until 1991. Since then it has been open to the public. The Bureau of Land Management (BLM) manages 2.7 million acres of public land along the road. Much of it is open to recreational prospecting. Road conditions vary depending on weather, maintenance and time of year. The highway is one of the most scenic in Alaska. Wildlife and recreational opportunities abound. For a detailed description of the Dalton Highway refer to The Milepost.

Acknowledgements

A book of this scope is necessarily a compilation of the work of others. A list of principal publications appear in the reference section at the end of this book. Acknowledgments are given in the text for the use of direct quotations or specially organized material; here is acknowledged the debt to all who have contributed to the great pool of printed knowledge without which this book could not have been compiled.

There are, in addition, those who have contributed in other ways. Ray & Jimmy Wilkes, and Tommy Rongstad, three of the most dedicated mining partners anywhere, for the many days of hard work and endless years of friendship. To my friends on Canyon Creek: Marshall Ronnie, Al & Patti Adams, Ken May, and Flemming Cochran for a place to hang out. To Steve Herschbach, of Alaska Mining and Diving Supply, for the continued encouragement and assistance. And finally, to my wife Bonnie and my two children Brian and Sheri. Without there support none of this would have been possible.

It remains only to state that by acknowledging the debt to an individual does not imply that the individual agrees with all the statements contained in this book. Many indeed, of those herein mentioned, did not even know that this book was in preparation. Any shortcomings that exist are the responsibility of the author.

R.L.

Contents

Beaver Quadrangle

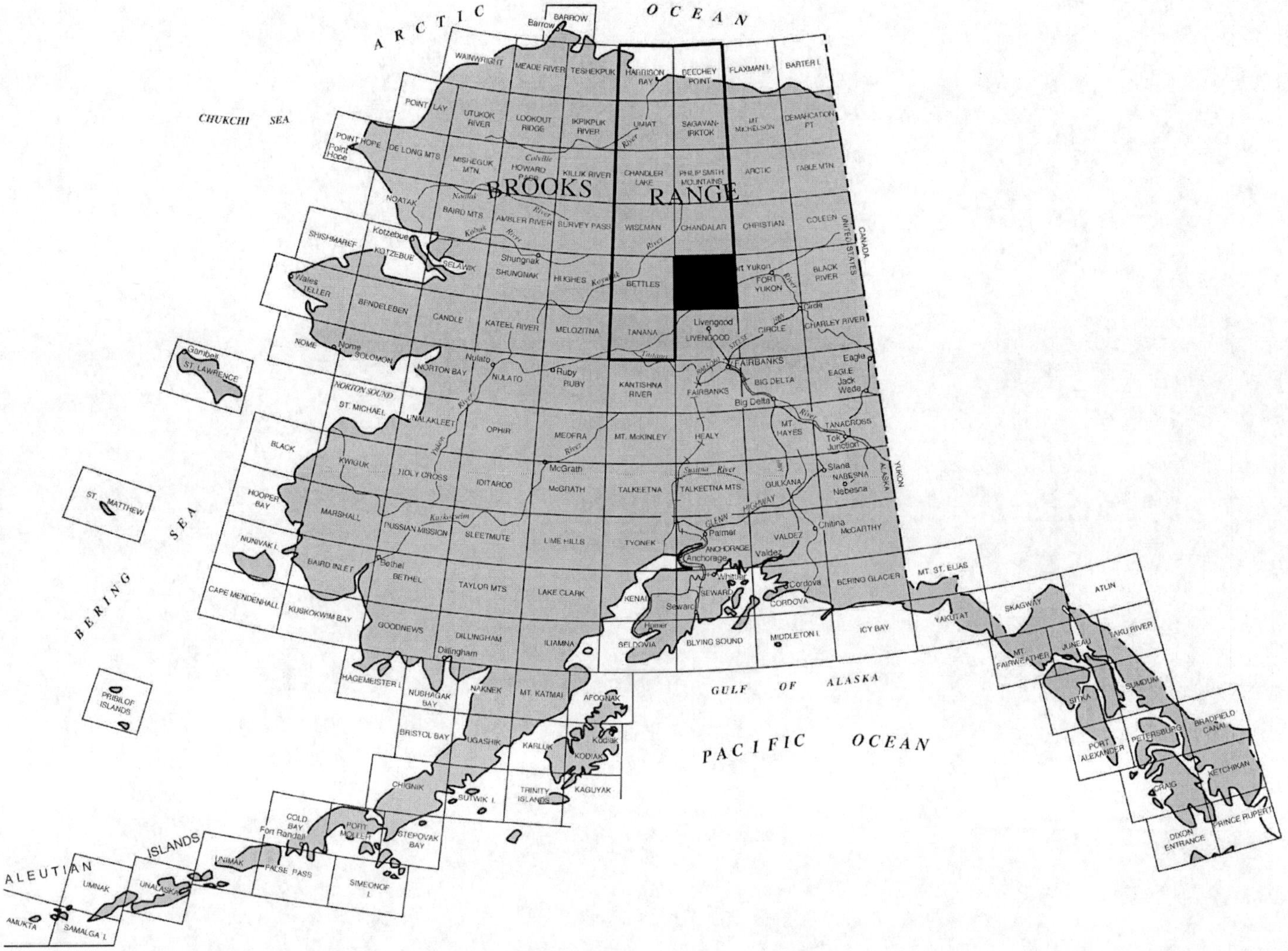

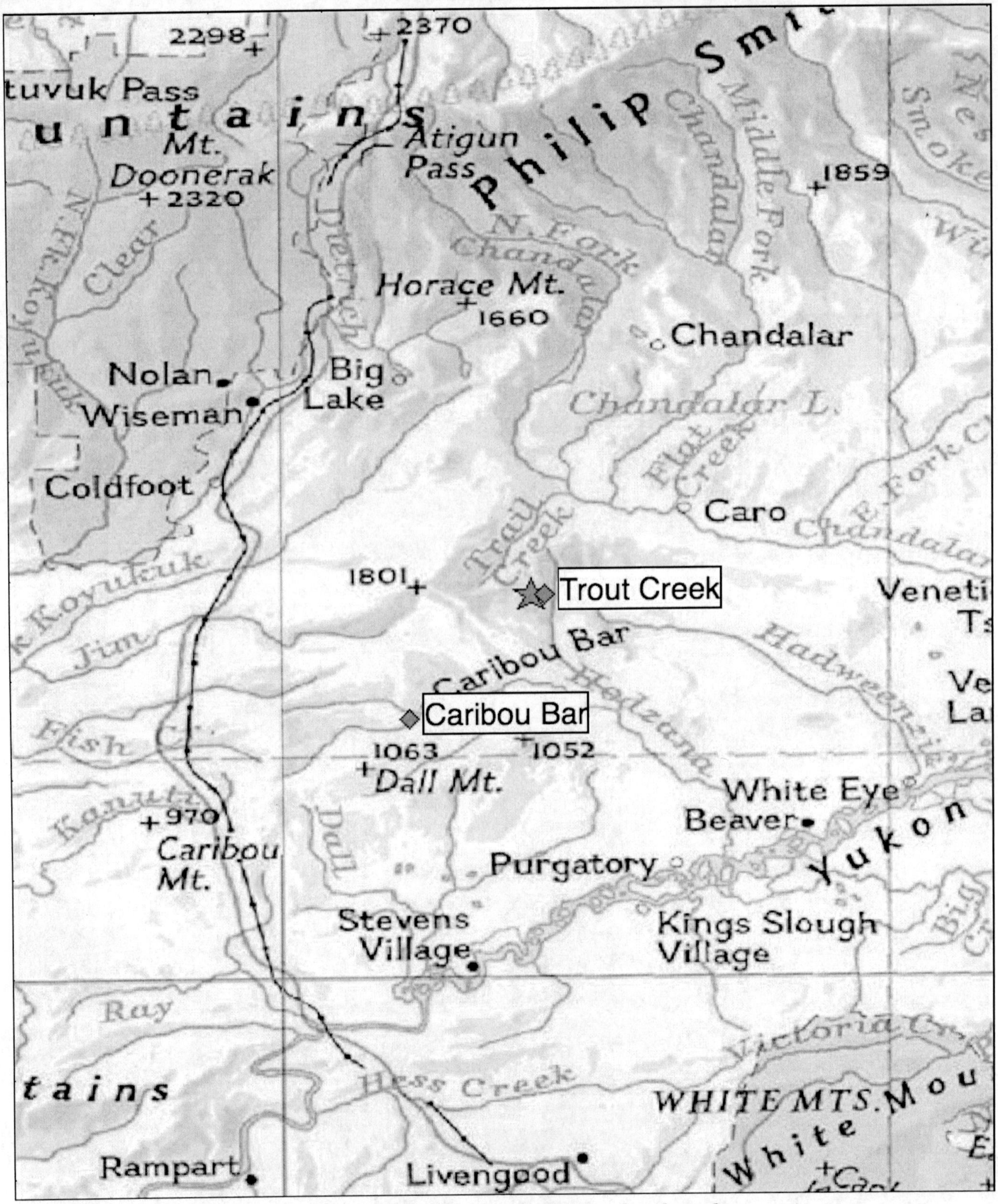

Gold Placer Locations in the Beaver Quadrangle

♦ Diamonds shows placer gold locations. ★ Stars show gold-bearing hard rock locations. Hard rock occurrences are not identified by name nor is a description of the occurrence included in this book. They are provided so the reader can see where the hard rock sources for the placer deposits occur. Areas are located by latitude and longitude WGS 84 coordinates.

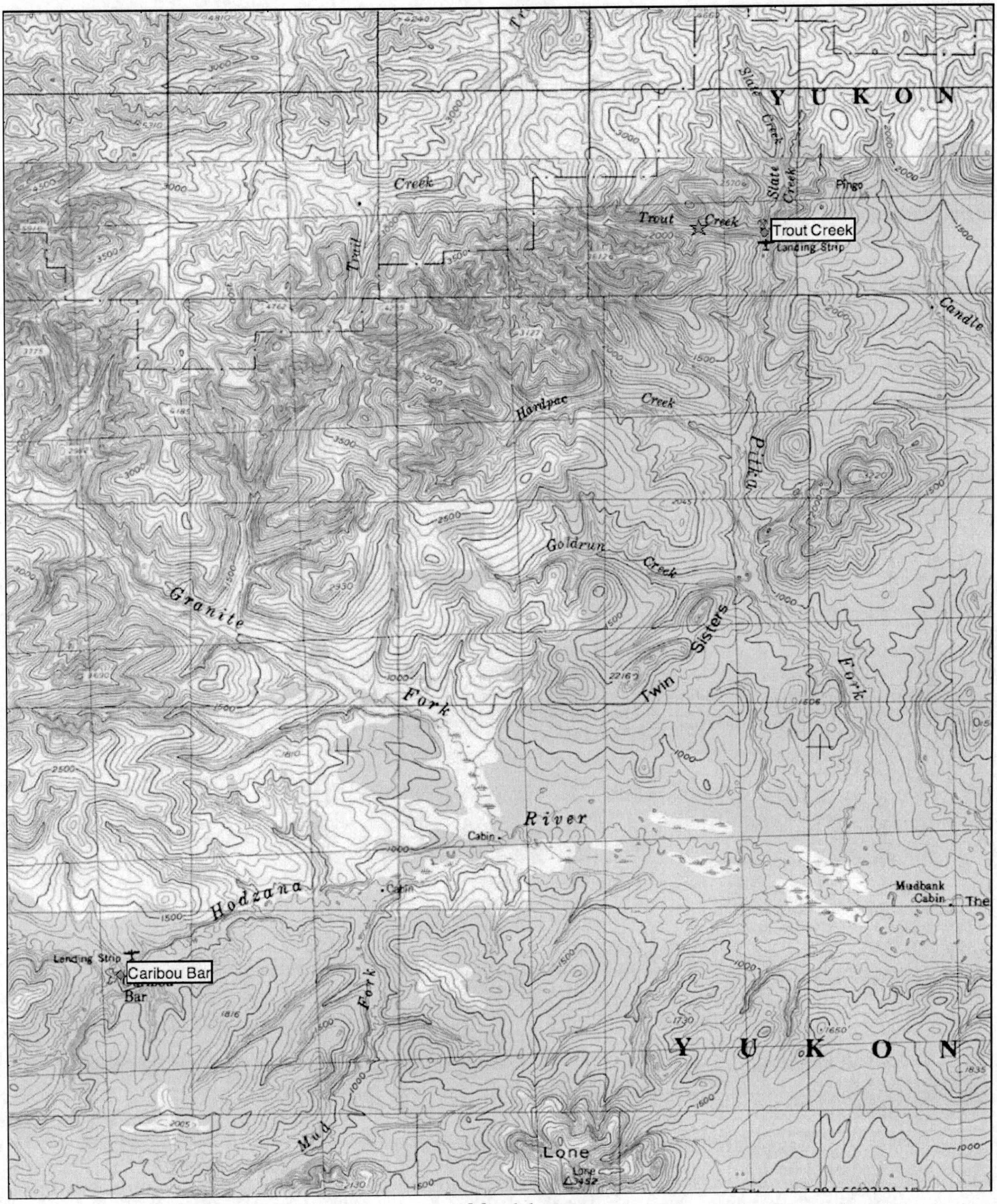
YUKON
Slate Creek
Slate Creek
Creek
Trail
Pingo
Trout Creek
Trout Creek
Landing Strip
Candle
Hardpac Creek
Pitka
Goldrun Creek
Granite
Twin Sisters
Fork
Fork
River
Cabin
Hodzana
Cabin
Mudbank Cabin
The
Landing Strip
Caribou Bar
Bar
Fork
YUKON
Mud
Lone

Map 1-1

Hodzana River

(Gold) **Quadrangle:** Beaver (C5); **Coordinates:** 66.6551°N-149.2430°W; Map: 1-1; **Worked:** 1910- ; **Access:** Trails from landing strip at 66.6637°N, 149.2372°W. **Summary:** Tributary of Yukon River. Good Prospects reported in 1910 where 50 men were prospecting. Hosiana Creek was the old name of the Hodzana River. The prospecting may well have been at or near Caribou Bar, as this is the only place in the Hodzana drainage basin where placer gold mining has been reported (Cobb, 1978, OF 78-94 p. 5).

Trout Creek

(Gold) **Quadrangle:** Beaver (D4); **Coordinates:** 66.9675°N-148.5608°W; **Map:** 1-1; **Worked:** 1950s-1960s; **Access:** Trails from landing strip at 66.9628°N, 148.5608°W. **Summary:** Trout Creek and Slate Creek combine to form the Pitka Fork of the Hodzana River. Hosiana Creek was the old name of the Hodzana River. Specimen of quartz, pyrite, spalerite, and molybdenite. Small-scale placer gold mining in the 1950's and 1960's. The only mining reported in the Yukon Flats district. Stream gradient of Trout Creek is 100 ft./mi. (Cobb, 1978, OF 78-94 p. 6).

Bettles Quadrangle

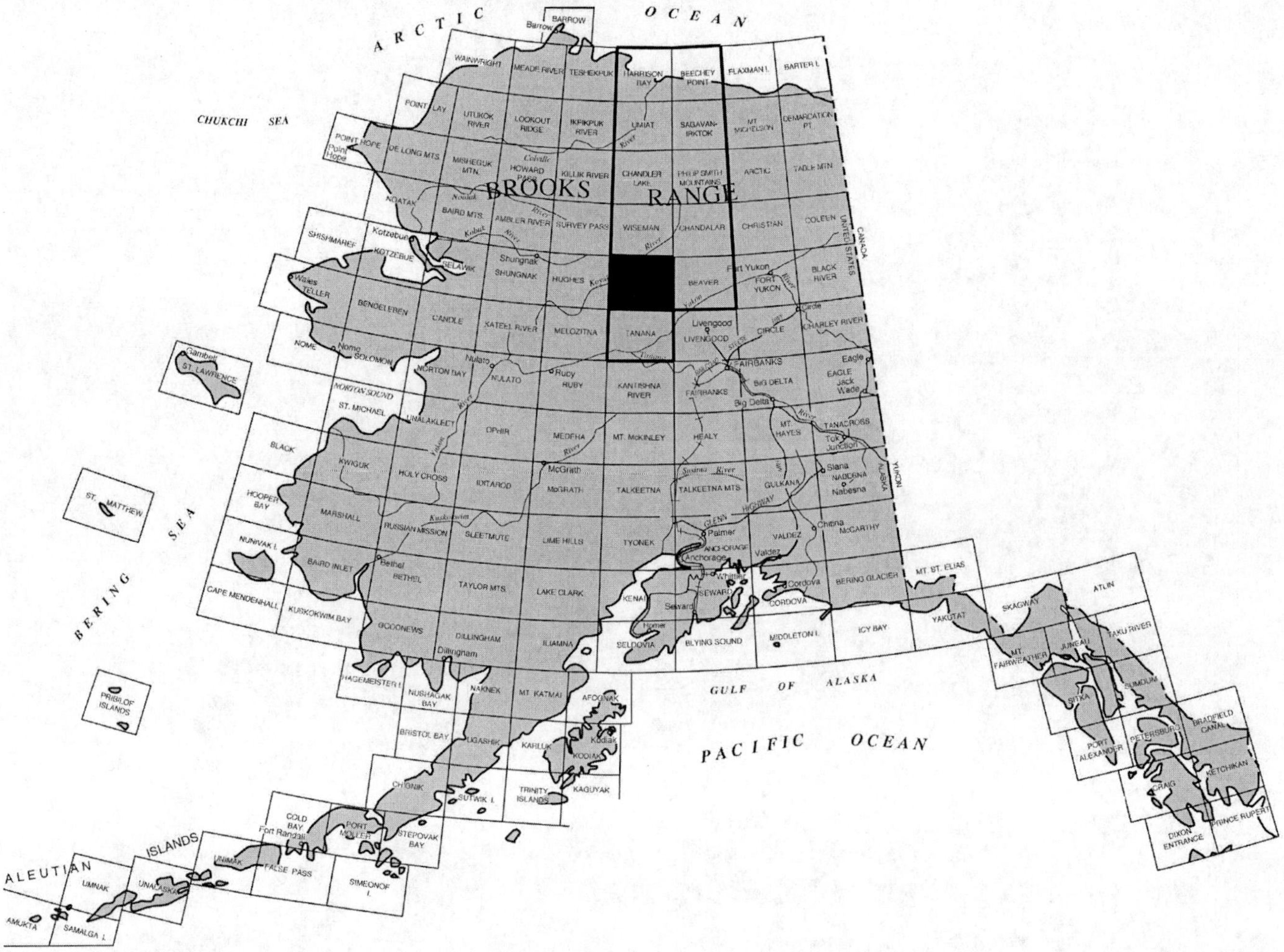

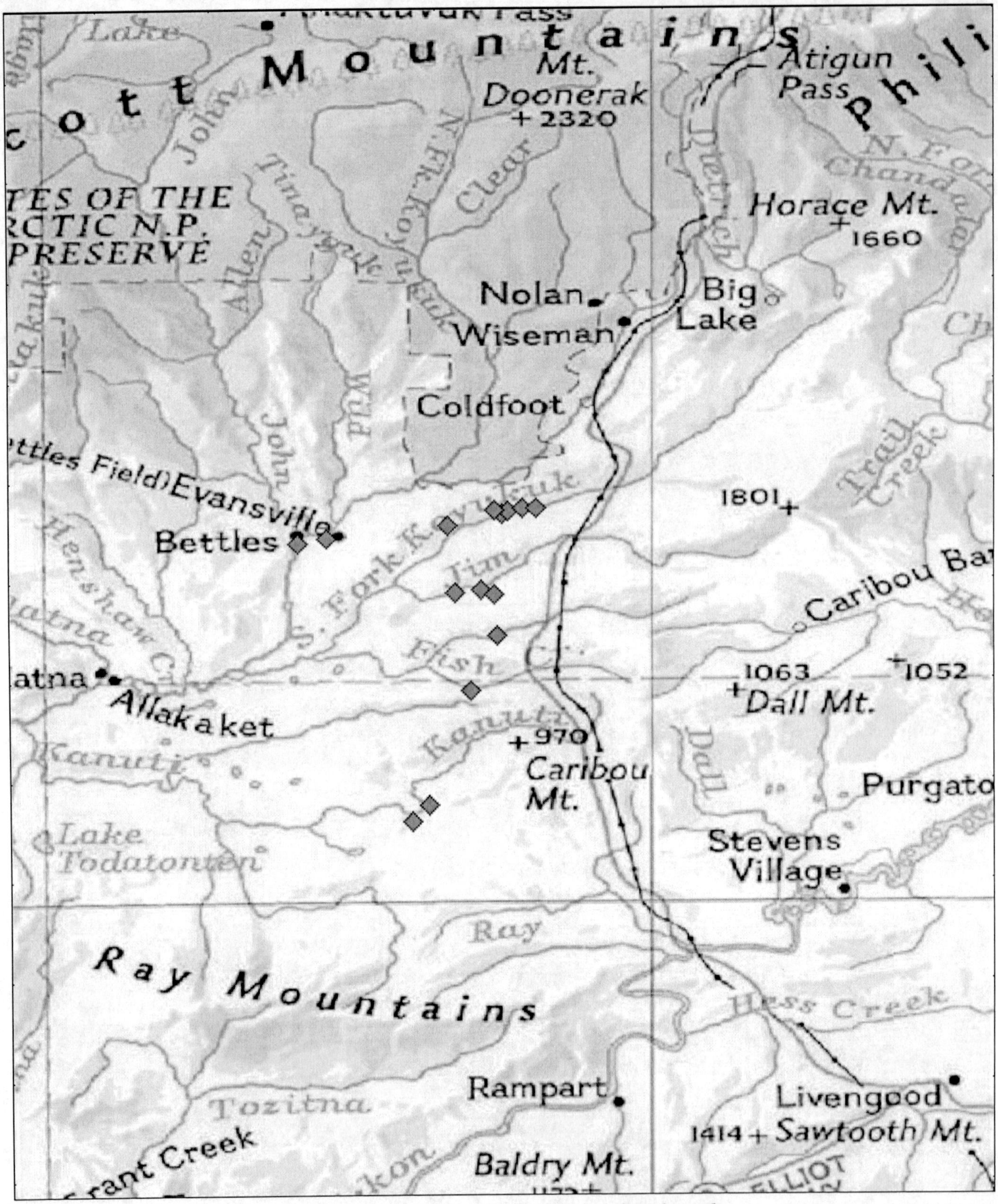

Gold Placer Locations in the Bettles Quadrangle

♦ Diamonds shows placer gold locations. ★ Stars show gold-bearing hard rock locations. Hard rock occurrences are not identified by name nor is a description of the occurrence included in this book. They are provided so the reader can see where the hard rock sources for the placer deposits occur. Areas are located by latitude and longitude WGS 84 coordinates.

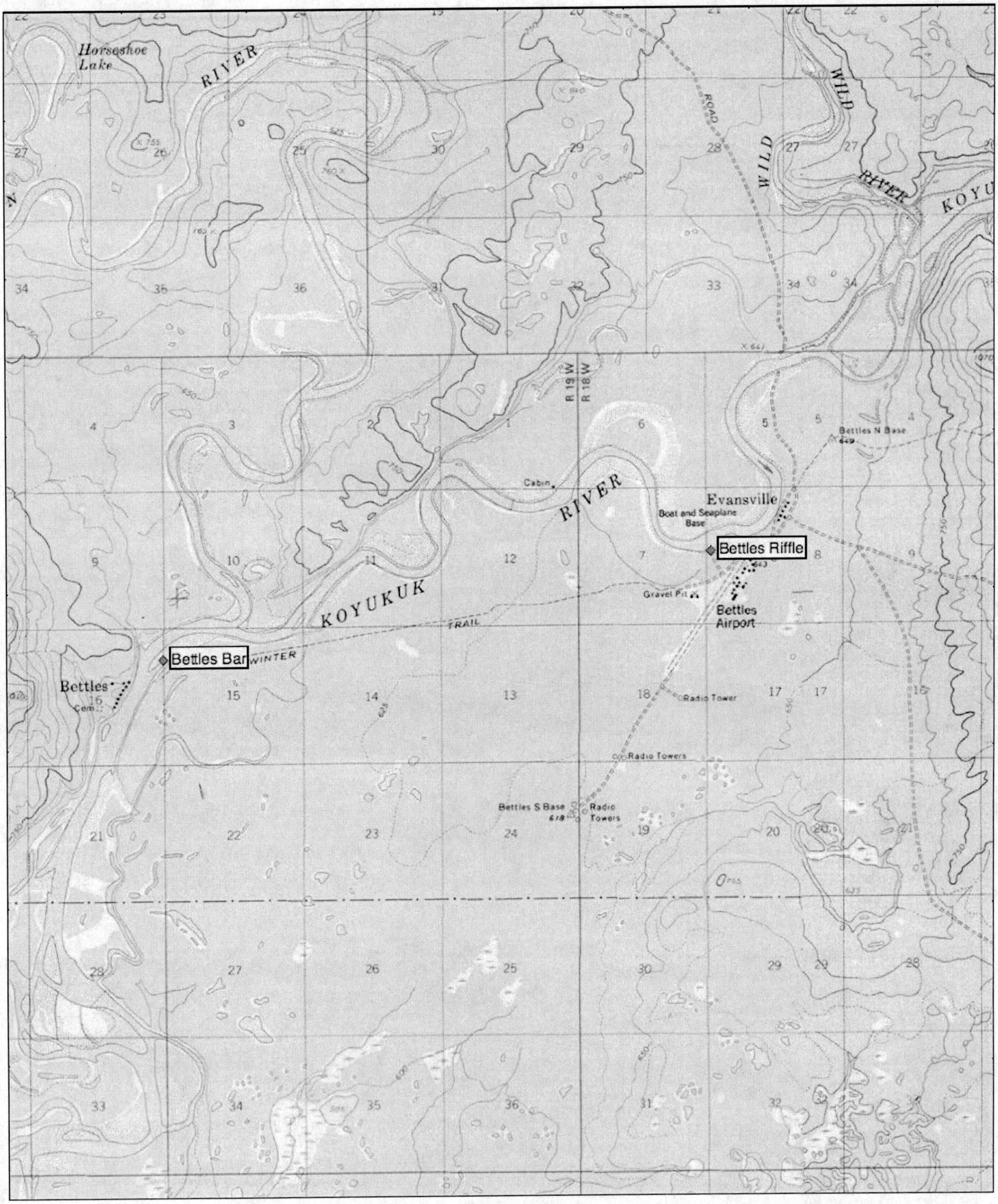
Horseshoe Lake
RIVER
ROAD
WILD
WILD
RIVER
KOYU
R 19 W
R 18 W
Bettles N Base
Cabin
RIVER
Evansville
Boat and Seaplane Base
Bettles Riffle
Gravel Pit
Bettles Airport
KOYUKUK
TRAIL
WINTER
Bettles Bar
Bettles
Cem.
Radio Tower
Radio Towers
Bettles S Base
Radio Towers

Map 2-1

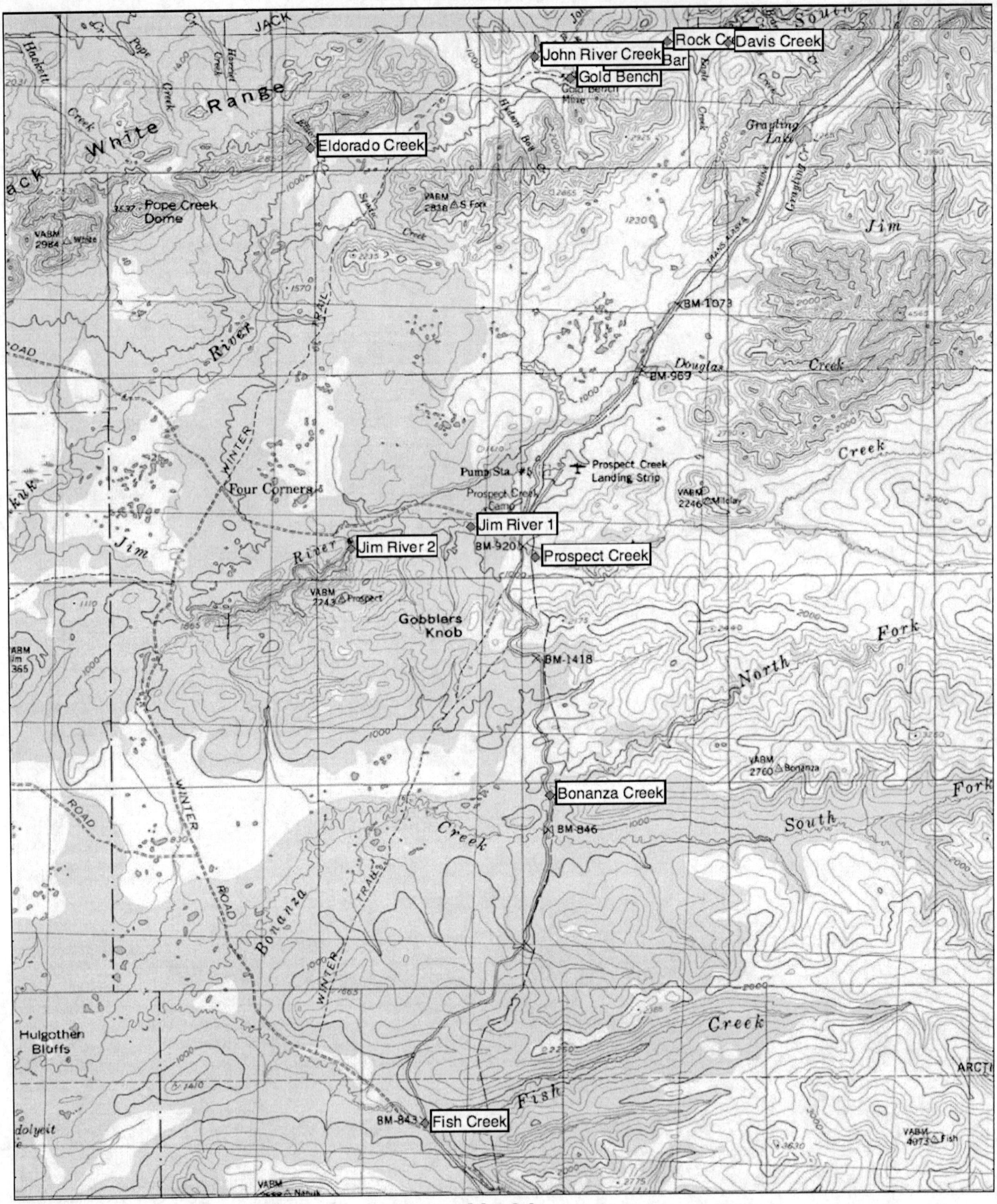
Rock C
Davis Creek
John River Creek
Bar
Gold Bench
Eldorado Creek
White Range
Pope Creek Dome
Grayling Lake
Jim
BM-1073
Douglas Creek
BM-969
Pump Sta. #5
Prospect Creek Landing Strip
Prospect Creek Camp
Four Corners
Jim River 1
Jim River 2
BM-920
Prospect Creek
Gobblers Knob
BM-1418
North Fork
Bonanza Creek
BM-846
South Fork
Hulgothen Bluffs
Fish Creek
BM-843

Map 2-2

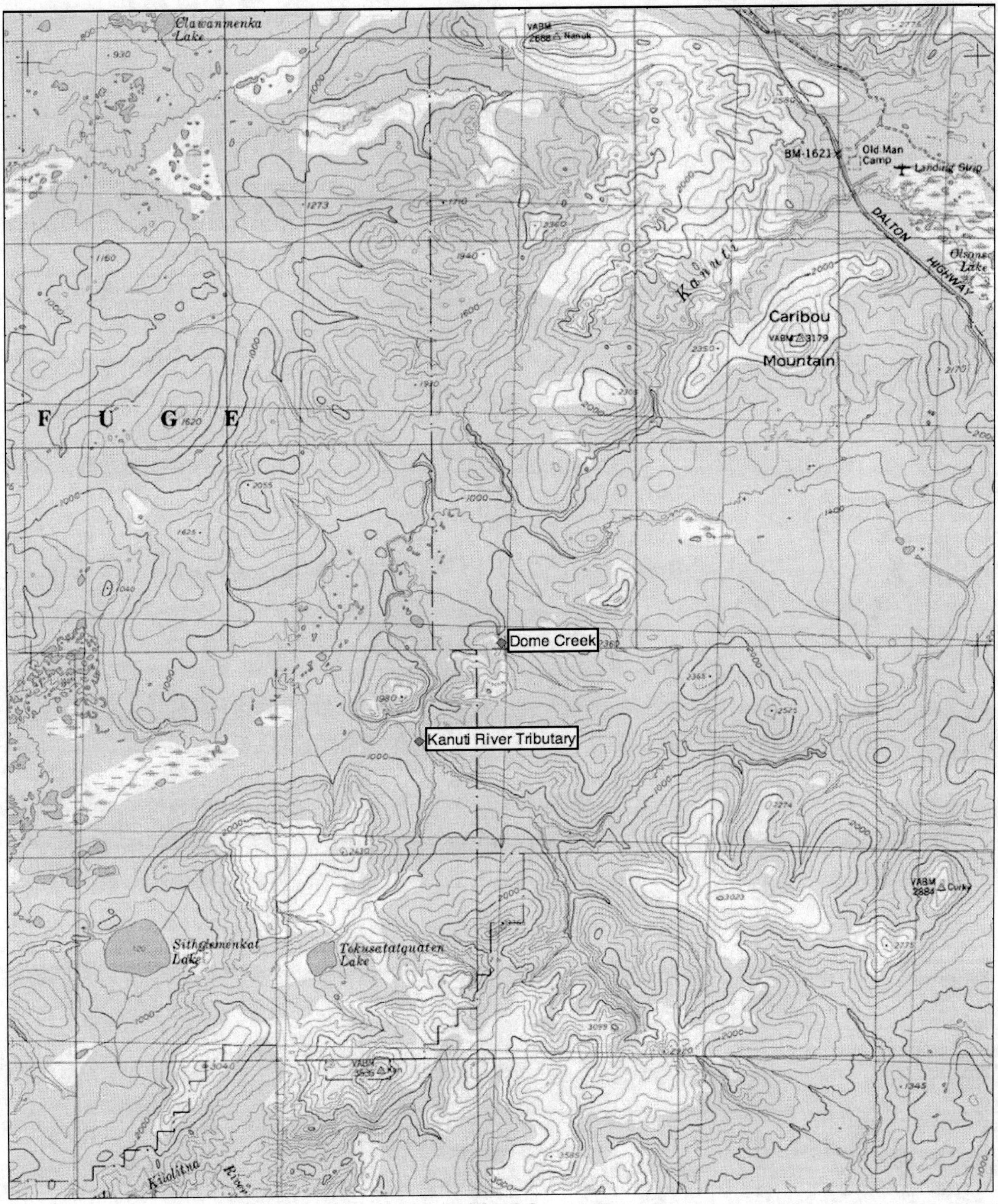
Clawanmenka Lake
VABM 2688 Nanuk
BM-1621
Old Man Camp
Landing Strip
DALTON HIGHWAY
Olsons Lake
Kanuti
Caribou Mountain
VABM 3179
F U G E
Dome Creek
Kanuti River Tributary
Sithylemenkat Lake
Tokusatatquaten Lake
VABM 2884 Curky
Kilolitna River

Map 2-3

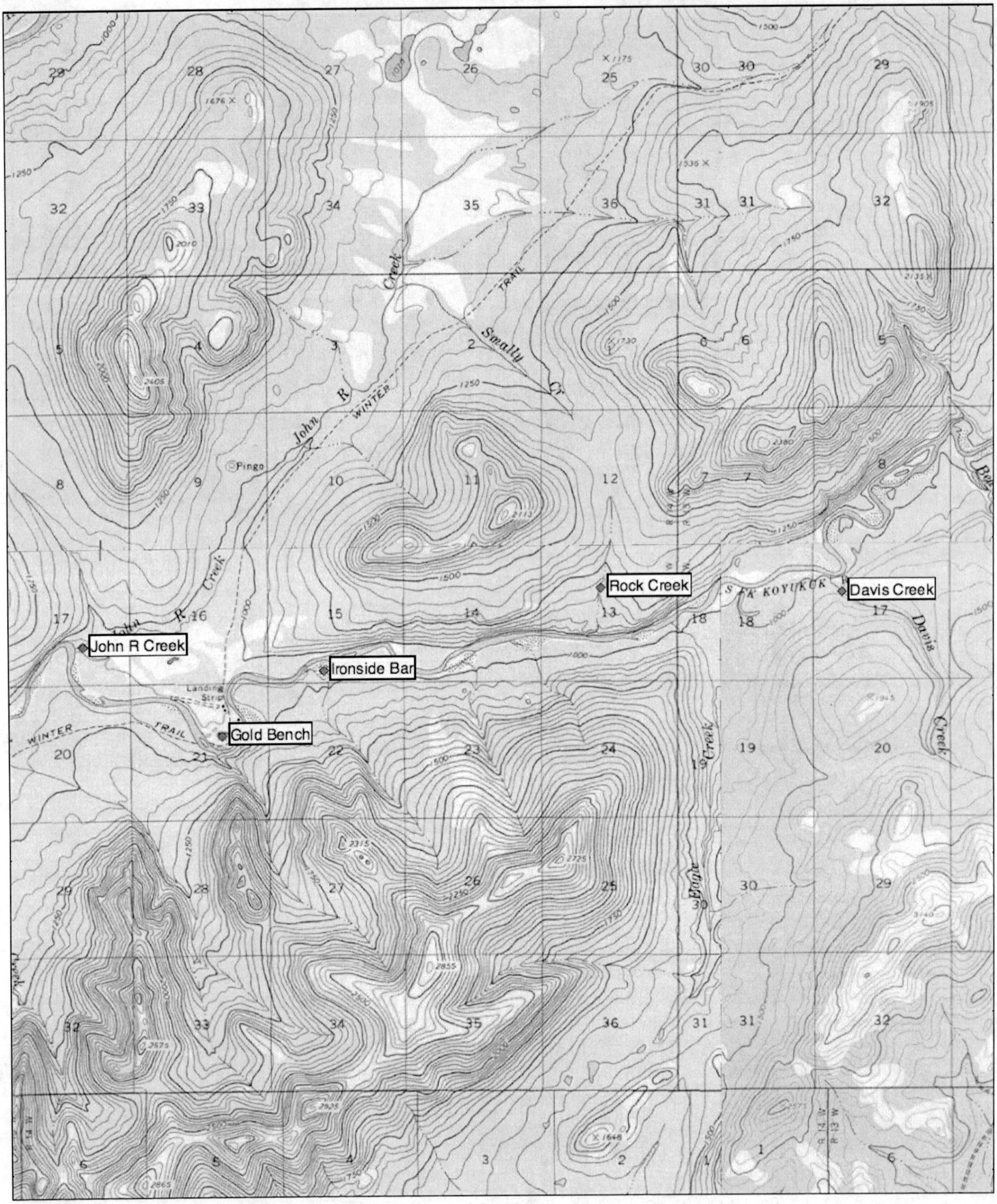
John R Creek
Ironside Bar
Gold Bench
Rock Creek
Davis Creek
Landing Strip
WINTER TRAIL
Smally Cr
John R Creek
Pingo
S FK KOYUKUK
Davis Creek
Eagle Creek

Map 2-4

Bettles Bar

(Gold) **Quadrangle:** Bettles (D4); **Coordinates:** 66.9096°N-151.6730°W; **Map:** 2-1; **Worked:** 1935-Present; **Access:** Trail from Bettles Airport at 66.9140°N, 151.5286°W. **Summary:** Bettles Bar is on the east bank of the Koyukuk River, opposite the old site of Bettles, and about 0.5 mile downstream from the junction of John and Koyukuk Rivers. The mine is in NE1/4 of section 16, T. 24 N., R. 19 W. of the Fairbanks Meridian. **Geology:** The gold in the Bettles Bar deposit is in a gravel layer composed of rounded cobbles embedded in fine sand, which in turn is mostly covered by a layer of fine river sand and silt (Reed, 1938). The gold is immediately below the sand and silt in the upper three feet of the cobble-and-sand layer, and is distributed throughout the sand fraction. The gold is very fine; so fine that when dry it is said to float. Although the locations of the bar deposits are relatively fixed, the individual spots where the river concentrates the gold are said to change from year to year (Reed, 1938). The deposits can be as much as 15 feet above the low-water level of the river (Reed, 1938). Similar deposits also occur along the Koyukuk River in the Bettles area, particularly at the so-called the Bettles Riffle (66.9206°N-151.5273°W) near the Bettles Airport, at Evansville (Kurtak and others, 2002). **Workings:** According to Reed (1938), with some effort, a person could generally recover as much as $10 (0.286 ozs) per day (gold at $35 per ounce) working these bar deposits. In 1937, as much as $600 (17 ozs) in gold was said to have been recovered in two weeks by a man working a short distance above Bettles. Grybeck (1977) indicated mining activity through 1975. Bettles Bar is one of a number of gold placers in the gravel bars along the Koyukuk River from Bettles upstream to the vicinity of Tramway Bar (Reed, 1938). See also Tramway Bar, Hanshaw Bar, and Grubstake Bar in the Wiseman quadrangle. (The Alaska Resource Data File., Bettles Quadrangle).

Bonanza Creek

(Gold) **Quadrangle:** Bettles (C2); **Coordinates** 66.6763°N, 150.6614°W; **Map:** 2-2; **Yield:** 0.0025 oz/yd^3; **Summary:** Tributary of Fish Creek. Bonanza Creek has been designated as a **Recreation Mining Area** by the State of Alaska upstream for 2.5 miles from the North Fork Bridge or downstream any distance except for the pipeline right-of-way. Placer mining claims begin 2.5 miles upstream on the North Fork of Bonanza Creek from the Dalton Highway bridge.

Davis Creek

(Gold, Bismuth) **Quadrangle:** Bettles (D1); **Coordinates:** 66.9944°N-150.4705°W; **Map:** 2-2, 2-4; **Worked:** 1900-1990s; **Production:** 1900-1909=240 ozs; **Access:** Boat from the Dalton Highway Bridge at 67.0181°N, 150.2884°W downstream 6 miles to the mouth of Davis Creek. **Summary:** Tributary of the South Fork of the Koyukuk River. **Geology:** Thick, washed gravel above the mouth of Davis Creek is cut by the creek channel and contains small amounts of gold (Maddren, 1913). The rocks near the mouth of the creek are Paleozoic pelitic schist with subordinate quartzite (Patton and Miller, 1973). A panned-concentrate sample of gravel collected about 0.8 mile above the mouth of Davis Creek contains 6 coarse and 6 fine gold particles; the sample was analyzed and contains 220 ppm gold and 351 ppm bismuth (Kurtak and others, 2002). A sample of a 3-inch-wide quartz vein that cuts micaceous schist on Davis Creek contains 150 ppb gold (Kurtak and others, 2002). **Workings:** Maddren (1913) reports that a little mining occurred on Davis Creek in the early 1900's. Reed (1938), however, indicates that considerable mining was done then. Davis Creek may have been mined as late as the 1990's (Kurtak and others, 2002). Little mining was done near the mouth as the bars near the mouth are typically dry. Placer mining has occurred along the lower 0.8 mile of the creek; most of the mining was within a few hundred feet of its mouth. (The Alaska Resource Data File., Bettles Quadrangle).

Dome Creek

(Gold, Chromite) **Quadrangle:** Bettles (B3); **Coordinates:** 66.5379°N, 150.7958°W; **Map:** 2-3; **Worked:** 1911-; **Summary:** Tributary of the Kanuti River. The coordinates above are based on U.S. Bureau of Mines and Bureau of Land Management records that locate the occurrence, but do not indicate the basis on which the location was determined (Bottge, 1986; Kurtak and others, 2002). **Geology:** Brooks (1912) reported that placer gold prospects were found on Dome Creek in 1911. There are no other published reports of mining or prospecting on this creek. The U.S. Bureau of Mines investigated the area in the 1980's, apparently for its chrome potential (Foley and McDermott, 1983; Kurtak and others, 2002). Two panned-concentrate samples from Dome Creek, about three miles east of the Kanuti River, contain 2 very fine gold flakes, and 425 and 508 ppm chromium (Kurtak and others, 2002). (The Alaska Resource Data File., Bettles Quadrangle).

Eldorado Creek

(Gold) **Quadrangle:** Bettles (D2); **Coordinates:** 66.5379°N, 150.7958°W; **Map:** 2-2; **Access:** Boat from the Dalton Highway Bridge at 67.0181°N, 150.2884°W downstream 18 miles to the mouth of Davis Creek at 66.9339°N, 150.9029°W. **Summary:** South-flowing tributary to the South Fork, Koyukuk River. The junction of Eldorado Creek and South Fork Koyukuk River is opposite Squaw Creek, about 8 miles west-southwest of Gold Bench. The occurrence is about 1 mile upstream from the mouth of Eldorado Creek. **Geology:** A panned-concentrate sample collected during a Bureau of Land Management examination of the area contained a single particle of coarse gold (Kurtak and others, 2002). The concentrate contained 8.79 ppm gold. Several other panned-concentrate samples, as well as a stream-sediment sample, were barren. The rocks in the area are Cretaceous graywacke interbedded with mudstone (Patton and Miller, 1973). No other information is available. (The Alaska Resource Data File., Bettles Quadrangle).

Fish Creek

(Gold) **Quadrangle:** Bettles (C1, C2); **Coordinates:** 66.5379°N, 150.7958°W; **Map:** 2-2; **Yield:** 0.0025 oz/yd^3; **Summary:** Tributary of the South Fork of the Koyukuk River. Fish Creek has been designated as a **Recreation Mining Area** by the State of Alaska 1 mile upstream and 1 mile downstream from the Dalton Highway bridge.

Gold Bench

(Gold, Bismuth, Copper, FM, Lead, Mercury, Monazite, Tin, Tungsten) **Quadrangle:** Bettles (D2); **Coordinates:** 66°58'43"N, 150°38'20°W (Mine Symbol); **Map:** 2-2, 2-4; **Worked:** 1900s-Present; **Production:** 1900s=6,000 ozs. **Access:** Landing strip at 66.9832°N, 150.6445°W or by boat from the Dalton Highway Bridge at 67.0181°N, 150.2884°W downstream 10 miles or by winter trail from Dalton Highway at 67.0797°N, 150.3509°W. **Summary:** Location on the South Fork of the Koyukuk River. **Minerals:** Magnetite, hematite, garnet, and small amounts of pyrite, chalcopyrite, cinnabar, rutile, cassiterite, scheelite, monazite, uranothorianite, galena, sphene, and bismuthinite. Source of heavy minerals not known but probably a reconcentration from glacial deposits. Some may have been derived from mountains between South Fork and Jim River (Cobb, 1978, OF 78-94 p. 10-11). **Geology:** Gravel-capped bench on unconsolidated material; about 30 - 60 feet above present grade of South Fork of Koyukuk River. The bench is about 1,000 feet wide and 3/4 mile long, slopes gently toward river, covering an area of about 60 to 100 acres. It was most likely formed when the stream flowed at a higher level. Gold is flattened and scattered through top 8 - 10 feet of gravel, but most is on false bedrock of reddish sand 18 - 24 inches below surface of fine washed stream gravels composed largely of schist and quartz pebbles. Small amounts of gold occurred throughout the gravel section (Maddren, 1913). Bedrock is decomposed to blue clay that probably was derived from shale (Reed, 1938). The depth to bedrock was estimated by Reed (1938) to be about 6 feet. Production averaged 3.9 to 4.35 ounces per day using a rocker box. Gold at foot of bench is angular, rough, and honeycombed. In early 1900's production by simple hand methods was about 6,000 ounces of gold. Placer concentrates contain as much as 0.18 percent equivalent uranium. The source of the gold is not known. Maddren (1913) speculated that it might have come from the Tramway Bar area on the Middle Fork, Koyukuk or, more likely, from the hills to the south. Cobb (1973 [B1374]) thought that at least some of the gold was probably from reworked glacial deposits. **Workings:** In the 1940's heavy equipment was used with no production data during that time. Heiner and Wolff (1968) noted mining in the 1940's using heavy equipment. Grybeck (1977) indicated mining activity through 1975. (The Alaska Resource Data File., Bettles Quadrangle).

Ironside Bar

(Gold) **Quadrangle:** Bettles (D2); **Coordinates:** 66°59'07"N, 150°36'06°W Mine Symbol); **Map:** 2-4; **Worked:** 1900s-Present; **Purity:** 907.5Au; **Yield:** 0.11 - 0.22 oz/yd^2; **Production:** Through 1909=100 ozs. **Access:** Landing strip at 66.9832°N, 150.6445°W or by boat from the Dalton Highway Bridge at 67.0181°N, 150.2884°W downstream 9 miles or by winter trail from Dalton Highway at 67.0797°N, 150.3509°W. **Summary:** Location on the South Fork of the Koyukuk River. **Geology:** The high channel which constitutes the Ironside Bench placer is about 30 feet higher than the current level of the South Fork Koyukuk River. It is about at the same level as the nearby Gold Bench channel. Reed (1938) suggests that the channels at both locations are the same. River-washed gravel at Ironside Bench is heavy, coarse and waterworn, with few large boulders. The gravel lies on a layer of blue clay, which Reed (1938) suspected is decomposed shale. The depth to bedrock varies from 6 to 30 feet; the upper 5 to 6 feet is muck and ice. The gold is waterworn, fine and flaky; it occurs throughout the gravel but is found mostly at the top of the clay. The only production figures available are Maddren's (1913), who indicates that a total of $2,000 (about 100 oz.) in gold had been recovered by 1909. No figures are available for later production. Reed (1938) reported that the value of the ground ranged from 25 to 50 cents per square foot of bedrock and that the average fineness of the gold is 907.5. **Workings:** Mining was generally by hand methods. Mining by ground sluicing and shoveling into sluice boxes was being done in 1937; about 40 acres were said to have been mined by that time (Reed, 1938). Grybeck (1977) indicated mining activity through 1975. Small-scale mining in the early 1900's and the 1930's. Total production of Ironside Bench and Eagle Bluff in 1900 was about 50 ounces. During 1937 the ground that was worked averaged 0.11 - 0.22 oz/yd^2 of false bedrock (Cobb, 1978, OF 78-94 p. 12). (The Alaska Resource Data File., Bettles Quadrangle).

Jim River

(Gold) **Quadrangle:** Bettles (C2, C3, D1, D2); **Coordinates:** 66°47'24"N, 151°11'09°W; **Map:** 2-2; **Purity:** Au; **Yield:** 0.0003-0.0025 oz/yd^3; **Access:** Unimproved road from the Dalton Highway at 66.7915°N, 150.6876°W, west 1.2 miles to Prospect Creek Camp. **Summary:** Tributary of the South Fork of the Koyukuk River. The Jim River has been designated as a **Recreation Mining Area** by the State of Alaska from the end of the road in old Prospect Creek camp downstream for 2 miles. There are two placer-gold occurrences are along the Jim River. One is at the mouth of Prospect Creek. The second is in the canyon of Jim Creek about 6 miles downstream. **Geology:** Fine placer gold, identified in panned concentrates occurs at two sites along Jim River (Kurtak and others, 2002). At the 'Jim Prospect Confluence', a panned concentrate contained a single, very fine grained, gold particle, with abundant magnetite. An analysis of this concentrate indicated that it contained 1,590 ppb gold. Another placer sample from the same site contained 13 very fine gold particles and an analysis showed that it has a value of 0.0003 ounce of gold per cubic yard. At a site on Jim River about 6 miles downstream from the mouth of Prospect Creek, very fine gold flakes were observed in a panned-concentrate; an analysis showed a value of 1,230 ppb gold (Kurtak and others, 1999). Kurtak and others (1999) suspect that the gold is derived from the Jim River pluton, which crops out north of the Jim River canyon. The rocks in the region are Triassic to Jurassic mafic volcanic and intrusive rocks, and mid-Cretaceous granitic, plutonic rocks (Patton and Miller, 1973). The rocks in the vicinity of the occúrrences on the Jim River are altered volcanics, including pillow basalts (Kurtak and others, 2002). No bedrock was observed at the 'Jim Prospect Confluence' occurrence. (The Alaska Resource Data File., Bettles Quadrangle).

John R Creek

(Gold) **Quadrangle:**Bettles (D2); **Coordinates:** 66.9891°N-150.6744°W; **Map:** 2-4; **Worked:** 1930s; **Access:** Landing strip at 66.9832°N, 150.6445°W or by boat from the Dalton Highway Bridge at 67.0181°N, 150.2884°W downstream 11 miles or by winter trail from Dalton Highway at 67.0797°N, 150.3509°W. **Summary:** Tributary of the South Fork of the Koyukuk River. The old name for John River Creek was Jean D'Arc Creek. Gold reported to have been recovered from 17 prospect holes drilled in 1930. Ground probably too lean for profitable mining (Cobb, 1978, OF 78-94 p. 13). **Geology:** Probably similar to Gold Bench and Ironside.

Kanuti River Tributary

(Gold) **Quadrangle:** Bettles (D1, D2); **Coordinates:** 66.2103°N-151.0936W; **Map:** 2-3; **Worked:** 1960s-1970s; **Summary:** This placer occurrence is on an unnamed, north-flowing tributary to the Kanuti River about 0.5 mile upstream from its mouth. **Geology:** Two very fine gold flakes were noted in eight panned samples collected from this unnamed creek. The panned samples contain 2,119 ppb gold and 252 ppm chromium (Kurtak and others, 2002). The rocks in the area are chlorite schist and interbedded metagraywacke cut by quartz veinlets (Patton and Miller, 1973). Upstream, the creek passes over the contact of the Sithylemenkat Pluton. **Workings:** Placer claims were located in the 1960's and 1970's. (The Alaska Resource Data File., Bettles Quadrangle).

Prospect Creek

(Gold) **Quadrangle:** Bettles (D2); **Coordinates:** 66.7768°N-150.6759°W; **Map:** 2-2; **Worked:** 1909-1993; **Yield:** 0.0025 oz/yd^3; **Access:** Unimproved road from the Dalton Highway at 66.7915°N, 150.6876°W, west 1.2 miles to Prospect Creek Camp. **Summary:** Tributary Jim River. Prospect Creek has been designated as a **Recreation Mining Area** by the State of Alaska upstream 1 mile and downstream any distance from Dalton Highway Bridge at 66.7816°N-150.6891°W, except for the pipeline right of way. **Geology:** Maddren (1913) reported that in 1909 a Geological Survey party in 1909 found gold colors while panning gravels from the upper part of Prospect Creek. Reed (1938), however, noted that considerable later prospecting was unable to locate the deposit reported by Maddren. Bundtzen and others (1984; 1994) noted claim staking and prospecting on Prospect Creek in 1983, and that a mining operation was said to be active in 1993. (The Alaska Resource Data File., Bettles Quadrangle).

Rock Creek

(Gold) **Quadrangle:** Bettles (D1, D2); **Coordinates:** 66.9950°N-150.5349°W; **Map:** 2-4; **Production:** 1900-1910=40 ozs; **Access:** Boat from the Dalton Highway Bridge at 67.0181°N, 150.2884°W downstream 7 miles to the mouth of Rock Creek at 66.9897°N, 150.5349°W. **Summary:** It is a short, south-flowing tributary to South Fork Koyukuk River. Its mouth is about 2 miles east (upstream) from Ironsides Bar. The coordinates are for a site on the lower potion of Rock Creek but the exact location of placer mining on the creek is uncertain. Creek descends very steeply into the valley of South Fork. **Geology:** The rocks in the area are predominantly Cretaceous conglomerate (Patton and Miller, 1973). Reed (1938) reported that $800 in gold (about 40 ounces) was produced in from 1900 to 1910.

Chandalar Quadrangle

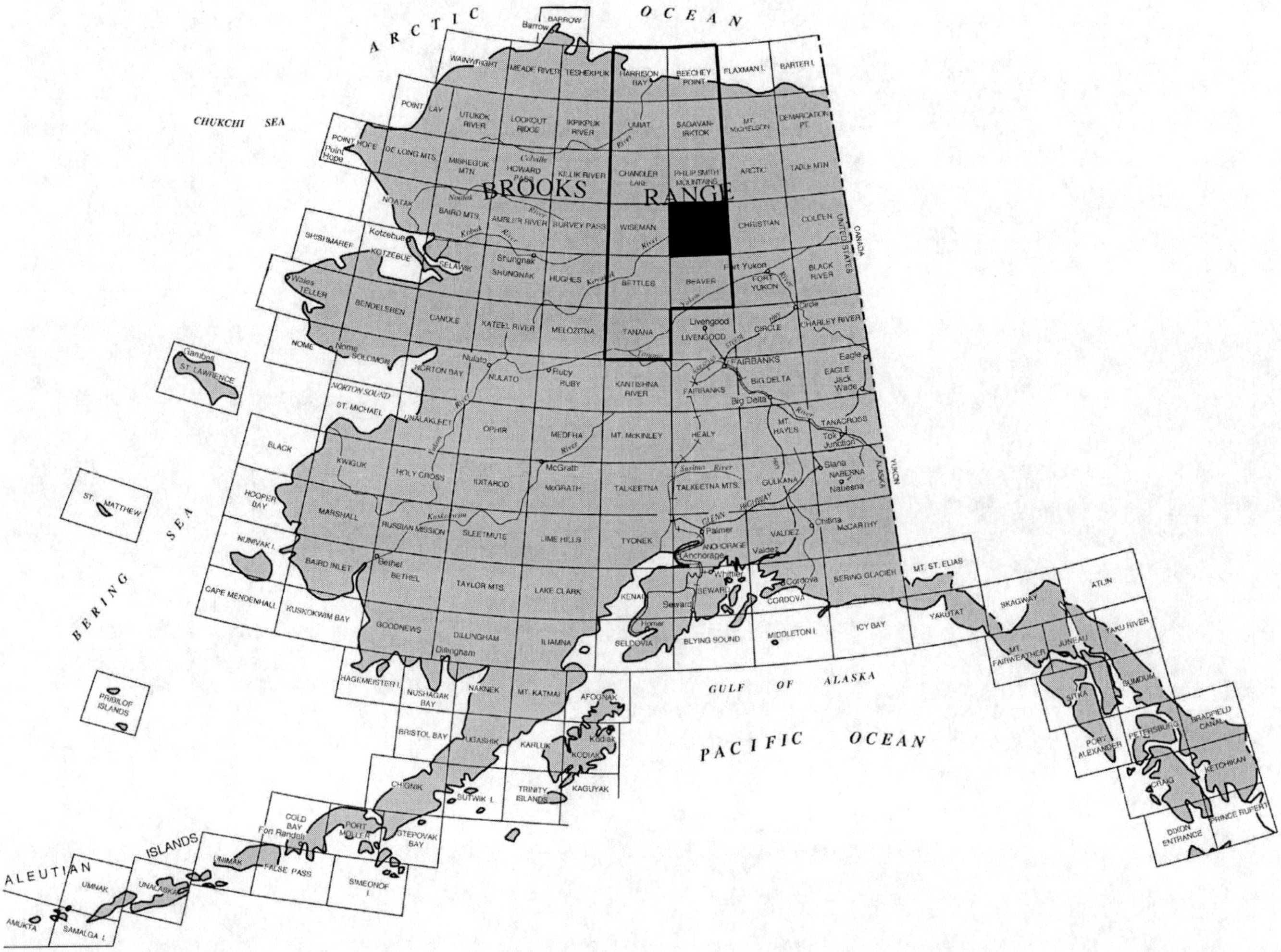

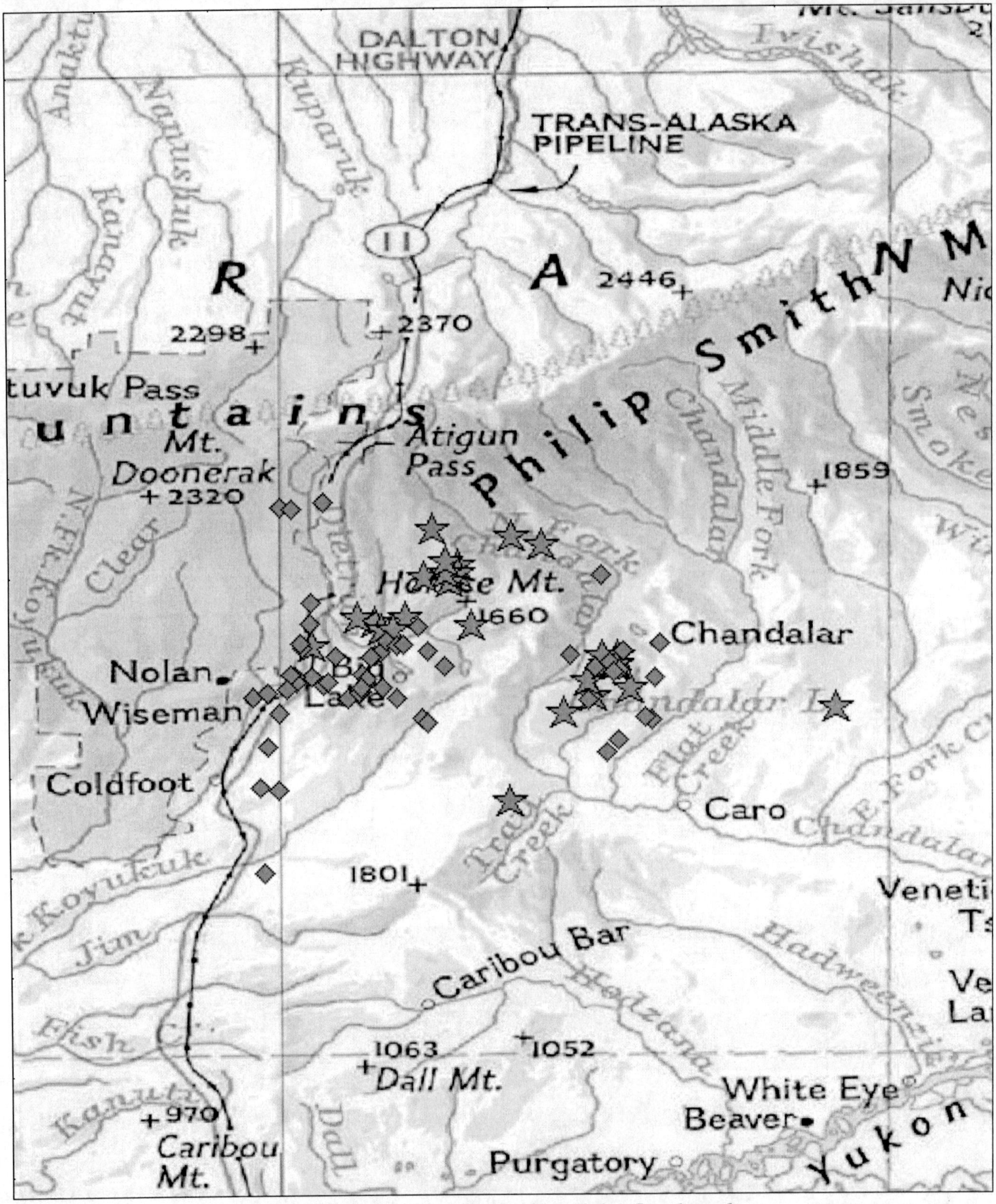

Gold Placer Locations in the Chandalar Quadrangle

♦ Diamonds shows placer gold locations. ★ Stars show gold-bearing hard rock locations. Hard rock occurrences are not identified by name nor is a description of the occurrence included in this book. They are provided so the reader can see where the hard rock sources for the placer deposits occur. Areas are located by latitude and longitude WGS 84 coordinates.

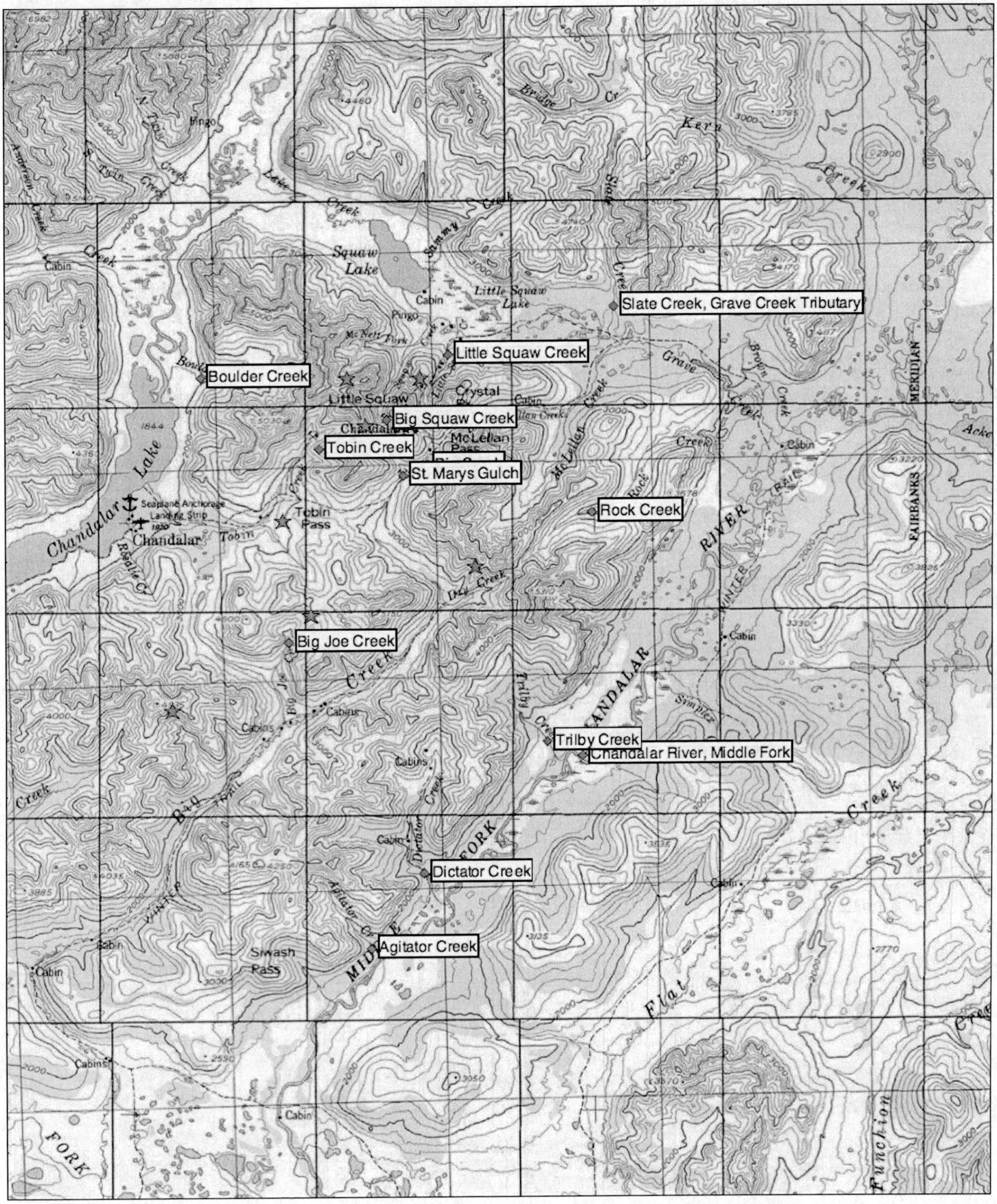
Slate Creek, Grave Creek Tributary
Little Squaw Creek
Boulder Creek
Big Squaw Creek
Tobin Creek
St. Marys Gulch
Rock Creek
Big Joe Creek
Trilby Creek
Chandalar River, Middle Fork
Dictator Creek
Agitator Creek
Squaw Lake
Little Squaw Lake
Chandalar Lake
Chandalar
Tobin Pass
McLellan Pass
Siwash Pass
Seaplane Anchorage
Landing Strip
FAIRBANKS MERIDIAN

Map 3-1

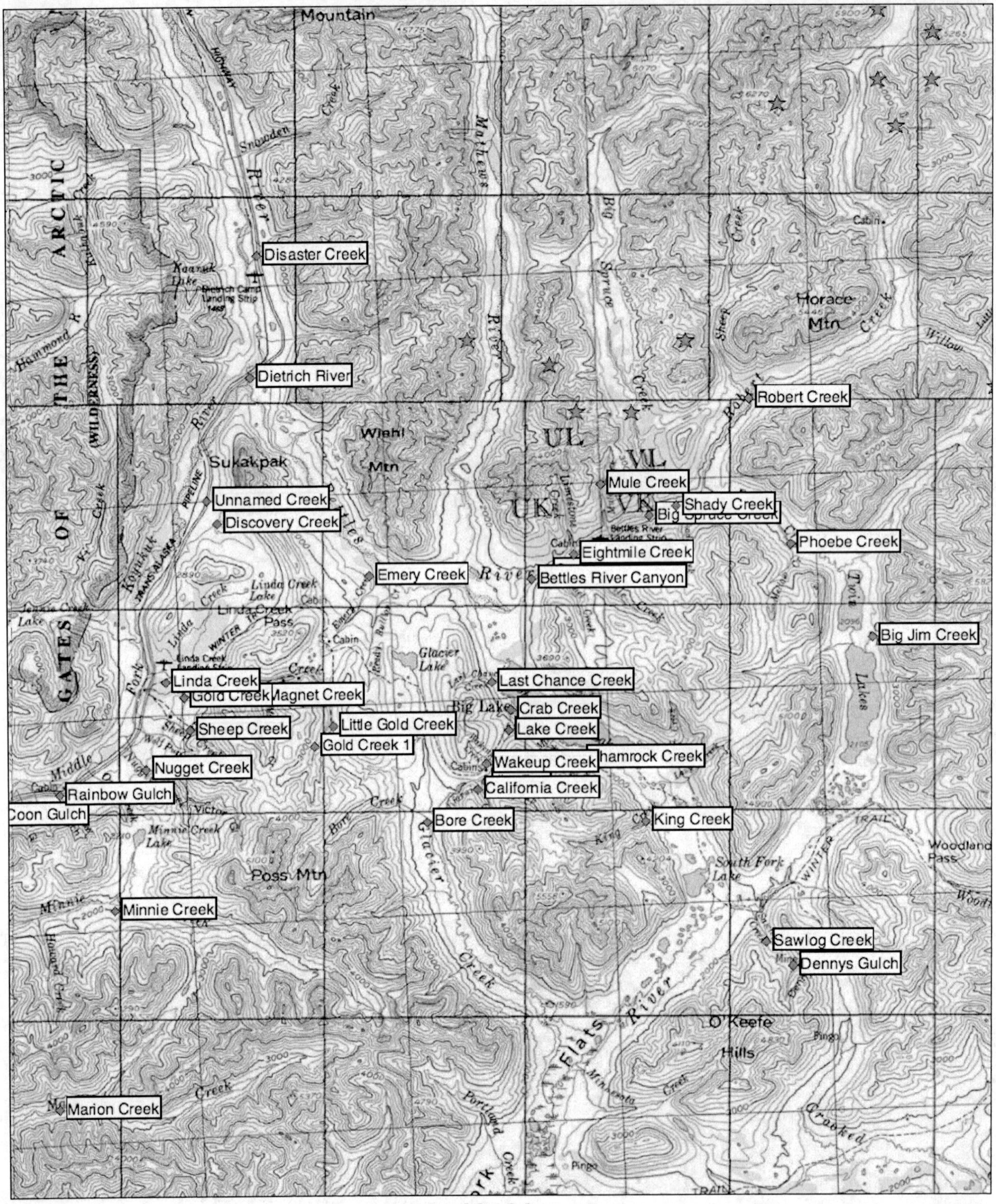
Disaster Creek
Dietrich River
Robert Creek
Unnamed Creek
Discovery Creek
Mule Creek
Shady Creek
Big Spruce Creek
Phoebe Creek
Eightmile Creek
Emery Creek
Bettles River Canyon
Big Jim Creek
Linda Creek
Gold Creek
Magnet Creek
Last Chance Creek
Crab Creek
Lake Creek
Sheep Creek
Little Gold Creek
Gold Creek 1
Wakeup Creek
Shamrock Creek
Nugget Creek
California Creek
Rainbow Gulch
Coon Gulch
Bore Creek
King Creek
Minnie Creek
Sawlog Creek
Dennys Gulch
Marion Creek
Mountain
Sukakpak
Wiehl Mtn
Horace Mtn
Poss Mtn
O'Keefe Hills
Woodland Pass
Big Lake
Glacier Lake
Linda Creek Pass
GATES OF THE ARCTIC
(WILDERNESS)

Map 3-2

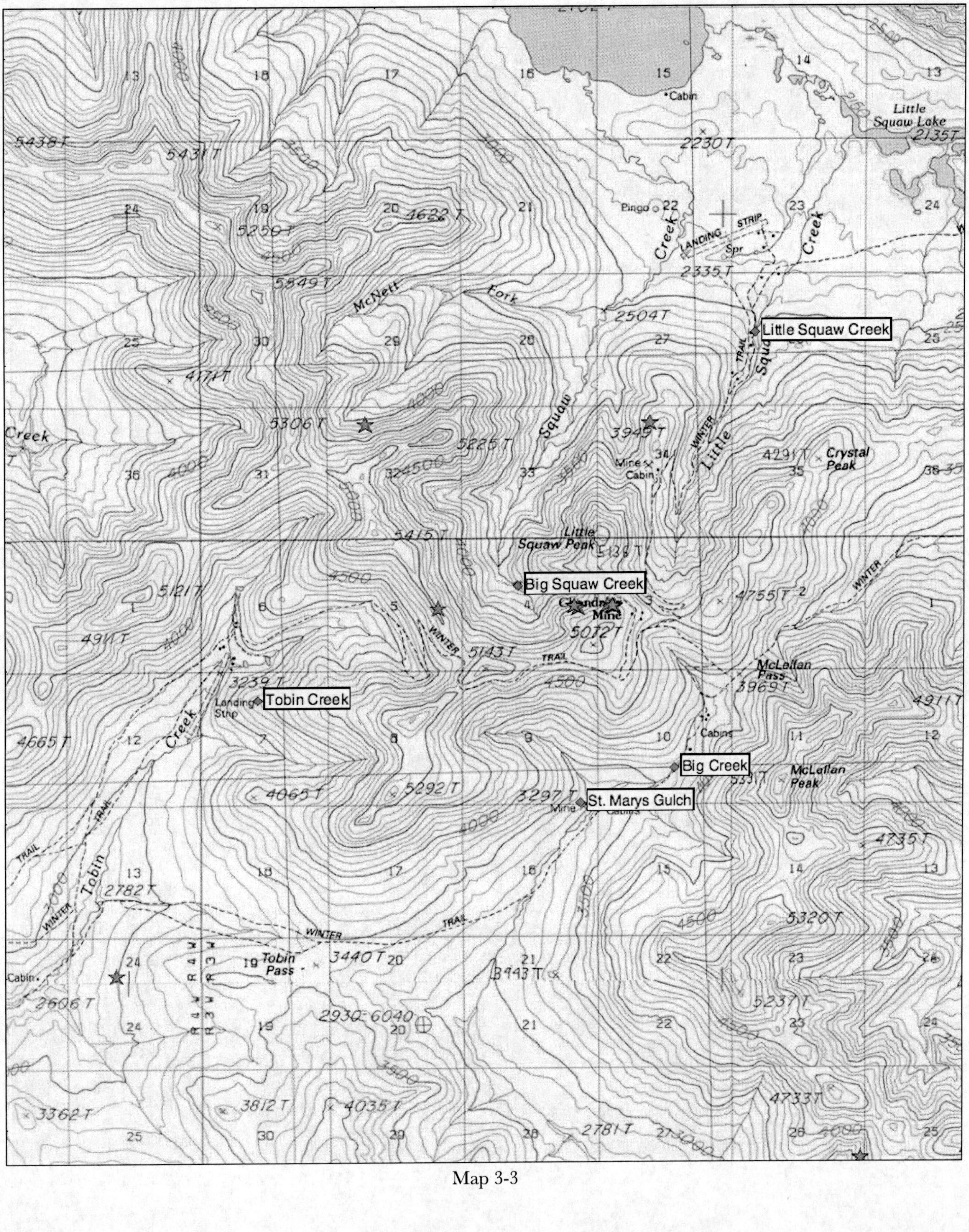
Little Squaw Lake
Little Squaw Creek
Big Squaw Creek
Tobin Creek
Big Creek
St. Marys Gulch
Little Squaw Peak
Crystal Peak
McLellan Pass
McLellan Peak
Tobin Pass
McNett Fork
LANDING STRIP
WINTER TRAIL
Mine Cabin
Cabins
Landing Strip

Map 3-3

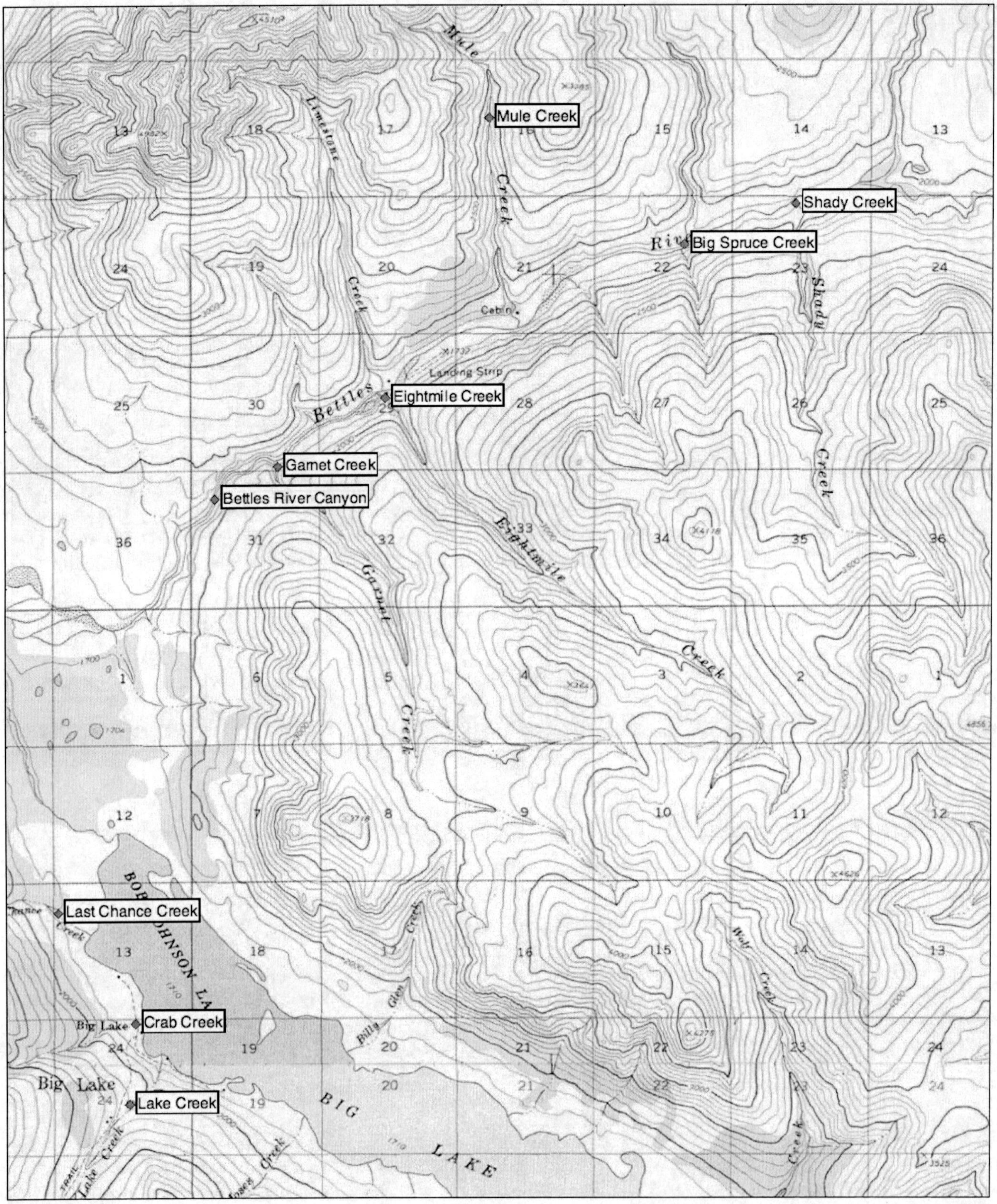
Mule Creek
Shady Creek
Big Spruce Creek
Eightmile Creek
Garnet Creek
Bettles River Canyon
Last Chance Creek
Crab Creek
Lake Creek
Mule Creek
Limestone Creek
Bettles
Cabin
Landing Strip
Shady Creek
Eightmile Creek
Garnet Creek
Wolf Creek
Billy Glen Creek
BOB JOHNSON LAKE
Big Lake
BIG LAKE

Map 3-4

Jennie Creek Lake
Jennie Creek Lake
MIDDLE
Linda
Landing Strip
Linda Creek
Gold Creek
THE TRANS-ALASKA PIPELINE
WINTER TRAIL
Landing Strip
Sheep Creek
Sheep Creek
Wolf
Sheep
Pup
Jennie Creek
RIVER
KOYUKUK
UNDER CONSTRUCTION
Nugget Creek
Nugget
MIDDLE FORK
Over Creek
Cabin
Rainbow Gulch
THE TRANS-ALASKA PIPELINE
Rainbow Gulch
Creek
Victor
River
Coon Gulch
Coon Gulch
Montana Gulch
Minnie Creek Lake
Minnie

Map 3-5

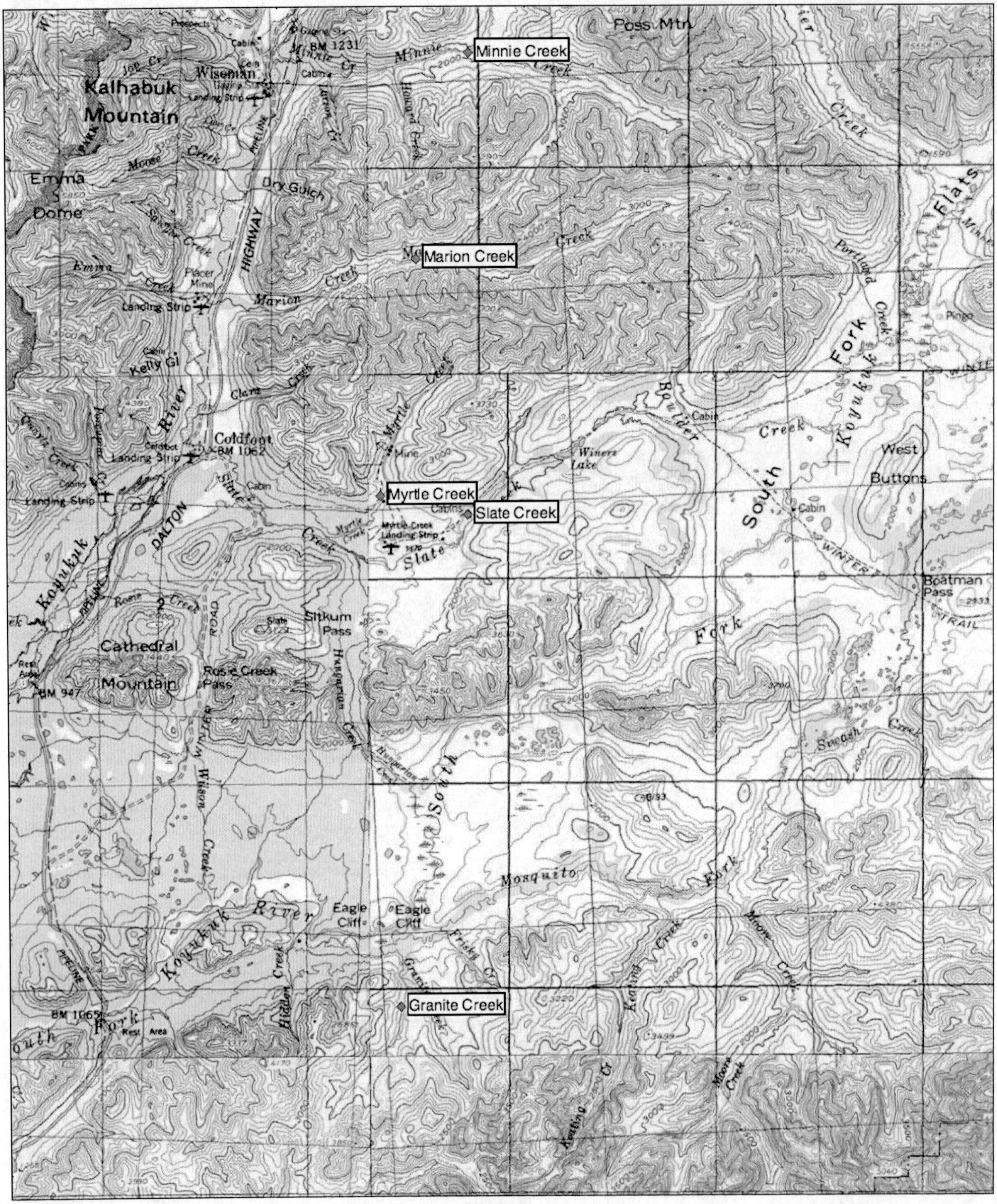
Minnie Creek
Marion Creek
Myrtle Creek
Slate Creek
Granite Creek
Poss Mtn
Kalhabuk Mountain
Wiseman
Emma Dome
Dry Gulch
HIGHWAY
Placer Mine
Landing Strip
Kelly Gl
Coldfoot
BM 1062
DALTON
ROAD
Cathedral Mountain
Rosie Creek Pass
Sitkum Pass
West Buttons
Boatman Pass
Wiseman Lake
South Fork Koyukuk
Koyukuk River
Eagle Cliff
Pingo
BM 947
BM 1065
BM 1231

Map 3-6

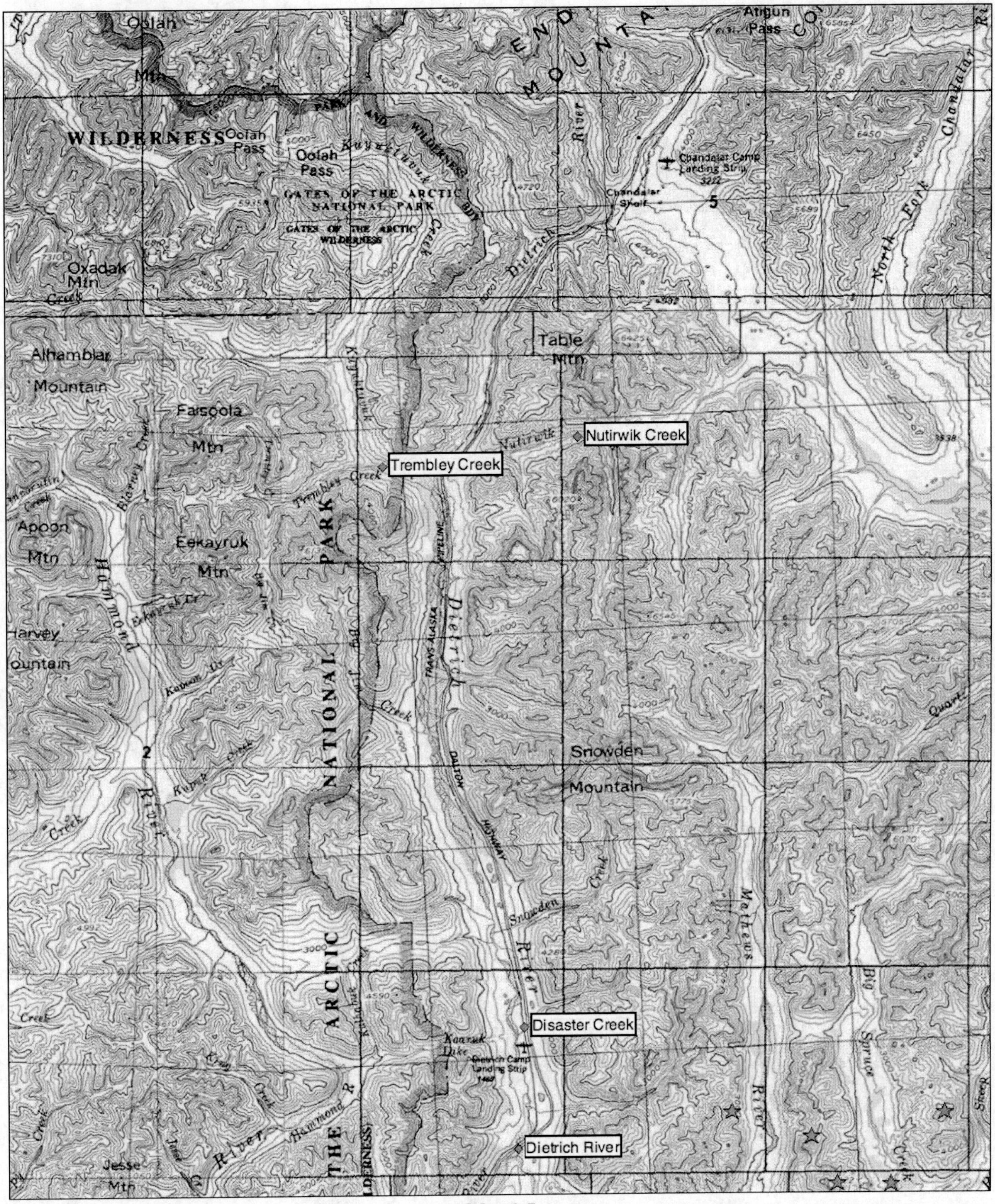

Oolah
Mtn
WILDERNESS
Oolah Pass
Atigun Pass
GATES OF THE ARCTIC NATIONAL PARK
GATES OF THE ARCTIC WILDERNESS
Chandalar Camp Landing Strip
Chandalar Shelf
North Fork Chandalar
Oxadak Mtn
Dietrich
Table Mtn
Alhambra Mountain
Faiscola Mtn
Nutirwik Creek
Trembley Creek
Apoon Mtn
Eekayruk Mtn
Hammond
Harvey Mountain
TRANS ALASKA PIPELINE
Dietrich
Snowden Mountain
DALTON HIGHWAY
Snowden
Mathews River
Big Spruce Creek
Disaster Creek
Dietrich Camp Landing Strip
Dietrich River
Jesse Mtn
Hammond R
THE ARCTIC NATIONAL PARK

Map 3-7

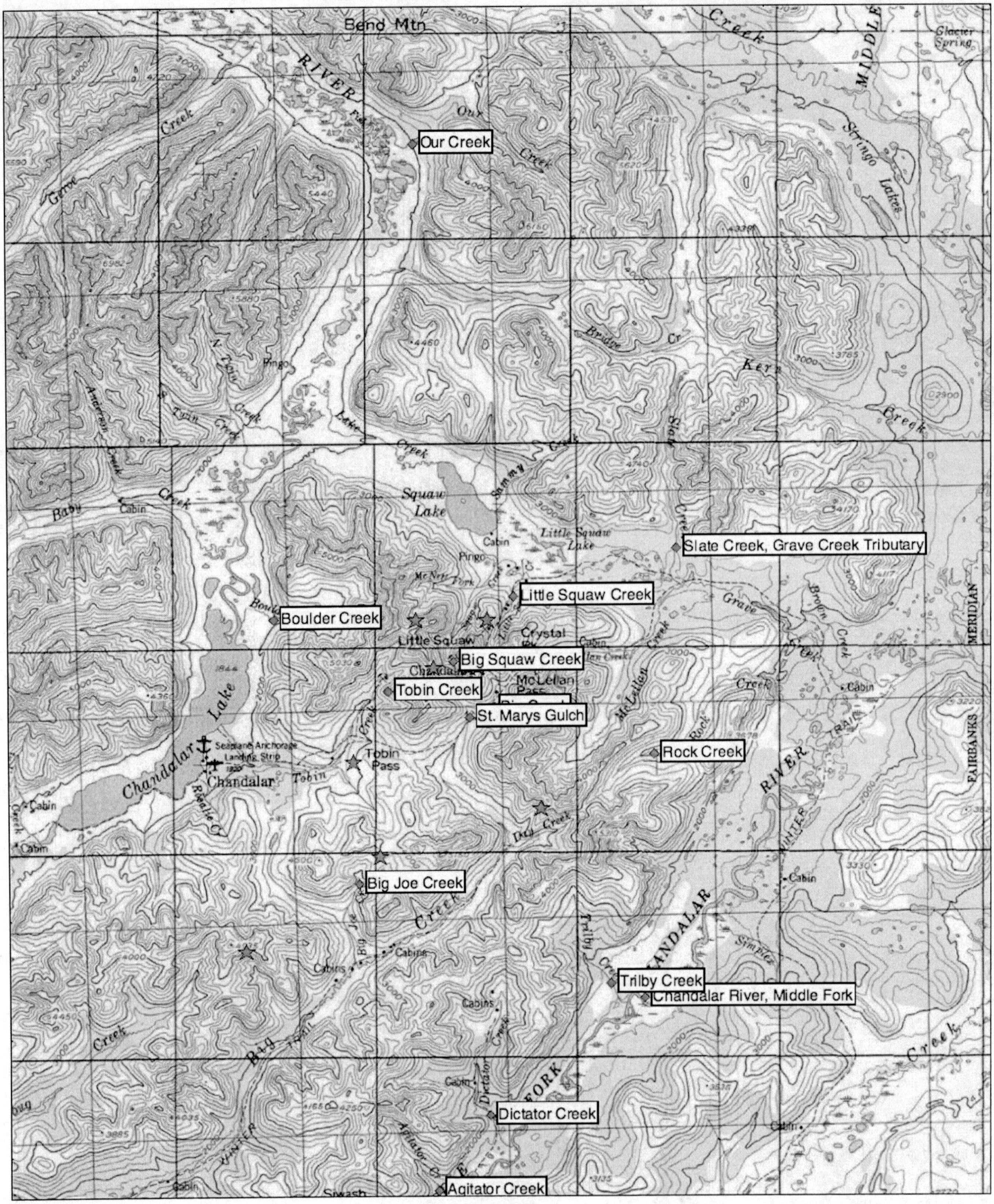
Bend Mtn
Our Creek
Slate Creek, Grave Creek Tributary
Little Squaw Creek
Boulder Creek
Big Squaw Creek
Tobin Creek
St. Marys Gulch
Rock Creek
Big Joe Creek
Trilby Creek
Chandalar River, Middle Fork
Dictator Creek
Agitator Creek
Squaw Lake
Little Squaw Lake
Chandalar Lake
Chandalar
Tobin Pass
McLellan Pass
Stringo Lakes
MERIDIAN
FAIRBANKS

Map 3-8

Agitator Creek

(Gold) **Quadrangle:** Chandalar (C6); **Coordinates:** 67.3183°N, 148.2443°W; **Map**: 3-1, 3-8; **Worked:** 1910-; **Access:** Boat from Middle Fork of the Chandalar River at 67.3146°N, 148.2352°W. **Summary:** Southeast flowing tributary to the Middle Fork Chandalar River. **Geology:** Reed (1930, MR 31-4, p. 10) noted that encouraging prospects had reportedly been found in the early days (about 1910) but that no mining had occurred and no prospecting was taking place at the time of Reed's report. No other information is available on this occurrence. (The Alaska Resource Data File., Chandalar Quadrangle).

Bettles River Canyon

(Gold) **Quadrangle:** Chandalar (C5); **Coordinates:** 67.5592°N, 149.4290°W; **Map**: 3-2, 3-4; **Worked:** 1927-1937; **Yield:** 0.216 oz/ft^2; **Summary:** Tributary of Middle Fork of the Koyukuk River. Bettles River was named in 1899 by the USGS after an old Yukon, Gordon Bettles. **Geology:** Most of the placer mining along the Bettles River has been on its tributaries, Reed (1938) documented drift mining along the Bettles River from just below Garnet Creek to the mouth of Mule Creek. **Workings:** Several claims across from mouth of Garnet Creek were drift mined. Considerable gold mined during period of very low water just before freeze-up in winter of 1927-28. Results of prospecting bench deposits on top of south canyon wall and flood-plain deposits below canyon are not known. An area mined out in 1936-37 was 14 -16 feet deep and contained gold worth about 0.216 oz/ft^2 of bedrock cleaned. Most of mining reported as on Bettles River was on tributaries. See also Eightmile Creek, Emory Creek, Mule Creek, Phoebe Creek, Robert Creek, Spruce Creek (Cobb, 1976, OF 76-340. p. 5). The Bettles River has been designated as a **Recreation Mining Area** by the State of Alaska upstream two miles from the Dalton Highway at 67.6449°N-149.7398°W. Average yield of gravel is estimated to be 0.0025 oz/yd^3 of fine gold.

Big Creek

(Antimony, Copper, Gold, Lead, Monazite, Silver, Tungsten, Zinc) **Quadrangle:** Chandalar (C3); **Coordinates:** 67.5185°N, 148.2087°W; **Map**: 3-3; **Worked:** 1906-1990s; **Production:** 10,000 ozs; **Access:** Trail from Chandalar landing strip at 67.5025°N, 148.4914°W. **Summary:** Tributary of the North Fork of the Chandalar River. Gold discovered in 1906 followed by stampede in 1907. Big Creek is one of the major placer gold producers in the Chandalar district. Most of production from head of Big Creek and St. Marys Gulch. **Minerals:** Gold, pyrite, hematite, limonite, monazite, scheelite, rutile, arsenopyrite, chalcopyrite, galena, ilmenite. **Geology:** It drains directly from the area of several of the Chandalar gold quartz properties. The gravels in the stream are relatively shallow, averaging about 12 feet deep on the uppermost portion of the creek and deepening to about 20 to 22 feet on the lower portion of the mined ground (Reed, 1930, MR 31-4; MR 195-13). The gold occurs in the lower 3 to 5 feet, and there is very little gold on bedrock. The gravel is fairly coarse with numerous greenstone boulders as much as 3 feet in diameter. There is only one generation of placers in the drainage (they are not separable into pre- and post-glacial deposits). The gold is reported to be bright yellow, generally irregular, and shot-like to flattened, but some occurs as crystals. Its size averages 1 mm with many 2- to 3-mm nuggets. A few nuggets contain included quartz crystals, limonite, and goethite. It is also reported that the concentrates contain a large suite of heavy minerals in addition to the gold (Mertie, 1925; White, 1952). A pan evidence that mining has occurred in this lower area. There are at least two reports of a quartz vein in the creek bed on the upper reaches of Big Creek (Chipp, 1970; Heiner and Wolff, 1968). Heiner and Wolff (1968) described it as a gold-bearing mineralized zone that may extend from Big Creek to upper Tobin Creek. concentrate sample had eU of 0.05 percent (White, 1952, p. 11). Coarse gold was reported several miles downstream from the mined area, but there is no evidence that mining has occurred in this lower area. There are at least two reports of a quartz vein in the creek bed on the upper reaches of Big Creek (Chipp, 1970; Heiner and Wolff, 1968). Heiner and Wolff (1968) described it as a gold-bearing mineralized zone that may extend from Big Creek to upper Tobin Creek. **Workings:** Mining was by hand methods until after World War II, when mechanized equipment was introduced; total production was about 15,000 fine ounces, of which two-thirds was after 1950. Placers were derived from lodes near head of creek. One man placer mining and 2 men developing quartz vein near head in 1952. Production from district has been 40,000 ounces of which about 15,000 ounces was from Big Creek, and 10,000 ounces since 1950. (The Alaska Resource Data File., Chandalar Quadrangle).

Big Jim Creek

(Gold, Copper) **Quadrangle:** Chandalar (B3); **Coordinates:** 67.5342°N-149.0542°W; **Map**: 3-2; **Access:** Float plane on Twin Lake at 67.5269°N, 149.0605°W. **Summary:** Big Jim Creek is a west-flowing tributary into the southern lake of Twin Lakes, which are approximately 15 miles west of Chandalar. **Workings:** Small placer mine produced native copper nuggets as well as gold (Reed, 1938). The amount of production is unknown, but it was probably quite small. (The Alaska Resource Data File., Chandalar Quadrangle).

Big Joe Creek

(Gold) **Quadrangle:** Chandalar (B3); **Coordinates:** 67.4484°N-148.3336°W; **Map**: 3-, 3-81; **Access:** Winter Trail from

Coldfoot, Alaska at 67.2533°N, 150.1794°W. **Summary:** Tributary of Big Creek. Gold in quartz vein in quartz-mica schist near diorite intrusives or andesitic flows (Cobb, 1976, OF 76-340. p. 11). Presence of placer gold below quartz vein possible.

Big Spruce Creek

(Gold) **Quadrangle:** Chandalar (C5, D5); **Coordinates:** 67.5875°N-149.2989°W; **Map:** 3-4; **Access:** Trail from landing strip at 67.5733°N, 149.3704°W. **Summary:** Tributary of the Bettles River. Heads in limestone; creek has cut down into underlying schist. Two shafts about 45 feet deep did not find good prospects where stream enters valley of Bettles River. Good prospects said to have been found further upstream in early days probably just below the lode gold deposit at 67.6300°N-149.3193°W. Spruce Creek is the old name for Big Spruce Creek (Cobb, 1976, OF 76-340. p. 68). The creek traverses the Horace Mountain plutons and the Chandalar copper belt.

Big Squaw Creek

(Gold, Silver, Molybdenum, Lead, Thorium, Antimony) **Quadrangle:** Chandalar (B3); **Coordinates:** 67.4484°N-148.3336°W; **Map**: 3-1, 3-3, 3-8; **Worked:** 1923-1928; **Access:** Trail from Chandalar landing strip at 67.5025°N, 148.4914°W. **Summary:** Big Squaw Creek (Squaw Creek on the modern topographic map) drains the area of the Chandalar gold lodes and terminates at Squaw Lake. **Minerals:** Gold, pyrite, arsenopyrite, stibnite, monazite, uranothorianite, galena, molybdenite, and zircon. **Geology:** Squaw Creek cuts through the Chandalar lode gold district, and the placer gold occurrences in this area are generally attributed to direct weathering of the gold-bearing veins. In spite of the proximity of Squaw Creek to the gold-bearing veins, most of the placer mining activity on this creek appears to have been limited to an area near the head of the creek (Chipp, 1970). This area on the upper creek, just below the Jupiter claim on the Eneveloe prospect, was mined, although there was little gravel in the creek bed and in places the creek was running on bedrock. On the lower creek, early efforts at mining the deeper gravels were thwarted as the gravels were thawed and the shafts were flooded. The gold that has been produced was in thin gravels along an irregular bedrock surface. The gravels are coarse and subangular with numerous greenstone boulders and slabs of schist. Glacial damming and disruption of drainage has created both pre- and post-glacial generations of placers. **Workings:** Small-scale surface workings, by hand methods only. Mining in 1923 was in gravel 3 to 4 feet deep. Mining was reported as late as 1928, but there is no information on any later activity. (The Alaska Resource Data File., Chandalar Quadrangle).

Bore Creek

(Gold) **Quadrangle:** Chandalar (B6); **Coordinates:** 67.4570°N-149.5443°W; **Map**: 3-2, 3-6; **Worked:** 1901-; **Access:** Winter Trail from Dalton Highway at 67.5159°N, 149.8485°W. **Summary:** Tributary of Glacier Creek. Bedrock is schist. Maddren reported that gold was found in 1901 and that (possibly) there was some mining. Gold recoveries were said to be poor, earning only enough to cover wages. Reed reported that prospect shaft failed to find pay. Gold may have been present, but certainly was not payable quantities. (Cobb, 1976, OF 76-340. p. 13).

Boulder Creek

(Gold) **Quadrangle:** Chandalar (B5, B6); **Coordinates:** 67.5606°N-148.4274°W; **Map**: 3-1, 3-8; **Worked:** 1991-; **Access:** From Chandalar landing strip at 67.5025°N, 148.4914°W. **Summary:** This occurrence is on Boulder Creek, a tributary to the North Fork Chandalar River, about a mile east of the north end of Chandalar Lake (sec. 34, T. 32 N., R. 4 W., of the Fairbanks Meridian). **Geology:** There is no description of this placer occurrence other than a reference to gold being produced from the creek in 1991 (Bundtzen and others, 1992, p. 28). It drains an area immediately west of the Chandalar gold quartz vein deposits. (The Alaska Resource Data File., Chandalar Quadrangle).

California Creek

(Gold) **Quadrangle:** Chandalar (B5); **Coordinates:** 67.4706°N-149.4885°W; **Map**: 3-2, 3-6; **Worked:** 1901-1990; **Production:** 1901-1909=145 ozs; **Access:** Winter Trail from Dalton Highway at 67.5159°N, 149.8485°W. **Summary:** Tributary of Glacier creek. Gold was discovered in 1901 and mined on a small scale, intermittently, until 1931. Some deep mining from shafts about 18 feet deep; some mining of gravel of present stream channel. In 1929 mining was reported at and below the mouth of Jim Pup, both from shafts about 18 feet deep and in present creek channel where gravels are 2 - 5 feet thick. No mining or prospecting after 1931. Forty-eight fine ounces of gold mined in 1901. Production from 1901-09 was 145 fine ounces (Cobb, 1976, OF 76-340. p. 15).

Chandalar River, Middle Fork

(Gold) **Quadrangle:** Chandalar (B6); **Coordinates:** 67.3998°N-148.0150°W; **Map:** 3-1, 3-8; **Worked:** 1910-; **Summary:** The occurrence is on the Middle Fork Chandalar River approximately 1/2 mile above the mouth of Trilby Creek Reed (1930, MR 31-4, p. 10) reported that in the early days (about 1910), $6 to $7 in gold per day was picked up off bedrock cropping out in the stream bed but that no further work had been done up to that time. (The Alaska Resource Data

File., Chandalar Quadrangle).

Coon Gulch

(Gold) **Quadrangle:** Chandalar (B6); **Coordinates:** 67.4610°N-150.0207°W; **Map:** 3-5; **Access:** From Dalton Highway at 67.4625°N, 150.0238°W. **Summary: T**ributary of the Middle Fork of the Koyukuk River. Good prospects said to have been found in early days. (Cobb, 1976, OF 76-340. p. 52).

Crab Creek

(Gold) **Quadrangle:** Chandalar (C5); **Coordinates:** 67.5040°N-149.4511°W; **Map**: 3-2, 3-4, 3-6; **Access:** Winter Trail from Dalton Highway at 67.5159°N, 149.8485°W. **Summary:** Tributary of Big Lake which is the old name for Bob Johnson Lake. Only traces of gold. Three pieces of gold found during extensive prospecting (Cobb, 1976, OF 76-340. p. 18).

Dennys Gulch

(Gold) **Quadrangle:** Chandalar (B5); **Coordinates:** 67.3954°N-149.1407°W; **Map:** 3-2; **Worked:** 1955-1964; **Nugget Size:** 2.8 ozs; **Summary:** Tributary of Crooked Creek. Gold has been mined. Bedrock is mica schist cut by many thin quartz veins that contain pyrite. Recent placer mining reported in 1964 (Cobb, 1976, OF 76-340. p. 20). **Geology:** This area was not described by early workers, which suggests that the discovery of placer gold was considerably later than that in many other parts of the district. There are reports of active placer mining on this creek as late as the 1960s but little information after that. Heiner and Wolff (1968) reported that placer gold has been mined from gravels that range in thickness from 6 to 70 ft and that coarse nuggets (valued at as much as $100) have been recovered, although no pay streak was delineated. Bedrock is mapped as quartz-muscovite schist that is highly deformed and cut by many thin quartz veins that contain pyrite (Freeman, 1963). Brosgé and Reiser (1964) noted 50 ppm Pb in a stream sediment sample from this location. Anomalous radioactivity has been observed at the O'Keefe placer claim near the map site. **Workings:** Placer mining reported as late as 1964 (Brosgé and Reiser, 1964), and six lode gold claims were located in 1955 (Heiner and Wolff, 1968).

Dictator Creek

(Gold) **Quadrangle:** Chandalar (B3); **Coordinates:** 67.3499°N-148.1900°W; **Map:** 3-1, 3-8; **Worked:** 1928-1976; **Access:** Trail from Middle Fork of the Chandalar River at 67.3258°N, 148.2059°W. **Summary:** Tributary of Middle Fork of the Chandalar River. Mining in 1928 and 1933 76-340. p. 23). The actual location of mining along the 5-mile length of this creek is unknown. Two claims staked in 1976 are near the reference point. **Geology:** The area is mapped as quartz-muscovite schist; a few small outcrop areas of intercalated greenschist are in the hills above the creek (Brosgé and Reiser, 1964). **Workings:** Reed (1930) reported a 190-foot shaft that was flooded without reaching bedrock. (The Alaska Resource Data File., Chandalar Quadrangle).

Dietrich River

(Gold) **Quadrangle:** Chandalar (C6, D6); **Coordinates:** 67.6450°N-149.7401°W; **Map:** 3-2, 3-7; **Yield:** 0.0025 oz/yd^3; **Summary:** Tributary of the Middle Fork of the Koyukuk River. Dietrich River has been designated as a **Recreation Mining Area** by the State of Alaska from Dalton Highway Bridge, upstream 2 miles.

Disaster Creek

(Gold) **Quadrangle:** Chandalar (C6); **Coordinates:** 67.6969°N-149.7327W; **Map:** 3-2, 3-7; **Yield:** 0.0025 oz/yd^3; **Summary:** Tributary of the Dietrich River. Disaster Creek has been designated as a **Recreation Mining Area** by the State of Alaska from the Dalton Bridge upstream for 2 miles.

Discovery Creek

(Gold) **Quadrangle:** Chandalar (C6); **Coordinates:** 67.5837°N-149.7758°W; **Map:** 3-2; **Worked:** 1982-; **Access:** From Dalton Highway at 67.5901°N, 149.7880°W. **Summary:** Discovery Creek, which is not named on the topographic map, drains the south flank of Sukakpak Mountain. **Geology:** Discovery Creek drains the Sukakpak Sb-Au vein occurrence. The creek gravels contain clasts of stibnite-quartz vein material (Dillon, 1982). Although there is evidence of early workings on the creek (Dillon, 1982), there are no published descriptions of the deposit. The presence of gold in the creek is substantiated by reports of gold recovered with a suction dredge in 1982 (Dillon, 1982). **Workings:** Recent (about 1982) testing with suction dredge and evidence of drift mining in the early 1900s (Dillon, 1982).

Eightmile Creek

(Gold, Mercury) **Quadrangle:** Chandalar (C5); **Coordinates:** 67.5699°N-149.3818°W; **Map:** 3-2, 3-4; **Worked:** 1938-1950s; **Access:** Landing strip at 67.5733°N, 149.3704°W. **Summary:** Tributary of the Bettles River. **Geologic description:** Eightmile Creek is one of several tributaries to the Bettles River in which placer gold has been reported. Little descriptive information is available for this creek, but Reed (1938) reported that very good prospects had been found and that mining had occurred near the mouth in the early 1900s. Joesting (1943) reported placer cinnabar on this creek. Reed (1938) reported schist bedrock near the mouth of the creek, but Dillon and others (1996) mapped Ordovician black phyllite and marble near the mouth and Ordovician to Cambrian(?) quartz-feldspar schist farther upstream. **Workings:** Heiner and Wolff (1968) reported that the creek had been worked in the 1950s. (The Alaska Resource Data File., Chandalar Quadrangle).

Emery Creek

(Gold) **Quadrangle:** Chandalar (C6); **Coordinates:** 67.5607°N-149.6073°W; **Map:** 3-2; **Worked:** 1900-1937; **Production:** 1901-1909=483 ozs; **Access:** Trail from Dalton Highway at 67.5159°N, 149.8486°W. **Summary:** Tributary of Bettles River. Only tributary of the Bettles River where more than prospecting occurred. Mining at mouth of creek in early 1900's by one man. Gold found on bedrock. Production through 1909 was 483 fine ounces. No one prospecting or mining reported after 1937 (Cobb, 1976, OF 76-340. p. 24). **Geology:** According Maddren (1913), Emery Creek had produced $10,000 in gold by 1909. The gold occurred on bedrock amidst boulders which had to be moved to recover the gold. Reed (1938) saw no evidence of the mining that Maddren had earlier reported, but he noted prospecting near the junction of Emory Creek with an east-flowing tributary (67.5494°N-149.620°W). The results of this prospecting were unknown. Bedrock in the area is mapped as Ordovician black phyllite and intercalated marble (Dillon and Reifenstuhl, 1990). (The Alaska Resource Data File., Chandalar Quadrangle).

Garnet Creek

(Gold) **Quadrangle:** Chandalar (C5); **Coordinates:** 67.5626°N-149.4119°W; **Map:** 3-4; **Worked:** 1900-1959; **Purity:** 872Au; **Yield:** 0.198 oz/ft^2; **Production:** 1900-1909=48 ozs; **Access:** Trail from landing strip at 67.5733°N, 149.3697°W. **Summary:** Tributary of Bettles River. Lower part of stream course in narrow canyon; upper part above forks in shallow basin. All mining below forks. Bedrock is schist; gravels are coarse and up to 8 feet thick. Gold is on bedrock or in lower part of gravel. Production, 1900-09, was 48 fine ounces. Said to have yielded 0.34 -0.44 ounces per man per day through 1909. A high channel had not been prospected in 1937. Ground near mouth in 0.198 oz/ft^2 of bedrock (Cobb, 1976, OF 76-340. p. 27). **Geologic description:** Garnet Creek is one of several tributaries to the Bettles River in which placer gold has been reported. The gold is in the present channel of the creek from the mouth to the forks, mostly on schist bedrock, but also in the lower few feet of the gravel (Reed, 1938). The gravel is coarse and waterworn, with many large boulders. The gravel is about 8 feet thick near the mouth of the creek and thins to 1 to 3 feet thick 1/2 mile from the mouth. Ground near the mouth of the creek was mined in 1937 and said to run $0.77 per square foot of bedrock (0.022 oz Au) (Reed, 1938). The gold was fairly fine, with a few nuggets. Reed (1938) reported the presence of a high channel, but it had not been prospected to any extent at that time. Dillon (1996) mapped Ordovician black phyllite and marble near the mouth and Ordovician to Cambrian(?) quartz-feldspar and calcareous schists farther upstream. **Workings:** Surface mining reported in 1937. Claims were staked in 1959. Gold fineness reported to be 872. (The Alaska Resource Data File., Chandalar Quadrangle).

Gold Creek

(Antimony, Gold) **Quadrangle:** Chandalar (C6); **Coordinates:** 67.5100°N-149.8200°W; **Map:**3-2; **Worked:** 1900-Present; **Purity:** NOT; **Yield:** 0.063 to 0.324 oz/ft^2; **Production:** Through 1909=11,223 ozs; **Access:** Trail from landing strip at 67.4913°N, 149.6548°W. **Summary:** Tributary of Middle Fork of the Koyukuk River. **Geology:** Bedrock is schist intruded by a diorite dike through which the stream has cut. Gold Creek originally flowed to the Middle Fork of the Koyukuk River where Linda Creek now is, but glacial drainage derangement or stream capture diverted the lower mile of its course. Gold Creek has bench placers about 8 feet above the present stream, stream placers of the present channel, and placers in a deeply buried channel (50 - 100 feet below present stream); have all been mined. Angular fragments of stibnite with quartz stringers in them have been found in stream gravels. The source of the gold is uncertain, but there has been some pyritic mineralization of the schist. The richest placers were between constrictions (caused by diorite dike and more resistant parts of schist) in valley. Some of gold is rough, but most is smooth and shotlike. Just downstream from the diorite dike gold is fairly coarse. **Workings:** Gold discovered in 1900 and mined sporadically until as recently as 1974; mining not reported between 1916 and 1934. In 1938 the value ground being worked ranged from 0.063 to 0.324 oz/ft^2. Production through 1909 was about 11,223 fine ounces (Cobb, 1976, OF 76-340. p. 28-29). Gold Creek flows through a narrow bedrock canyon, 50 to 250 feet deep, in the lower 1.5 miles of its course. Gold was originally discovered in 1900. The richest paystreaks averaged 5.3 ounces per man per day. Production in 1901 was reported to have been 5,700 ounces via hand methods. During the 1903-04 mining year only 100 ounces of gold was recovered. Gold is located on bedrock and in bedrock crevices. Gold Creek has been designated as a **Recreation Mining Area** by the State of Alaska upstream 2 miles upstream of the Dalton Highway at 67.5122°N-149.8514°W.

Granite Creek

(Gold) **Quadrangle:** Chandalar (A6); **Coordinates:** 67.01999°N-149.9700°W; **Map:** 3-6; **Worked:** 1967-1969; **Summary:** This site is at an elevation of about 1,350 ft on a principal left-limit tributary of Granite Creek, (SE1/4 sec. 4, T. 25 N., R. 11 W., of the Fairbanks Meridian). **Geologic description:** Mapping in the area shows Devonian(?) andesitic volcanic rocks and/or dioritic intrusives and discontinuous chert adjacent to a large mass of upper Paleozoic or Mesozoic granitic rocks (Brosgé and Reiser, 1964). **Workings/exploration:** Five claims staked in 1967 and suction-dredge work in 1969 (DeYoung, 1978).

Jim Pup (Gulch, Creek)

(Gold) **Quadrangle:** Chandalar (B5); **Coordinates:** 67.4739°N-149.4795°W; **Map:** 3-6; **Worked:** 1901-1990s; **Purity:** NOT; **Nugget Size:** 1.4 ozs; **Yield: Production:** 1900-1909=145 ozs; **Access:** Winter Trail from Dalton Highway at 67.5159°N, 149.8485°W. **Summary:** Tributary of California Creek. The creek has been mined from its mouth to 3/4 miles above confluence with California Creek. Gold is coarse, but the deposit is hard to mine because of rough bedrock surface. Largest nugget reported in 1938 weighed 1.4 ounces. **Geology:** The deep channel on Jim Pup is considered to be a continuation of the one on nearby Wakeup Creek and appears to generally extend downstream from the mouth of Wakeup Creek along the right limit of Jim Pup. Near the confluence of Jim Pup and Wakeup Creek the deep channel is about 55 feet deep and said to be thawed but dry (Reed, 1938). There were also reports of an 'upper' deep channel running on false bedrock on lower Wakeup Creek that extends to the confluence with Jim Pup, where it was said to be about 12 feet below the surface (Reed, 1938). It was thought that this upper deep channel might extend along the right limit of Jim Pup. There is a close spatial relationship among California, Jim Pup and Wakeup creeks, and their descriptions are commonly intermixed. **Workings:** Gold was discovered in 1901; mining in 1909, 1934, and 1938 was reported. Production, 1900-1909, was 145 fine ounces. (Cobb, 1976, OF 76-340. p. 35). Mining during those years was said to be small scale and to pay no better than wages. Mining was also reported in 1934 and 1938. Heiner and Wolff (1968) reported that the property was worked by hand methods into the 1940s and that work continued into the 1950s. Production was also noted in 1989 and 1990 (Bundtzen and others, 1990; Swainbank and others, 1991). (The Alaska Resource Data File., Chandalar Quadrangle).

King Creek

(Gold) **Quadrangle:** Chandalar (B5); **Coordinates:** 67.4566°N-149.3042°W; **Map:** 3-2; **Summary:** There is little descriptive information on this occurrence. Reed (1938) reported prospecting which reportedly found only small amounts of gold in a 70- to 75-foot shaft. A second shaft about 600 feet farther downstream from the first shaft reportedly found nothing. (The Alaska Resource Data File., Chandalar Quadrangle).

Lake Creek

(Gold) **Quadrangle:** Chandalar (B5, C5); **Coordinates:** 67.4954°N-149.4527°W; **Map:** 3-2, 3-4, 3-6; **Worked:** 1915-1990s; **Purity:** 906Au; **Nugget Size:** ; **Yield:** 0.01423 oz/ft^2; **Access:** Winter Trail from Dalton Highway at 67.5159°N, 149.8485°W. **Summary:** Tributary of Big Lake which is the old name for Bob Johnson Lake . Gold discovered in 1915. Lower part of stream is in a steep gully and has built a small delta into Big Lake. Gravel and sand mixed with schist slide rock. At head of creek, near a small lake in pass to Wakeup Creek. Gold in present channel is all coarse (Cobb, 1976, OF 76-340. p. 38). **Geology:** When Reed (1938, p. 38-40) visited the creek in 1937, the present stream had been worked from about 300 feet south of its mouth at Bob Johnson (Big) Lake to 1,500 ft south of the lakeshore. The stream in this area was 10 to 20 ft wide and the gravel was 9 to 12 ft deep. The gravels are reported to be coarse schist 'slide' rock mixed with fine sand and coarse waterworn gravel (Reed, 1938). The gold recovered from the present channel reportedly is all coarse. The values obtained were about $0.50 per square foot of bedrock (0.01423 oz/ft^2) and the gold's fineness was reported to be 906. Work on the old buried channel started in 1930 near the upper end of the creek. The depth to bedrock was 30 to 40 feet; 20 feet of this was a blue-gray mud and the rest gravel. The gold was finer than that recovered from the present channel workings and was on and in the upper 1 ft of bedrock. The bedrock was described as almost flat-lying gray schist (Reed, 1938). The gravel is coarse and waterworn with numerous erratic boulders and noticeable quantities of quartz. **Workings:** Surface mining on the present channel has been by ground sluicing and shoveling in open cuts (Reed, 1938). The deep channel was worked by drift mining. Two 38.5-ft shafts located the buried deep channel 150 to 180 feet north of the open-cut workings of the deep channel. One 76-foot shaft and one 40-foot shaft, 500 to 1000 feet from the open-cut workings on the deep channel, did not intersect that channel. Exploration activity on this creek was reported as late as 1990, but there were no details given. (The Alaska Resource Data File., Chandalar Quadrangle).

Last Chance Creek

(Gold) **Quadrangle:** Chandalar (C5); **Coordinates:** 67.5157°N-149.4721°W; **Map:** 3-2, 3-4; **Worked:** 1922-1969; **Access:** Winter Trail from Dalton Highway at 67.5159°N, 149.8485°W. **Summary:** Prospecting in 1922 yielded two grains of coarse gold at the bottom of a 30-foot-deep prospect shaft (Reed, 1938). Additional prospecting in 1932 yielded one grain of gold (said to be worth $0.05) from a 4-foot open cut near where the stream valley opens into the valley of Lake Creek. No other data are available for this occurrence. **Workings:** Last Chance Creek was prospected by at least three shafts to bedrock and an open cut. A 30-foot shaft was sunk in 1922 and 15-foot and 35-foot shafts were sunk in 1932. The 15- foot shaft reportedly was near the mouth of the creek and the 35-foot shaft was farther upstream 'at the forks'. Claims have been staked on Last Chance Creek since 1969, but there are no reports of any work. (The Alaska Resource Data File., Chandalar Quadrangle).

Linda Creek

(Gold) **Quadrangle:** Chandalar (C6); **Coordinates:** 67.5165°N-149.8327°W; **Map:** 3-2, 3-5; **Worked:** 1901-1990s; **Yield:** 0.414 oz/yd^2; **Production:** Through 1909=967=870 ozs; **Access:** Unimproved road from Dalton Highway at 67.5158°N, 149.8485°W of from landing strip at 67.5212°N, 149.8233°W. **Summary:** Tributary of the Middle Fork of the Koyukuk River. Linda Creek appears to have been a fairly consistent placer producer from the early 1900s until the present (Cobb, 1976, OFR 76-340; Bundtzen and others, 1996). Ground runs 0.414 oz/yd^2 of bedrock. Gold is in lower 3 feet of gravel and top 2 feet of bedrock. Gold was discovered in 1901 with small-scale mining reported sporadically since then. No data on total production. Production through 1909 was 967 fine ounces of which 870 ounces came from a claim about 1/2 mile above mouth. Hand and hydraulic mining in 1974 (Cobb, 1976, OF 76-340. p. 40). **Geology:** Most workers believe that the gold in Linda Creek was derived from a source other than the current Linda Creek drainage and that it most likely was the high channel on Gold Creek (Maddren, 1913). The high channel of Gold Creek probably originally drained into the Middle Fork along what is now lower Linda Creek before the current lower Gold Creek captured the drainage. The upper part of Linda Creek (above the map site where the creek abruptly turns to the west) runs in a wide, flat valley over glacial outwash material. Reed (1938) noted that no prospects had ever been found in that part of the drainage. The lower part of the creek (below where it turns to the west) has cut through the glacial fill and runs in a shallow cut in the underlying schist. Reed (1938) also noted that about 1/4 mile below where Linda Creek intersects the old Gold Creek channel the bedrock pitches off steeply to an unknown depth and that no values have been found below this point. In 1937 a small piece of the old channel that had not been found in the early days was being worked. This occurrence was described as 8 feet of gravel over a hard schist bedrock. The gravel was waterworn, fine material with few boulders but much heavy sediment in the sand. The gold occurred in the lower 3 feet of the gravel and upper 2 feet of the bedrock. The gold was said to be fairly fine. Maddren (1913) noted that the character of the gold was similar to that in Gold Creek. The fineness of the gold averaged about 945. **Workings:** There is little description of the early mining except that nearly all the gold was mined from the Discovery claim about 1/2 mile above the mouth, but only the lower half of this claim was very productive. Reed (1938) described a 'shoveling in' surface mining operation. Hand and hydraulic mining was reported in 1974, and in 1993 and 1996 underground drift mining was continuing 'as it has for many years' (Bundtzen and others, 1994; Bundtzen and others, 1996). Production through 1909 was estimated to have been approximately $20,000 (Maddren, 1913); almost all of it was in 1902. The ground in a portion of the old channel was reported to run $0.95 per square foot of bedrock (Reed, 1938). (The Alaska Resource Data File., Chandalar Quadrangle).

Little Gold Creek

(Gold) **Quadrangle:** Chandalar (B6); **Coordinates:** 67.4978°N-149.6490°W; **Map:** 3-2, 3-5; **Access:** Trail from landing strip at 67.4913°N, 149.6548°W or by trail from the Dalton Highway at 67.5159°N, 149.8488°W. **Summary:** Tributary of Gold Creek that heads against Eider Creek (Cobb, 1976, OF 76-340. p. 41). Reed (1938, p. 27-28) reported that a 76-foot shaft to bedrock was sunk about 200 feet from the mouth of Little Gold Creek, and although some gold was found, it was not in amounts sufficient to be workable. No other descriptive information is available on this occurrence. (The Alaska Resource Data File., Chandalar Quadrangle).

Little Squaw Creek

(Gold, Lead, Monazite, Tungsten) **Quadrangle:** Chandalar (C3); **Coordinates:** 67.5699°N-148.1600°W; **Map**: 3-1, 3-3, 3-8; **Worked:** 1905-Present; **Purity:** 848Au; **Access:** Trail from landing strip at 67.5798°N, 148.1687°W or winter trail from Middle Fork of the Chandalar River at 67.5266°N, 147.8058°W. **Summary:** Tributary of Little Squaw Lake. History similar to that of Squaw Creek. Complex glacial history. **Minerals:** Gold, pyrite, hematite, arsenopyrite, scheelite, galena, and monazite. **Geology:** Little Squaw Creek drains an area of auriferous quartz veins of the Chandalar district that are the presumed source of the placer gold. Mertie (1925, p. 254-259, 263) described the complex glacial history for the creek. The creek was dammed by ice in the North Fork Chandalar River valley during part of Pleistocene, and this damming resulted in both pre- and post-glacial channel and bench deposits along Little Squaw Creek. The creek and bench gravels are mixed, with no distinct boundaries. The upper part of the drainage is described as having gold on bedrock, while farther downstream the pay streak runs onto false bedrock in glacial gravels (Mertie, 1925). Post-glacial deposits, mainly along the lower course of the stream, are composed of 25 to 50 ft of glacial till overlain by 55 to 100 ft of gravel. The pre-glacial deposits, mainly along the upper stream course, are about 35 ft deep, with coarse gold in the upper 2 to 3 ft of bedrock. **Workings:** Surface and underground workings. Mostly small-scale drift mining in the old channels. Gold was discovered on Little Squaw Creek in 1905, and mining was reported in most years through 1940, with further activity noted as late as the mid-1980s. In 1983, bulk sampling indicated average grades of $11 to $16 per cubic yard; 0.0275-0.400 oz/yd^3; (gold at $400 per ounce) (Alaska Construction and Oil, 1984). Average fineness of the gold was reported as 848 (Mertie, 1925, p. 258). **Reserves:** Alaska Construction and Oil (1984) reported that Canadian Barranca Ltd., Inc., the operator in 1984, considered placer reserves in Tobin Creek, Little Squaw Creek, and two additional unnamed creeks in the area to be at least 100,000 ounces with potential for as much as 500,000 ounces. (The Alaska Resource Data File., Chandalar Quadrangle).

Magnet Creek

(Gold) **Quadrangle:** Chandalar (B6, C6); **Coordinates:** 67.5100°N-149.7299°W; **Map:** 3-2; **Worked:** -1974; **Access:** Trail from landing strip at 67.4913°N, 149.6548°W or by trail fron the Dalton Highway at 67.5159°N, 149.8488°W. **Summary:** Tributary of Gold Creek. Has been open-cut mining near mouth. High channel about 50 feet above creek. Creek bed has been mined for 500 - 600 feet between Gold and Magnet Creeks. No evidence of mining in 1974 (Cobb, 1976, OF 76-340. p. 45).

Marion Creek

(Gold) **Quadrangle:** Chandalar (B6); **Coordinates:** 67.3363°N-149.9522°W; **Map:** 3-2, 3-6; **Worked:** 1899-1909; **Production:** 1900-1909=48 ozs; **Summary:** Tributary of the Middle Fork of the Koyukuk River. Gold reported 1899. Good prospects said to have been found on bedrock at bottom of a 26-foot shaft 6 or 7 miles above mouth. Colors said to have been found near surface of gravels in valley. Production through 1909 was 48 fine ounces (Cobb, 1976, OF 76-340. p. 46).

Minnie Creek

(Gold) **Quadrangle:** Chandalar (B6); **Coordinates:** 67.4235°N-149.8952°W; **Map:** 3-2, 3-6; **Worked:** 1900s-1909; **Production:** 1909=48 ozs; **Access:** From Dalton Highway at 67.4201°N, 150.0768°W **Summary:** The prospect is approximately 2 miles south of Minnie Creek Lake. The reference point is at the junction of the tributary from Minnie Creek Lake and Minnie Creek (sec. 13, T. 30 N., R. 11 W., of the Fairbanks Meridian). This point is near the upper end of an area of placer activity extending downstream to the confluence of Minnie Creek and Howard Creek. A smaller stretch along Minnie Creek about 2.5 to 3 miles upstream from the reference point (sec. 21, T. 30 N., R. 11 W., of the Fairbanks Meridian) has also been placered. **Geology:** Minnie Creek was said to have been heavily prospected in the early 1900s, but water problems made reaching bedrock difficult and the early mining was not very successful (Reed, 1938). A shaft to bedrock in 1904 reportedly produced $500 of heavy 'shot' gold from a short drift before it was flooded. Drift mining was concentrated in a deep channel, although some gold was taken out at higher levels. No prospects were reported on the present channel or on a high channel, although several high benches indicate the presence of a high channel. **Workings:** Considerable prospecting and some drifting were reported in the early 1900s, but water on bedrock made prospecting and drift mining difficult. (The Alaska Resource Data File., Chandalar Quadrangle). Minnie Creek has been designated as a **Recreation Mining Area** by the State of Alaska upstream 1.5 miles of the Dalton Highway at 67.4203°N-150.0763°W.

Mule Creek

(Copper, Gold, Silver) **Quadrangle:** Chandalar (C5);**Coordinates:** 67.5994°N-149.3532°W; **Map:** 3-2, 3-4; **Worked:** 1900s-1938; **Purity:** 908Au; **Yield:** 0.00286 oz/ft2; **Production:** Through 1909=48 ozs; **Access:** Trail from landing strip at 67.5732°N, 149.3707°W. **Summary:** Tributary of the Bettles River. Shafts 75 - 80 feet deep in fan near mouth did not reach bedrock. Small nuggets of native silver and nuggets of native copper up to 7 lbs. Recorded production through 1909 was about 48 fine ounces (Cobb, 1976, OF 76-340. p. 49). **Geologic description:** Mule Creek is one of several tributaries to the Bettles River in this area that are reported either to have produced gold or to have good prospects. Maddren (1913) described good prospects 2 1/2 and 1 1/2 miles upstream from the mouth of Mule Creek. At 2 1/2 miles, there is a low bench in 8 feet of gravel; at 1 1/2 miles, gold was found in 2 feet of gravel which yielded $1 per man-hour (0.0483 oz/hr). Reed (1938) reported that the gravel was coarse and waterworn, with many large boulders, and that depth to bedrock was 3 to 8 feet. The gold is on bedrock and in the lowest foot of gravel; it is well worn and has a fineness of 908. Values of approximately 10 cents per square foot of bedrock were reported in 1938. Mule Creek heads in limestone. Dillon and others (1996) map this limestone as Devonian Skajit Limestone, which is in contact with Ordovician to Cambrian(?) calc-schist and Ordovician phyllite and marble. (The Alaska Resource Data File., Chandalar Quadrangle).

Myrtle Creek

(Gold) **Quadrangle:** Chandalar (A6, B6); **Coordinates:** 67.2360°N-149.9901°W; **Map:** 3-6; **Worked:** 1899-Present; **Purity:** 886-914Au; **Nugget Size:** 23 ozs; **Yield:** 0.0100-0.0343 oz/ft^2; **Access:** Trail from **Myrtle Creek** landing strip at 67.2151°N, 149.9862°W or by unimproved road and trail from **the Dalton Highway** at 67.2532°N, 150.1792°W. **Summary:** Tributary of Slate Creek. mined for approximately 6 miles above the confluence with Slate Creek. One of biggest producers in district. **Geology:** Gold was first discovered on Myrtle Creek in 1899 about 2 1/2 miles above the mouth (Maddren, 1913). The most productive ground was reported to be 2 to 3 miles above the mouth. Gold occurs in the Myrtle Creek drainage in the gravels of the present channel and in bench gravels along the lower 2 to 3 miles of the creek. The present channel was, according to Maddren (1913), 100 to 300 feet wide, although Reed (1938) puts the width at 30 to 150 feet. In general the gold was distributed across the present stream bed. The gravels were generally 2 to 4 feet thick and as much as 7 feet thick. Maddren (1913) described the gold as shot or wheat gold, although he speculated that much of the fine gold was not being recovered at the time. The gold was coarse, clean, and somewhat flattened (like melon seeds); occasional nuggets as much as about 1 oz were found. Smith (1942) reported recovery of a 23-oz nugget. The gold was said to be generally concentrated on bedrock and in crevices of the schist bedrock down to about 3 feet below the gravel/ bedrock contact. Reed (1938) reported active mining in 1937 from both the present channel and from benches. In at least two locations along the creek, Reed described the gold as having both as a fine and flaky component and coarse and well-worn pieces. At one of these locations, a bench about 2 3/4 miles from the mouth, Reed also noted fine and flaky gold was present all through the gravel and coarser, worn gold (with occasional small nuggets) was on or in the top 3 feet of bedrock. The gravels are generally reported to be slabby schist with varying numbers of greenstone boulders. The bedrock was reported to be graphitic to siliceous schist and slate with at least one observable greenstone dike; this is consistent with later mapping (Brosgé and Reiser, 1964). **Workings::** Mining began on Myrtle Creek in 1899. Mining in the early days was by open cut and later by drifting, principally on the benches. Hydraulicking was begun in 1909. Nearly all evidence of the early mining had disappeared by 1937. A dragline and bulldozers were in use in the early 1950s. Gravels are thawed near the present channel but frozen near the benches. Gold fineness from three locations along the creek was reported to be about 886 from one location and 914 from the other two (Reed, 1938). Exploration activity was noted in 1984 (Eakins and others, 1985) and development activity as late as 1992 (Swainbank and others, 1993). **Production:** Myrtle Creek has reportedly been one of the largest placer gold producers in the Koyukuk district. The total production is not known, but production through 1910 was estimated by Maddren (1913) to be $182,000 (8,805 ozs). Reed (1938) reported that at least $20,000 (967 ozs) was mined in the early days from a short section above Kelly's Pup in one summer. Values reported by Reed (1938) ranged from $1.20 (0.0343 oz/ft^2) to $0.35 (0.0100 oz/ft^2) per square foot of bedrock. The most productive part of the stream was reported by Maddren (1913) to be 2 to 3 miles above Slate Creek at the mountain front. The uppermost claims along the creek (above No. 15) were not profitable at that time. Mining has continued sporadically from 1899 until the present (1998). The most recent production from this property was noted in 1995 (Bundtzen and others, 1996). (The Alaska Resource Data File., Chandalar Quadrangle).

Nugget Creek

(Gold) **Quadrangle:** Chandalar (B6); **Coordinates:** 67.4800°N-149.8549°W; **Map:** 3-2, 3-5; **Worked:** 1901-; **Yield:** 0.0025 - 0.01 oz/yd^3 ; **Access:** From Dalton Highway at 67.4901°N-149.8756°W. **Summary:** Tributary of the Middle Fork of the Koyukuk River. Fair gold prospects have been found, but there has been very little mining (Cobb, 1976, OF 76-340. p. 53). Gold was discovered in Nugget Creek in 1901. Nugget Creek has been designated as a **Recreation Mining Area** by the State of Alaska upstream 1 mile on east side of the Dalton Highway at mile marker 67.4901°N-149.8756°W. Average yield of gravel is estimated to be 0.0025 - 0.01 oz/yd^3 of gold.

Nutirwik Creek

(Gold) **Quadrangle:** Chandalar (D6); **Coordinates:** 67.9467°N-149.6716°W; **Map:** 3-7; **Worked:** 1976-; **Access:** From Dalton Highway at 67.9345°N-149.8186°W. **Summary:** The occurrence is approximately 4 miles above the mouth of Nutirwik Creek (sec. 18, T. 36 N., R. 9 W., of the Fairbanks Meridian). Nutirwik Creek is a west-flowing tributary to Dietrich River. **Geology:** The country rocks in the area are shown by DeYoung (1978) to be principally Devonian Hunt Fork Shale and Devonian to Silurian Skajit Limestone with lesser Devonian slate, phyllite, limestone, and siltstone. No descriptive information is available. **Workings:** The only descriptions of this occurrence are references by the U.S. Bureau of Mines (1973) and Dillon (1987) to evidence of placer mining activity and by the location of claims in 1976. (The Alaska Resource Data File., Chandalar Quadrangle).

Our Creek

(Gold) **Quadrangle:** Chandalar (D3); **Coordinates:** 67.9467°N-149.6716°W; **Map:** 3-8; **Worked:** 1900s-; **Access:** Boat from the North Fork of the Chandalar River at 67.75213°N-148.3113°W. **Summary:** Tributary of the North Fork of the Chandalar River. Exact location of the occurrence is not known. **Geology:** Our Creek drains an area mapped as gneissic biotite granite (presumably part of the Baby Creek batholith), mica schist, calcareous schist and marble, and calc-silicate hornfels developed near the granitic rocks (Brosgé and Reiser, 1964). **Workings:** Reed (1930, MR 31-4) reported that very encouraging prospects had been found on Our Creek in the early days but that there was no report of any mining or any description of the deposit. (The Alaska Resource Data File., Chandalar Quadrangle).

Phoebe Creek

(Gold) **Quadrangle:** Chandalar (C5); **Coordinates:** 67.5738°N-149.1441°W; **Map:** 3-2; **Worked:** 1900s-; **Summary:** Tributary of the Bettles River. Reed (1938) reported that there is no record of prospecting on Phoebe Creek itself but that very good prospects were found on a right-limit tributary by Manual Mello in the early days 1900s. No other information is available on this occurrence. The location is probably near the cabin opposite Mellow creek named after Manual Mello. **Geology:** The occurrence is in an area underlain by Ordovician to Cambrian(?) feldspar-quartz schist, which has been intruded by biotite granite gneiss and aplite phases of the Devonian Baby Creek batholith (Dillon and others, 1996). Calcareous and siliceous hornfels have been developed near the intrusive contacts, and this entire assemblage is in thrust contact with Lower Paleozoic to Proterozoic units that are predominantly banded quartzite and graphitic schist. (The Alaska Resource Data File., Chandalar Quadrangle).

Rainbow Gulch

(Gold) **Quadrangle:** Chandalar (B6); **Coordinates:** 67.4695°N-149.9509°W; **Map:** 3-2, 3-5; **Worked:** 1900s; **Access:** From Dalton Highway at 67.4695°N-149.9509°W. **Summary:** Tributary of the Middle Fork of the Koyukuk River. Reed (1938, p. 30) reported that the creek had good prospects and speculated that the creek would be a producer, given improved freight rates to the area. **Geology:** Mapping shows the creek to drain an area underlain predominantly by Devonian siltstone and grit (Brosgé and Reiser, 1964). (The Alaska Resource Data File., Chandalar Quadrangle).

Robert Creek

(Gold) **Quadrangle:** Chandalar (C5, D5); **Coordinates:** 67.6355°N-149.1910°W; **Map:** 3-2; **Summary:** Robert Creek is one of the principal headwater tributaries to the Bettles River and Sheep Creek is a significant right-limit tributary to Robert Creek. The mouth of Sheep Creek is approximately 3 1/2 miles upstream from the Bettles River on Roberts Creek. The literature is unclear as to which (and how many) tributaries to Robert Creek contain gold. Because Sheep Creek is a tributary to Robert Creek and has similar gold placers, the reference point is arbitrarily placed at the confluence of Robert and Sheep creeks. **Geology:** Maddren (1913) reported that gold prospects had been found on Sheep Creek and 'especially those tributary to Robert Creek', along with a number of other creeks around the headwaters of the Bettles River. Maddren also noted that the gold was found on the lower parts of the tributaries to Roberts Creek, where miners had cut down to the schist underlying the limestones and that only prospect work had been done. No other descriptive information as to any further work, if any, is available. Recent mapping shows that lower Roberts Creek occupies a valley between significant granitic bodies that make up the Devonian Horace Mountain plutons to the north and the Baby Creek batholith to the south (Dillon and others, 1996). These plutons intrude Ordovician to Cambrian(?) schists and marbles. (The Alaska Resource Data File., Chandalar Quadrangle).

Rock Creek

(Gold) **Quadrangle:** Chandalar (C3); **Coordinates:** 67.5897°N-149.2677°W; **Map:** 3-1, 3-8; **Summary:** Rock Creek is a tributary to the Middle Fork Chandalar River. Rock Creek heads against the head of Day Creek, a tributary to Big Creek. **Geologiy:** Reed (1930, MR 31-4, p. 10) noted that encouraging prospects had been found in the early 1900s but that no mining had occurred and no prospecting was taking place at the time of his investigation. The area of Rock Creek is mapped as quartz-muscovite schist (Brosgé and Reiser, 1964). (The Alaska Resource Data File., Chandalar Quadrangle).

Sawlog Creek

(Gold) **Quadrangle:** Chandalar (B5); **Coordinates:** 67.4057°N-149.1700°W; **Map:** 3-2; **Worked:** 1964; **Summary:** Tributary of the South Fork of the Koyukuk River. Recent placer mining (as of 1964) Mining is just north of the low pass that separates Sawlog Creek from Dennys Gulch.(Cobb, 1976, OF 76-340. p. 60). **Geologyn:** Brosgé and Reiser (1964) reported recent placer mining in 1964, but there is no other information regarding this placer occurrence. The area is mapped as quartz-mica schist with intercalated greenschist and greenstone. Sawlog Creek heads against Dennys Gulch, in which there is documented placer activity, and both creeks are just south of the southern end of the Baby Creek batholith (DeYoung, 1978). **Workings:** Williams (1952) reported that the results of a magnetometer survey on Sawlog Creek looking for a pay streak were generally inconclusive, but because the survey was performed at the request of the property owners it suggests some activity at that time. Heiner and Wolff (1968) reported one claim located in 1960. (The Alaska Resource Data File., Chandalar Quadrangle).

Shady Creek

(Gold) **Quadrangle:** Chandalar (C5); **Coordinates:** 67.5904°N-149.2696°W; **Map:** 3-2, 3-4; **Worked:** 1930-; **Summary:** Tributary of the Bettles River. Lower part of stream course in limestone with schist farther upstream. Fair prospects said to have been found. Gus Creek was the old name for Shady Creek (Cobb, 1976, OF 76-340. p. 31). **Geology:** Reed (1938) reported only fair prospects. He described the creek as flowing through a limestone canyon on its lower end, but widening with bedrock changing to schist about 1/4 mile above the mouth. Dillon and others (1996) mapped the geology of the lower mile of the creek as Ordovician to Cambrian marble and two small bodies of Devonian biotite granite gneiss that are part of the Baby Creek batholith. A Devonian calcareous hornfels surrounds one of the granitic bodies. Rocks adjacent to the upper portion of the creek are mapped as Ordovician to Cambrian(?) feldspar-quartz schist. (The Alaska Resource Data File., Chandalar Quadrangle).

Shamrock Creek

(Gold) **Quadrangle:** Chandalar (B5); **Coordinates:** 67.4836°N-149.3745°W; **Map**: 3-2, 3-6; **Worked:** 1937-; **Summary:** Tributary of Bob Johnson (Big) Lake. Old name for Shamrock Creek was Butte Creek. Fair prospects found, 1937 (Cobb, 1976, OF 76-340. p. 14).

Sheep Creek (Gulch)

(Gold) **Quadrangle:** Chandalar (B6); **Coordinates:** 67.4963°N-149.8064°W; **Map:** 3-2, 3-5; **Worked:** 1900s-Present; **Yield:** 0.435-1.74 oz/yd^2; **Production:** NOT ozs; **Access:** Trail from landing strip at 67.5079°N, 149.8496°W or from the Dalton Highway at the same coordinates. **Summary:** Tributary of the Middle Fork of the Koyukuk River. Coarse gravel on schist bedrock was mined in the stream course and in a buried channel that extends into the valley of the Middle Fork of the Koyukuk. Mining was reported in early 1900's, 1930's, and 1950's or early 1960's. Production through 1909 was 96 fine ounces. One small area returned about 1.74 oz/yd^2 of bedrock. Mining in 1936 was confined to a deep channel, about 90 feet below surface in one place. It was drift mined for 7,000 feet upstream from Middle Fork valley. Average yield was 0.435 oz/yd^2 of bedrock. Coarse gold has been recovered from the present stream channel (Cobb, 1976, OF 76-340. p. 61). **Geology:** Recent mapping shows the area to be underlain by Devonian black slate, phyllite and phyllitic siltstone, chloritic siltstone and grit, and quartz-muscovite schist (Brosgé and Reiser, 1964). (The Alaska Resource Data File., Chandalar Quadrangle). Sheep Creek has been designated as a **Recreation Mining Area** by the State of Alaska upstream 2 miles on east side of the Dalton Highway at 67.5013°N-149.8609°W. Average yield of gravel is estimated to be 0.0025 - 0.01 oz/yd^3 of gold.

Slate Creek

(Gold) **Quadrangle:** Chandalar (B5); **Coordinates:** 67.4836°N-149.3745°W; **Map**: 3-6, 3-8; **Worked:** 1899-1993; **Production:** 1900-1909=145 ozs; **Access:** Trail from Myrtle Creek landing strip at 67.2152°N, 149.9856°W **Summary:** Tributary of the Middle Fork of the Koyukuk River. Mining in this area probably extended about 1 mile both up- and downstream from the reference point. There was also mining activity on lower Slate Creek below its confluence with Myrtle Creek in the Wiseman quadrangle. Mining began in 1899. Gold probably locally derived from quartz veinlets; large veins are barren. Most of the gold was from below Myrtle Creek in the Wiseman quadrangle, but there is some in bench gravel farther upstream. **Geology:** Much of the descriptive information available for Slate Creek does not distinguish between the lower part of the creek that lies in the Wiseman quadrangle and the upper part of the creek that lies in the Chandalar quadrangle. Reed (1938) reported that no deep channel was found on Slate Creek although placer gold was found on Slate Creek in the present channel and in high benches. On the upper creek (the part in the Chandalar quadrangle) the present channel was worked in the short canyon just below the reference point (about 2 miles east of the Myrtle Creek landing strip), and the old-timers were said to have rockered out $8 per day (0.387 oz/day) (gold at $20 per ounce). The present channel and a low left limit bench were also worked from a short distance above the canyon for a distance of 1 1/2 miles. The bench was said to have bedrock only a few feet above the bedrock in the present channel. Bedrock is described as mica schist and slate or phyllite with small bodies of quartz and cut by several altered diorite dikes (this description may apply principally to the lower two-thirds of the creek). Upper Slate Creek lies along the boundary between Devonian quartz-muscovite schist and a Devonian(?) sequence of graywacke, phyllite, and volcanic rocks and chert (Brosgé and Reiser, 1964). **Workings:** Mined mostly by simple hydraulic and hand methods; bulldozer and sluice plate in use in 1974. Stripping reported in preparation for mining in 1993. (The Alaska Resource Data File., Chandalar Quadrangle).

Slate Creek, Grave Creek Tributary

(Gold) **Quadrangle:** Chandalar (B5); **Coordinates:** 67.5899°N-147.9799°W; **Map**: 3-1; **Worked:** 1900s-; **Access:** Trail from Chandalar landing strip at 67.5029°N, 148.4899°W. Trail crosses Slate Creek at 67.5768°N, 147.9831°W. It can also be accessed by trail from the Middle Fork of the Chandalar River at 67.5268°N, 147.8063°W. **Summary:** The reference point for the occurrence is on the lower reaches of Slate Creek. **Geology:** Reed (1930, MR 31-4, p. 11) reported that very encouraging prospects had reportedly been found on Slate Creek in the early 1900s. No other information is available. The creek drains an area mapped as Devonian black slate, phyllite and phyllitic siltstone, and muscovite schist and chloritized amphibole schist (Brosgé and Reiser, 1964). (The Alaska Resource Data File., Chandalar Quadrangle).

St. Marys Creek (Gulch)

(Gold) **Quadrangle:** Chandalar (C3); **Coordinates:** 67.5187°N-148.2092°W; **Map:** 3-1, 3-3, 3-8; **Worked:** 1906-1986; **Yield:** 0.435-1.74 oz/yd^2; **Production:** NOT ozs; **Access:** Trail from landing strip at 67.5079°N, 149.8496°W **Summary:** St. Marys Creek drains the Chandalar mine area and joins Big Creek near the lower limit of mining on Big Creek. Has been one of the major producing creeks of the Chandalar area. Gold is bright with average grain size of gold is about 1 mm; many nuggets are 2-3 mm. Mining was reported for 1906-09, and 1914-15; there was probably some in other years also (Cobb, 1976, OF 76-340. p. 59). **Minerals:** Monazite(?), magnetite, hematite, rutile, pyrite, arsenopyrite, chalcopyrite, galena, stibnite, molybdenite, scheelite, and uranothorianite. **Geology:** Maddren (1913) noted that most of the gold that had been produced in the Chandalar district up to that time had come from upper Big Creek and St. Marys Gulch. Despite its production, there is little information describing St. Marys Creek, which drains the area of gold quartz lode deposits in the Chandalar district. Chipp (1970) noted that pan samples from Big Creek and St. Marys Gulch show the gold to be bright yellow and irregular shot-like to flattened; a few crystals occur. Some nuggets include quartz crystals and iron oxides. **Workings:** Most of the early mining was by drifting or by hand. Heiner and Wolff (1968) reported drift mining in the winter of 1915. Some exploration work, including drilling and sampling, was reported in 1986 on several properties in the area, including St. Marys Creek (Bundtzen and others, 1987). (The Alaska Resource Data File., Chandalar Quadrangle).

Tobin Creek

(Gold, Lead, Monazite, Tungsten) **Quadrangle:** Chandalar (B3, C3); **Coordinates:** 67.5299°N-148.3000°W; **Map:** 3-1, 3-3, 3-8; **Worked:** 1930s-1991; **Production: 1983=**2,129 ozs; **Access:** Trail from Chandalar landing strip at 67.5025°N, 148.4911°W or trail from the Middle Fork of the Chandalar River at 67.5268°N, 147.8063°W. **Summary:** Tributary of Chandalar Lake. Creek drains area where lode deposits have been developed. Prospecting was first reported in 1930 and mining in 1934. Mining continued until World War II and was reported since then in 1952 and 1969. Concentrates contain hematite, monazite, scheelite, gold, pyrite, magnetite, rutile, and galena. For general description of area see entries for Big Creek (Cobb, 1976, OF 76-340. p. 69). **Minerals:** Hematite, monazite, scheelite, pyrite, magnetite, rutile, and galena (White, 1952). **Geology:** Tobin Creek drains the area of the Mikado mine, one of several gold-quartz lode deposits in the Chandalar district and the Mikado fault, along which the Mikado vein occurs. The stream has probably been one of the principal producers in the Chandalar area, but there is only a modest amount of information about this mine. Initial prospecting began in 1930 and was encouraging. Mining was begun by at least 1934. The pay zone is reported to be more than 100 feet thick in thawed gravel. **Workings:** Prospecting with encouraging results began in 1930. Mining was reported sporadically from 1930 through as late as 1991. Bulldozer and hydraulic operations were set up in the 1950s. **Production notes:** Little production data are available, but the total has probably been significant. Heiner and Wolff (1968) reported that there had been moderate production from 1957 until 1968. Production of 2,129 ounces in 1983, said to be mostly from Tobin Creek but partially from Little Squaw Creek, was said to be more than double the 1982 production, and production in 1984 was expected to double again (Alaska Construction and Oil, 1984). Chandalar Development Corp. mined for two consecutive years on Tobin Creek and was reported to be the largest producer in the northern Alaska region (Bundtzen and others, 1992). Reserves: Alaska Construction and Oil (1984) reported that Canadian Barranca Ltd., Inc., the operator in 1984, considered placer reserves in Tobin Creek, Little Squaw Creek, and two additional unnamed creeks in the area to be at least 100,000 ounces with potential for as much as 500,000 ounces. (The Alaska Resource Data File., Chandalar Quadrangle).

Trembley Creek

(Gold) **Quadrangle:** Chandalar (D6); **Coordinates:** 67.9348°N-149.8890°W; **Map:** 3-7; **Worked:** 1963-; **Access:** West from the Dalton Highway at 67.9292°N, 149.8278°W **Summary:** The occurrence is near the confluence of Trembley Creek and Kuyuktuvuk Creek about 1 1/2 miles west of the Dietrich River. **Geology:** Two claims were reported to have been located in 1963, and there is evidence of placer mining activity, principally along Kuyuktuvuk Creek (U.S. Bureau of Mines, 1973; Dillon, 1987). Country rocks in the area are shown by DeYoung (1978) to be principally Devonian limestone and siltstone, Pennsylvanian and Mississippian limestone, dolomite, shale, and conglomerate of the Lisburne Group and Kayak Shale, and Devonian Hunt Fork Shale. No other information is available on this occurrence. **Workings:** Indications of placer mining activity noted along Kuyuktuvuk Creek above the mouth of Trembley Creek (U.S. Bureau of Mines, 1973; Dillon, 1987). Heiner and Wolff (1968) reported that claims were located at the mouth of Trembley Creek in 1963. (The Alaska Resource Data File., Chandalar Quadrangle).

Trilby Creek

(Gold) **Quadrangle:** Chandalar (D6); **Coordinates:** 67.4060°N-148.0522°W; **Map:** 3-1, 3-8; **Worked:** 1910-; **Access:**

Boat from the Middle Fork of the Chandalar River at 67.3977°N, 148.0346°W **Summary: Geologic description:** Reed (1930, MR 31-4, p. 10) noted that encouraging prospects had been found in the early days (about 1910) but that no mining had occurred and no prospecting was taking place at the time. (The Alaska Resource Data File., Chandalar Quadrangle).

Unnamed Creek

(Gold) **Quadrangle:** Wiseman (C6); **Coordinates:** 67.5935°N-149.7872; **Map:** 3-2; **Summary:** Tributary of the Middle Fork of the Koyukuk River. Designated a **Recreation Mining Area** by the State of Alaska upstream 2 miles on the east side of the Dalton Highway at mile marker 202.

Wakeup Creek

(Gold) **Quadrangle:** Chandalar (B5); **Coordinates:** 67.4815°N-149.4785°W; **Map:** 3-2, 3-6; **Worked:** 1930s-; **Yield:** 1.52 to 1.75 oz/yd^2; **Production:** NOT ozs; **Access:** Trail from lthe Dalton Highway at 67.5160°N, 149.8485°W **Summary:** Tributary of Jim Pup. Present channel does not cut bedrock; was mined in a small way in the early days, but is now covered with tailings from other operations. A deep channel is a continuation of the one in Jim Pup. It has been mined upstream from Jim Pup for 1,500 feet and deepens upstream. It is from 55 to 112 feet deep and 15 to 25 feet wide in hard schist. Gravel is about 5 feet thick and contains rough, but generally fine gold. The channel average between 1.52 to 1.75 oz/yd^2 of bedrock. High channel appears to have flowed to Big Lake (reverse of present direction) from Jim Pup. Gravel is coarse and subangular. No production by 1937, but active development underway. This channel averaged about 0.216 oz/ft^2 of bedrock. Reported mining in 1930's (Cobb, 1976, OF 76-340. p. 71).

Tanana Quadrangle

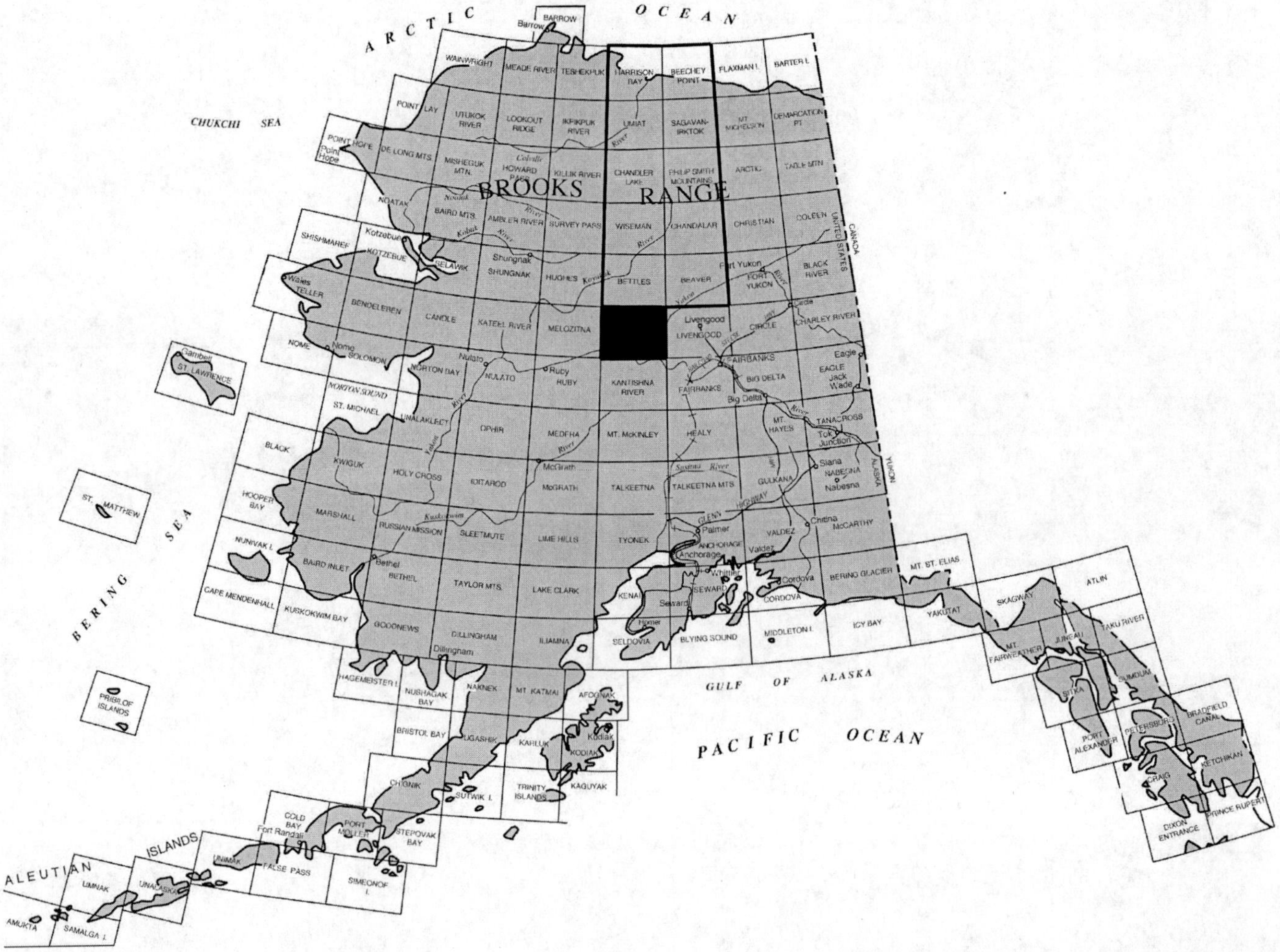

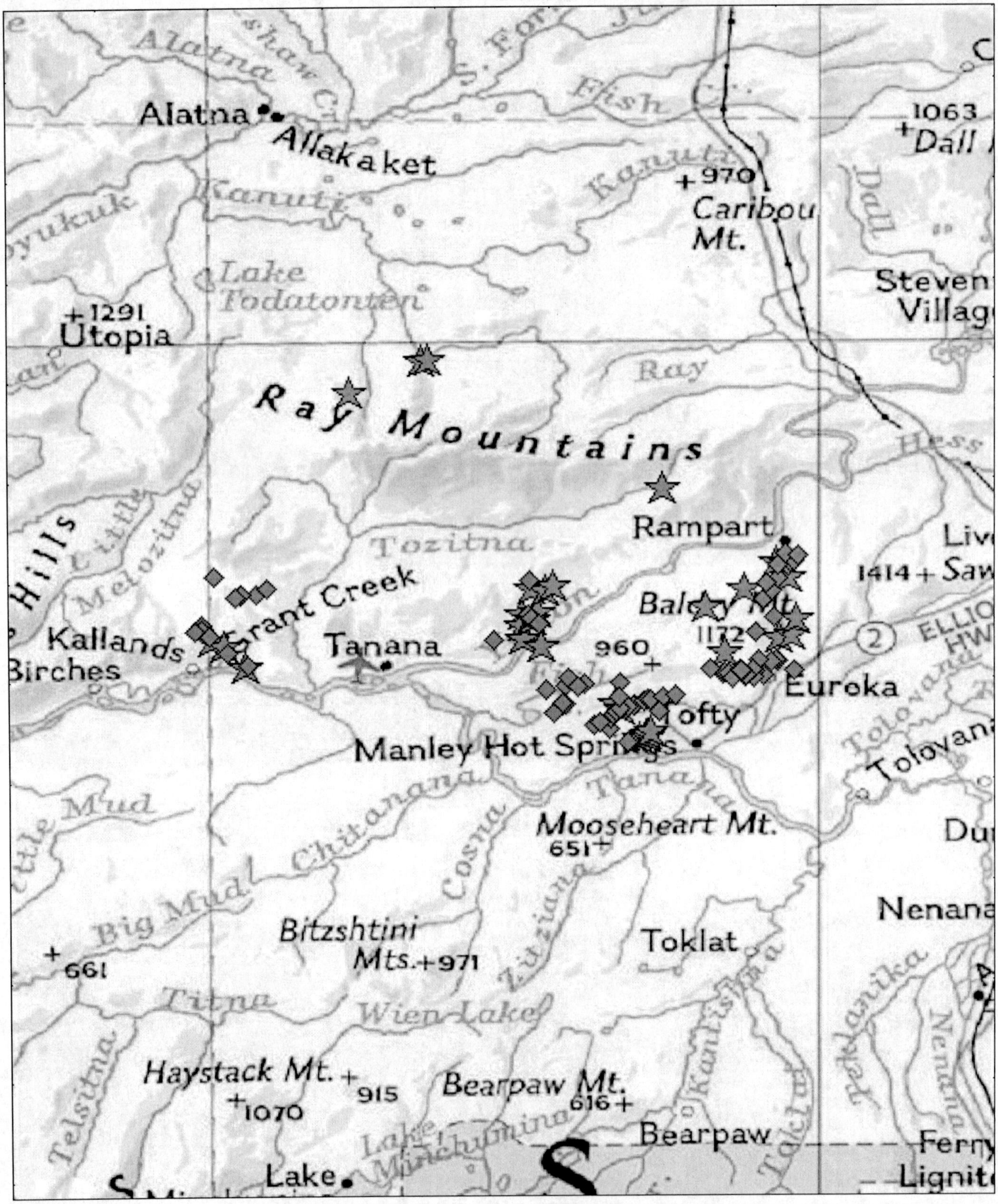

Gold Placer Locations in the Tanana Quadrangle

♦ Diamonds shows placer gold locations. ★ Stars show gold-bearing hard rock locations. Hard rock occurrences are not identified by name nor is a description of the occurrence included in this book. They are provided so the reader can see where the hard rock sources for the placer deposits occur. Areas are located by latitude and longitude WGS 84 coordinates.

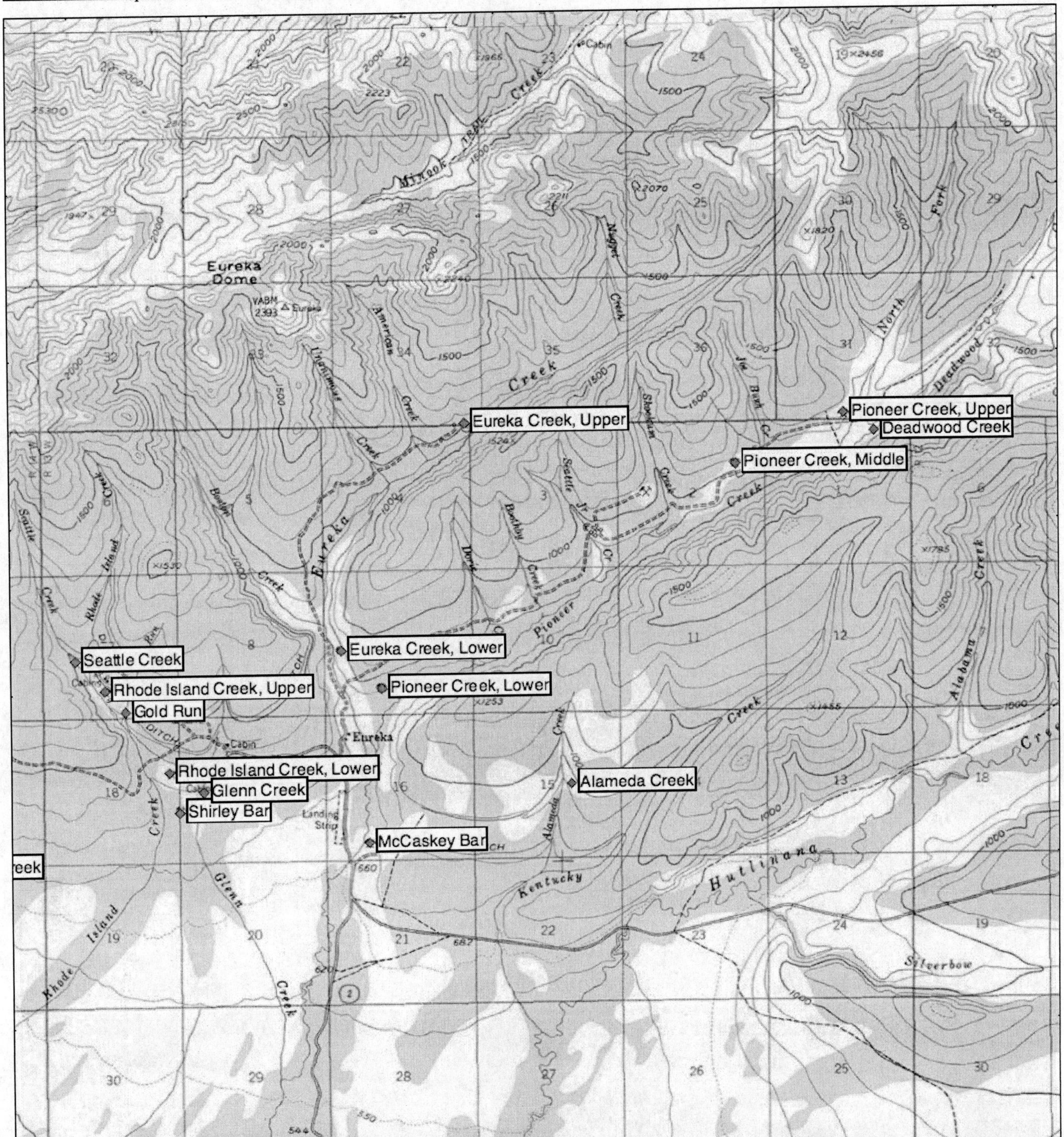
Eureka Dome
Eureka Creek, Upper
Pioneer Creek, Upper
Deadwood Creek
Pioneer Creek, Middle
Eureka Creek, Lower
Seattle Creek
Pioneer Creek, Lower
Rhode Island Creek, Upper
Gold Run
Rhode Island Creek, Lower
Alameda Creek
Glenn Creek
Shirley Bar
McCaskey Bar
reek

Map 4-1

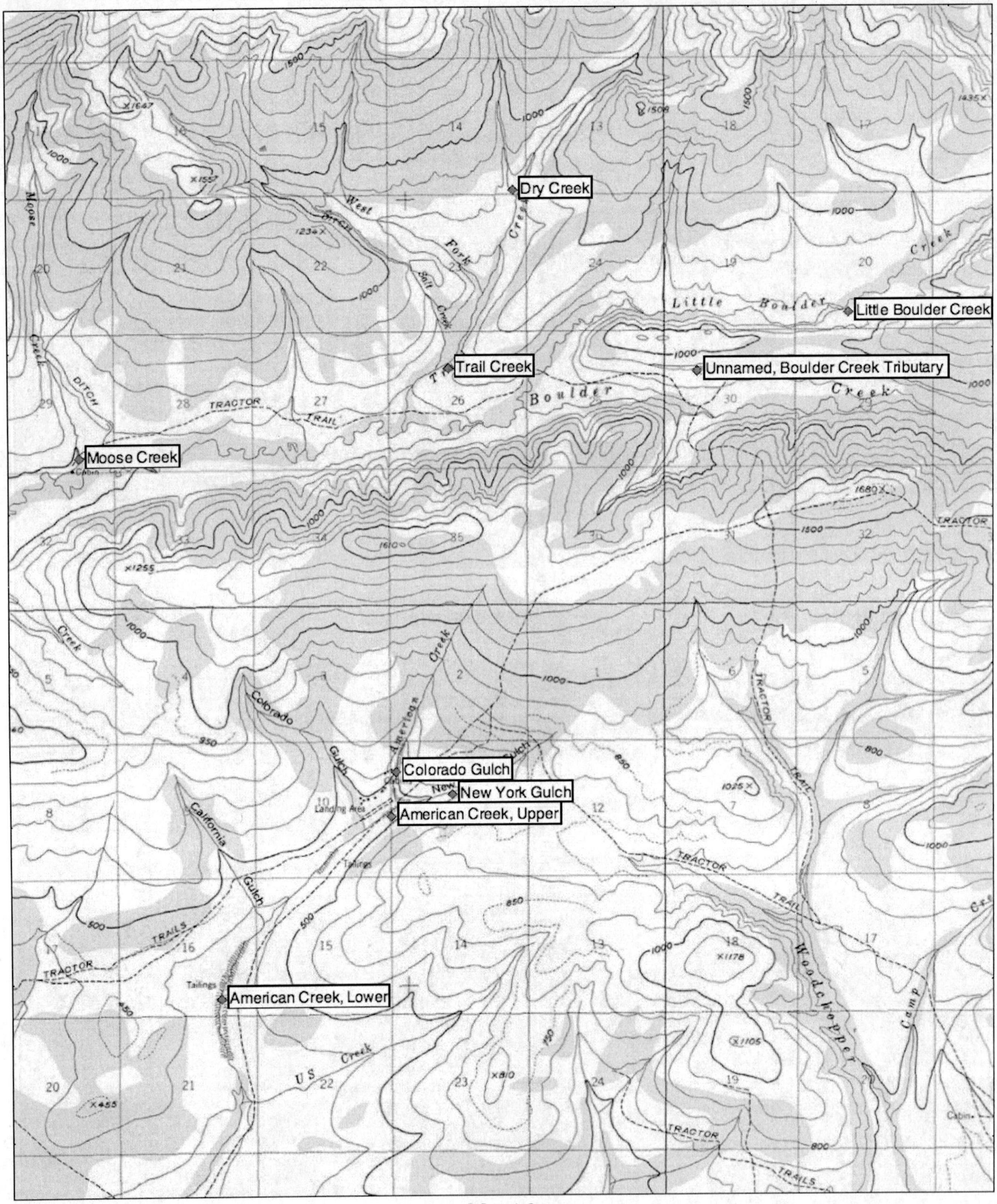
Dry Creek
Little Boulder Creek
Trail Creek
Unnamed, Boulder Creek Tributary
Moose Creek
Colorado Gulch
New York Gulch
American Creek, Upper
American Creek, Lower

Map 4-2

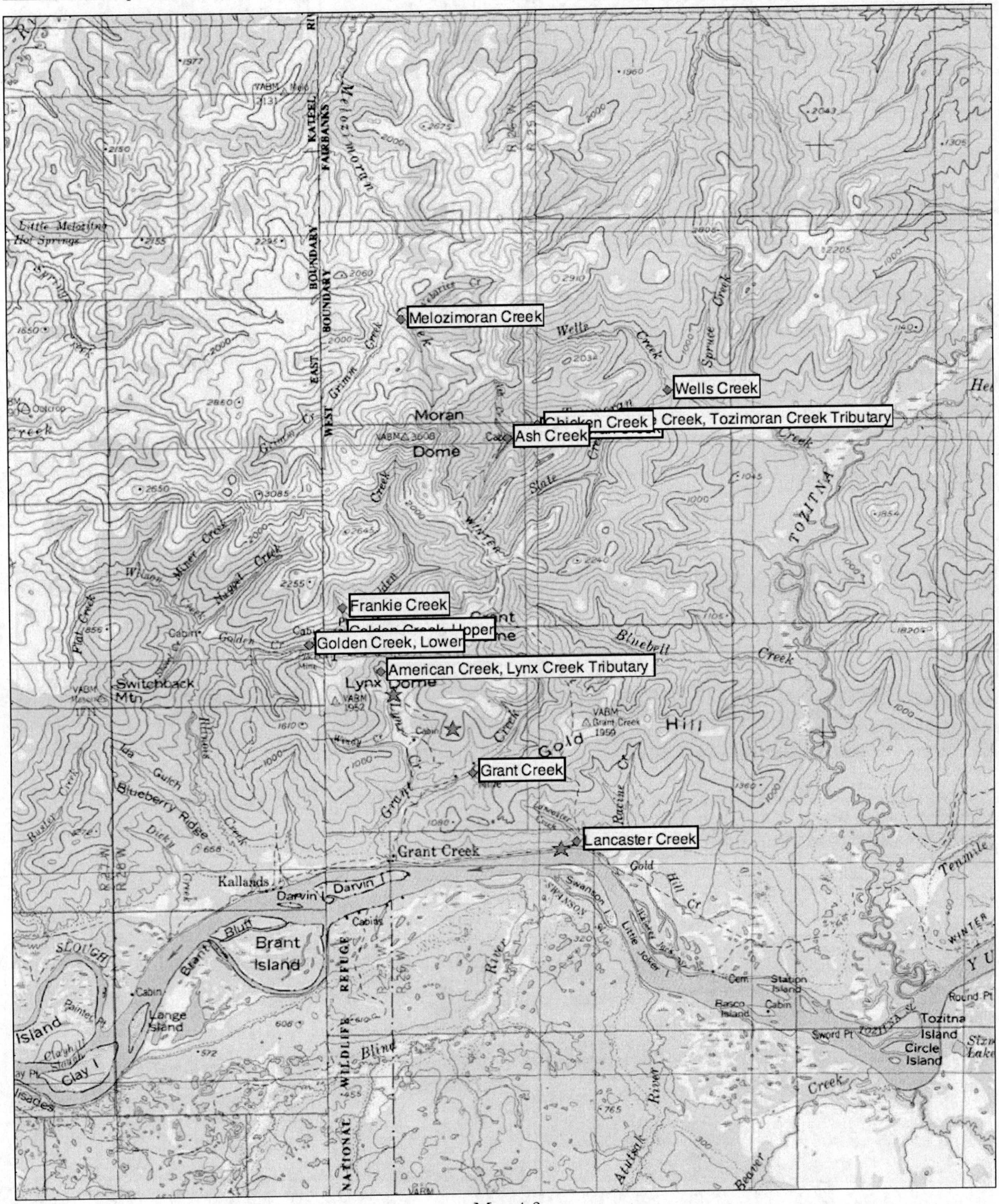
Melozimoran Creek
Wells Creek
Chicken Creek
Creek, Tozimoran Creek Tributary
Ash Creek
Frankie Creek
Golden Creek, Upper
Golden Creek, Lower
American Creek, Lynx Creek Tributary
Grant Creek
Lancaster Creek
Moran Dome
Lynx Dome
Switchback Mtn
Gold Hill
Brant Island
Lange Island
Clay I
Tozitna Island
Circle Island
Station Island
Kallands
TOZITNA
NATIONAL WILDLIFE REFUGE
WEST EAST BOUNDARY
KATEEL RIV
FAIRBANKS

Map 4-3

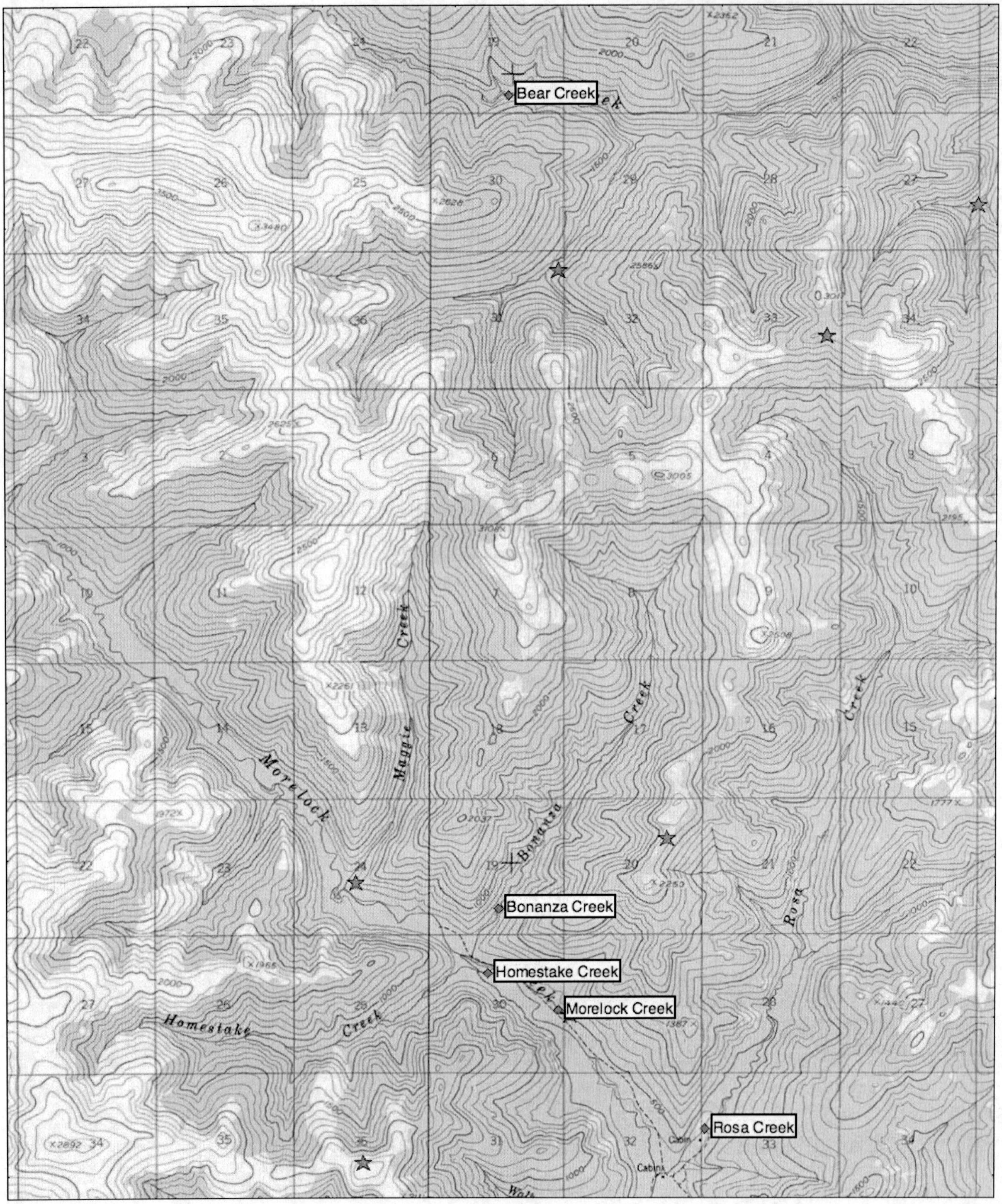

Bear Creek
Bonanza Creek
Homestake Creek
Morelock Creek
Rosa Creek
Morelock
Maggie
Creek
Bonanza
Creek
Rosa
Creek
Homestake
Creek
Cabin
Cabin

Map 4-4

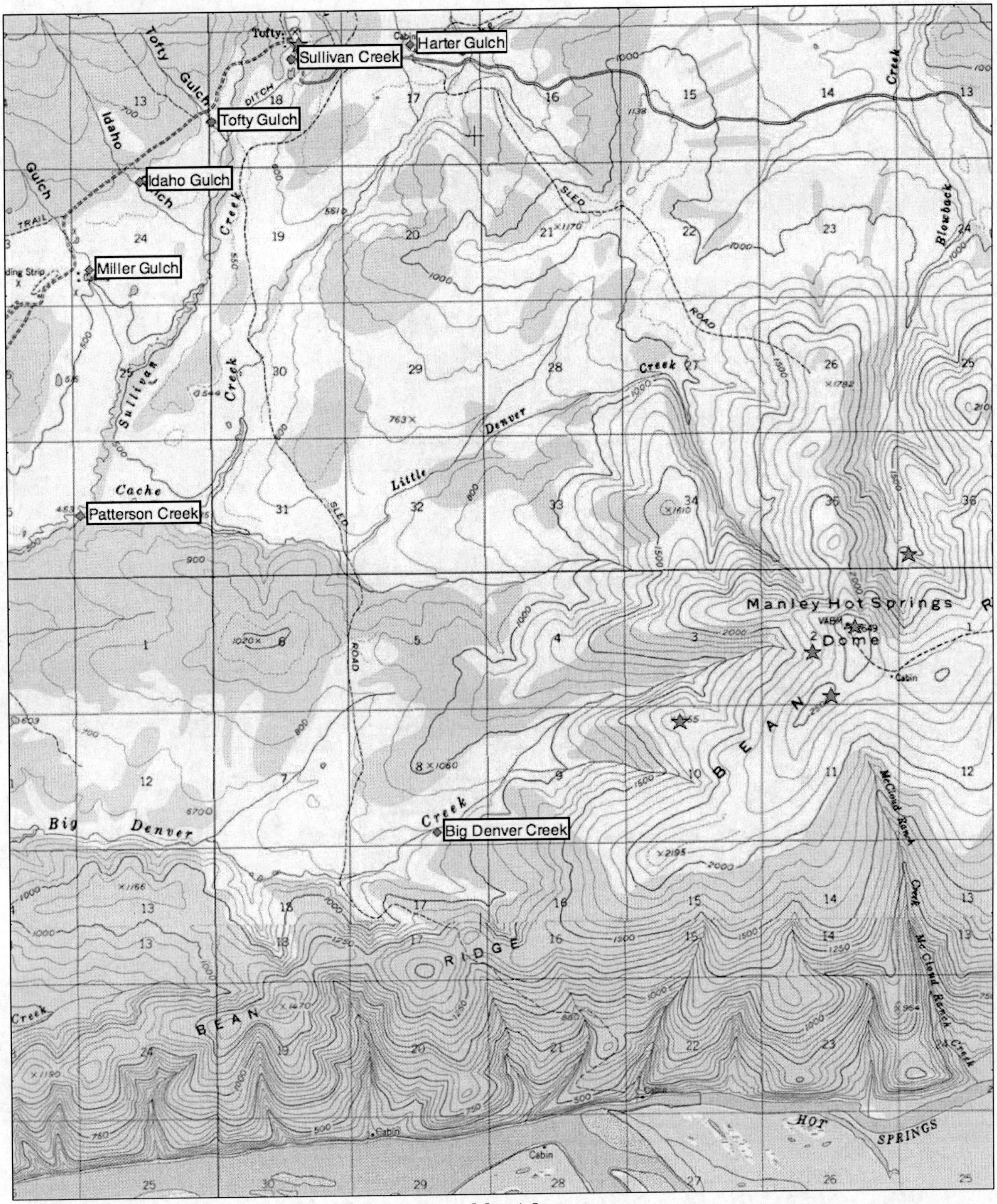
Tofty
Harter Gulch
Sullivan Creek
Tofty Gulch
Idaho Gulch
Miller Gulch
Patterson Creek
Big Denver Creek
Manley Hot Springs Dome
BEAN RIDGE
Little Denver Creek
Cache
Big Denver Creek
Blowback Creek
McCloud Ranch Creek
HOT SPRINGS
SLED ROAD
TRAIL
DITCH
Cabin

Map 4-5

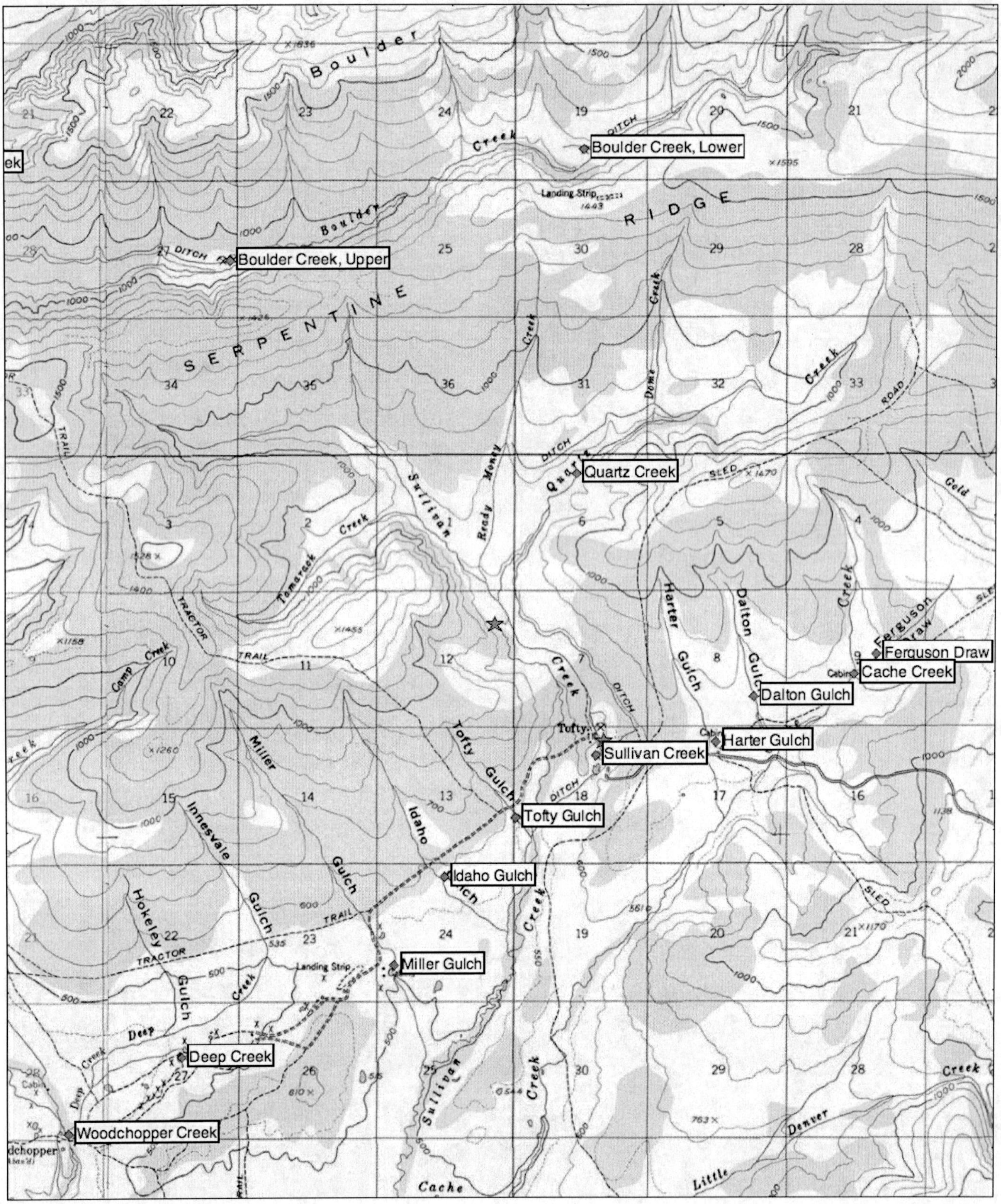
Boulder Creek, Lower
Boulder Creek, Upper
SERPENTINE
RIDGE
Quartz Creek
Ferguson Draw
Cache Creek
Dalton Gulch
Harter Gulch
Sullivan Creek
Tofty Gulch
Idaho Gulch
Miller Gulch
Deep Creek
Woodchopper Creek
Landing Strip
Tofty
Harter Gulch
Dalton Gulch
Tamarack
Camp Creek
Miller
Idaho
Innesvale
Hokeley
TRACTOR
TRAIL
DITCH
SLED
ROAD
Little
Denver
Cache

Map 4-6

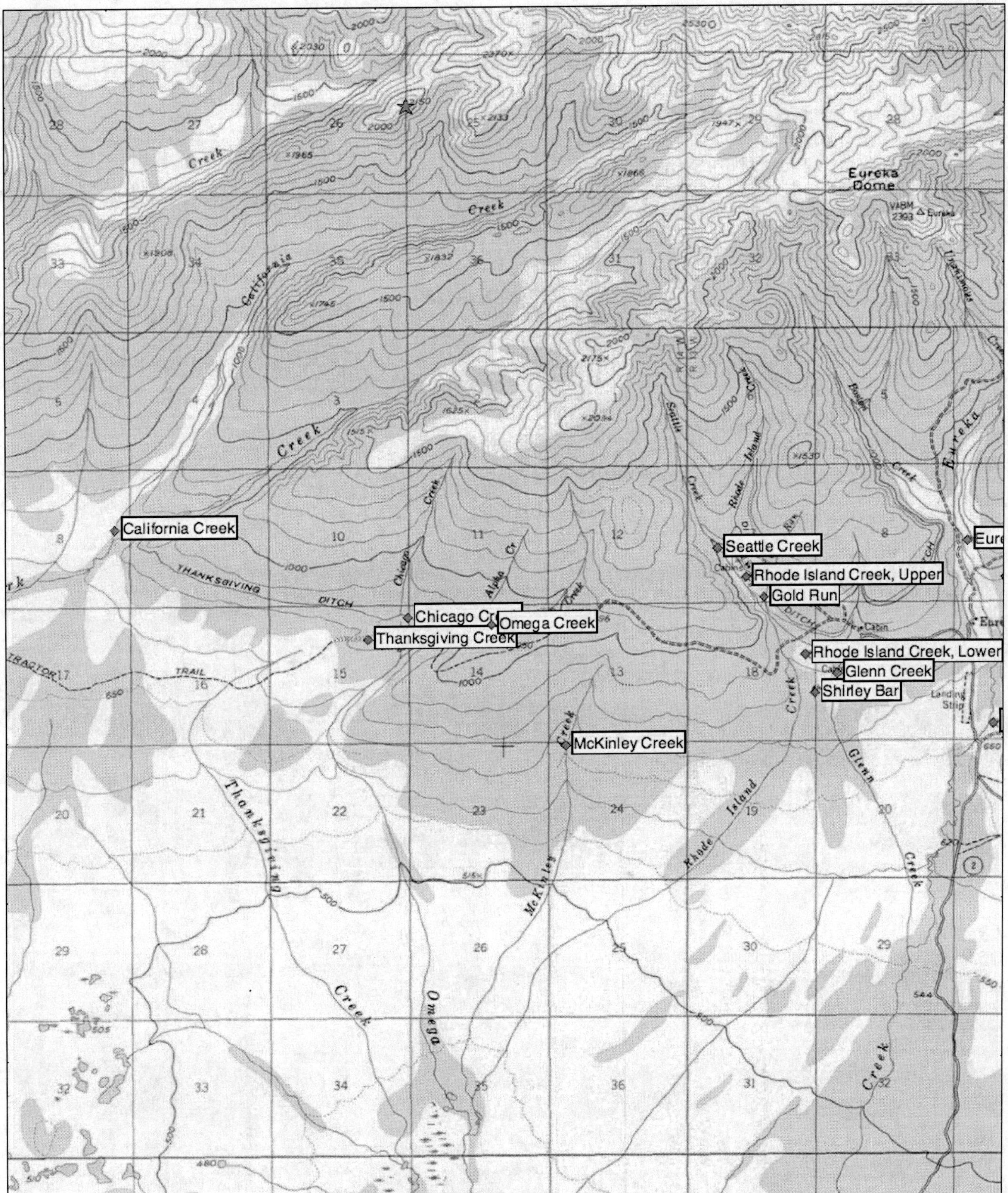
Eureka Dome
California Creek
Seattle Creek
Rhode Island Creek, Upper
Gold Run
Chicago Creek
Omega Creek
Thanksgiving Creek
Rhode Island Creek, Lower
Glenn Creek
Shirley Bar
McKinley Creek
THANKSGIVING DITCH
TRACTOR TRAIL
Thanksgiving Creek
Omega
McKinley
Rhode Island
Glenn Creek
Landing Strip

Map 4-7

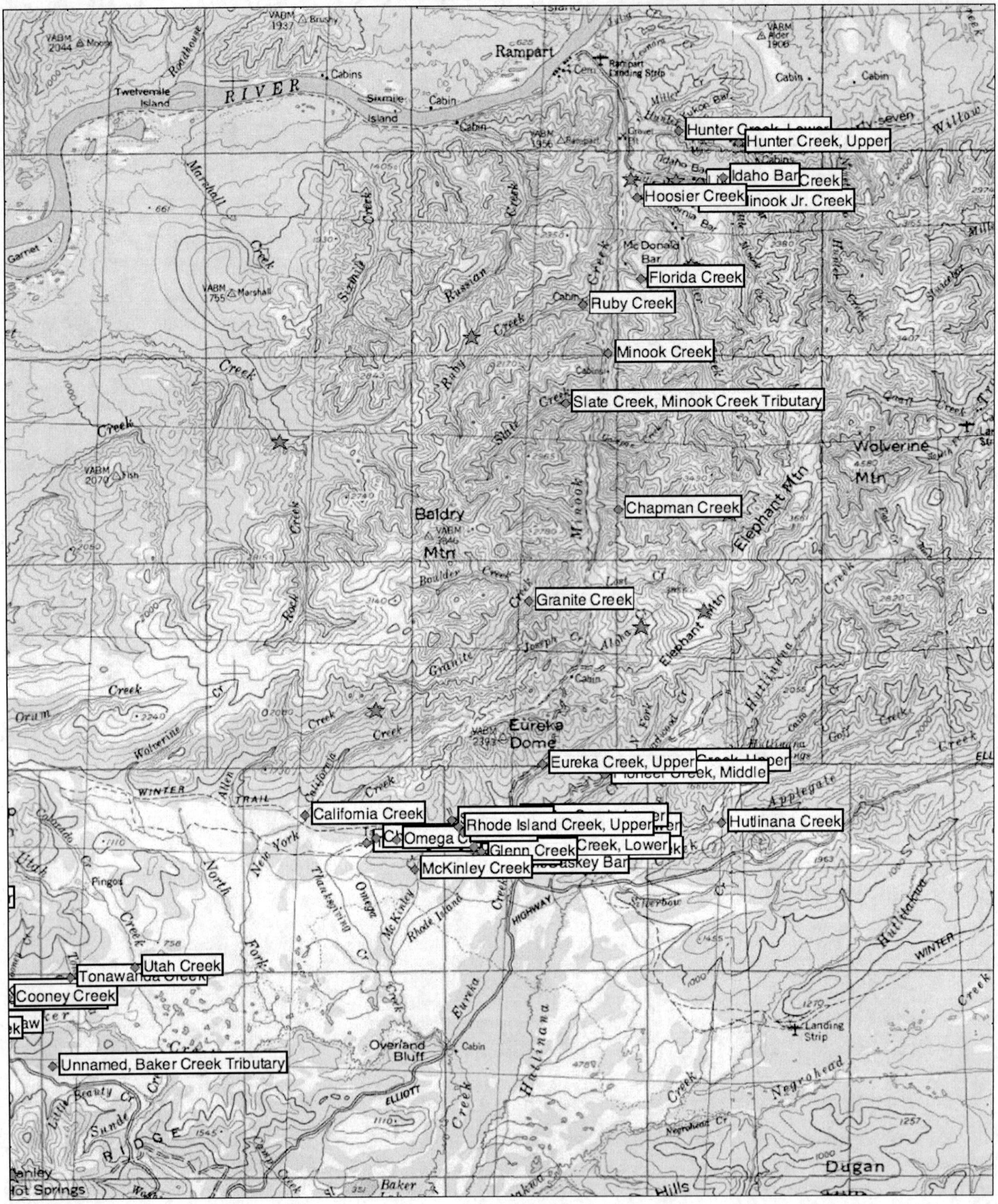
Rampart
RIVER
Twelvemile Island
Sixmile Island
Hunter Creek, Upper
Idaho Bar
Hoosier Creek
Florida Creek
Ruby Creek
Minook Creek
Slate Creek, Minook Creek Tributary
Wolverine Mtn
Chapman Creek
Baldry Mtn
Elephant Mtn
Granite Creek
Eureka Dome
Eureka Creek, Upper
California Creek
Rhode Island Creek, Upper
Glenn Creek
Hutlinana Creek
McKinley Creek
Utah Creek
Cooney Creek
Unnamed, Baker Creek Tributary
Overland Bluff
Landing Strip
Dugan
Hot Springs

Map 4-8

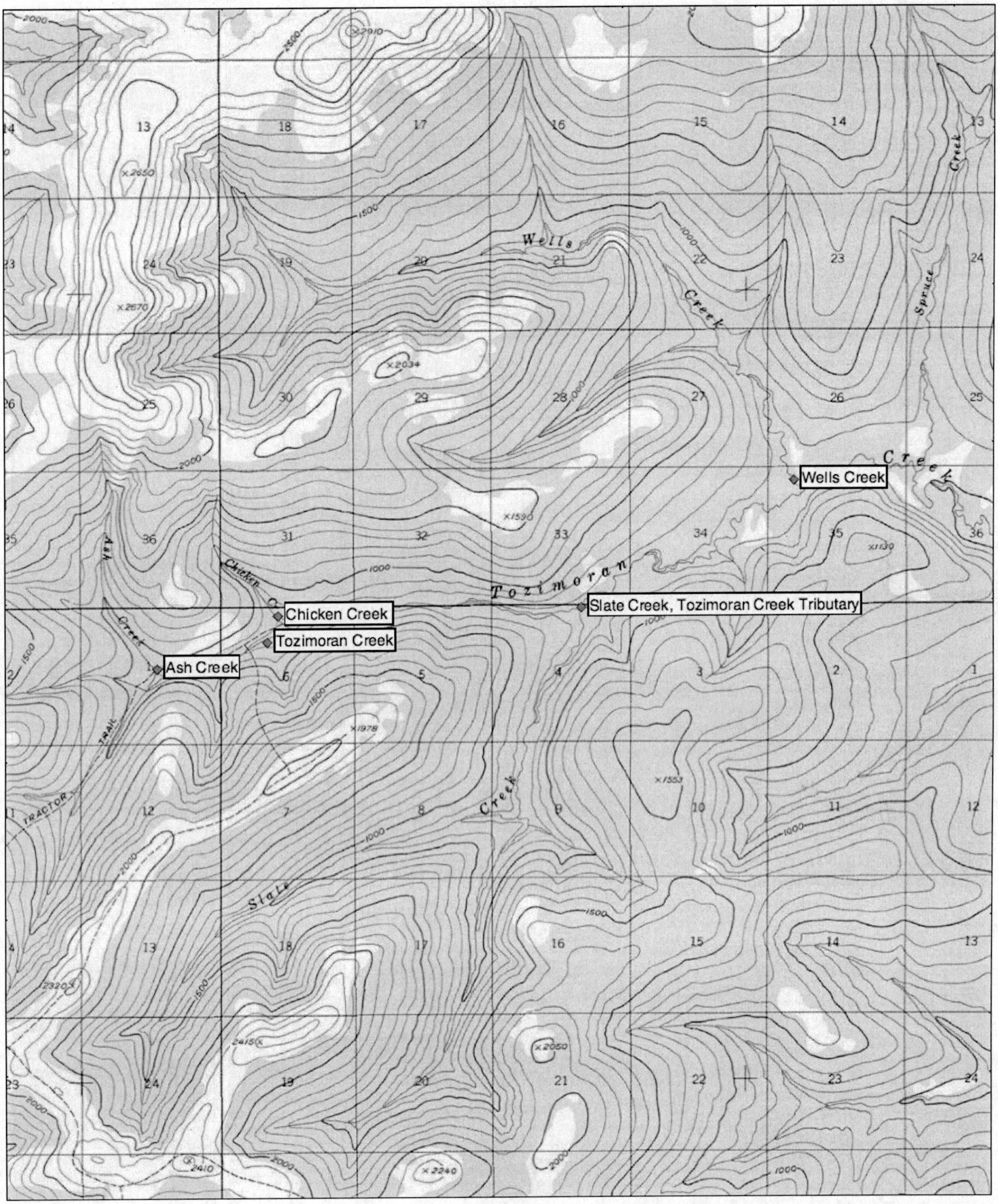
Wells Creek
Slate Creek, Tozimoran Creek Tributary
Chicken Creek
Tozimoran Creek
Ash Creek
Tozimoran
Wells Creek
Spruce Creek
Slate Creek
Ash Creek
Chicken Cr
Tractor Trail

Map 4-9

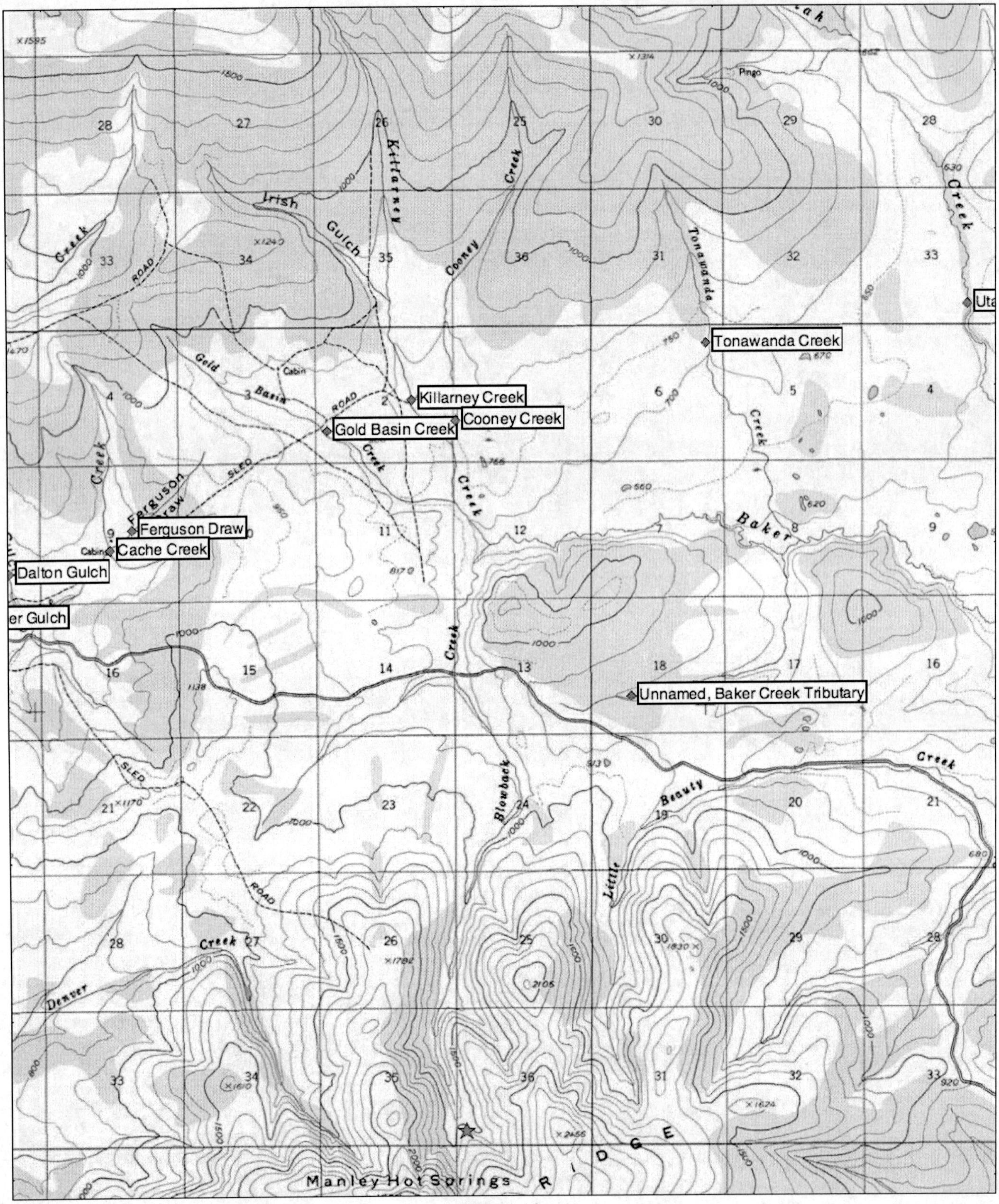
Tonawanda Creek
Killarney Creek
Cooney Creek
Gold Basin Creek
Ferguson Draw
Cache Creek
Dalton Gulch
er Gulch
Unnamed, Baker Creek Tributary
Uta
Irish Gulch
Killarney
Cooney Creek
Tonawanda
Gold Basin
Baker
Blowback
Little Beauty
Creek
Denver
Cabin
Pingo
ROAD
SLED
Manley Hot Springs RIDGE

Map 4-10

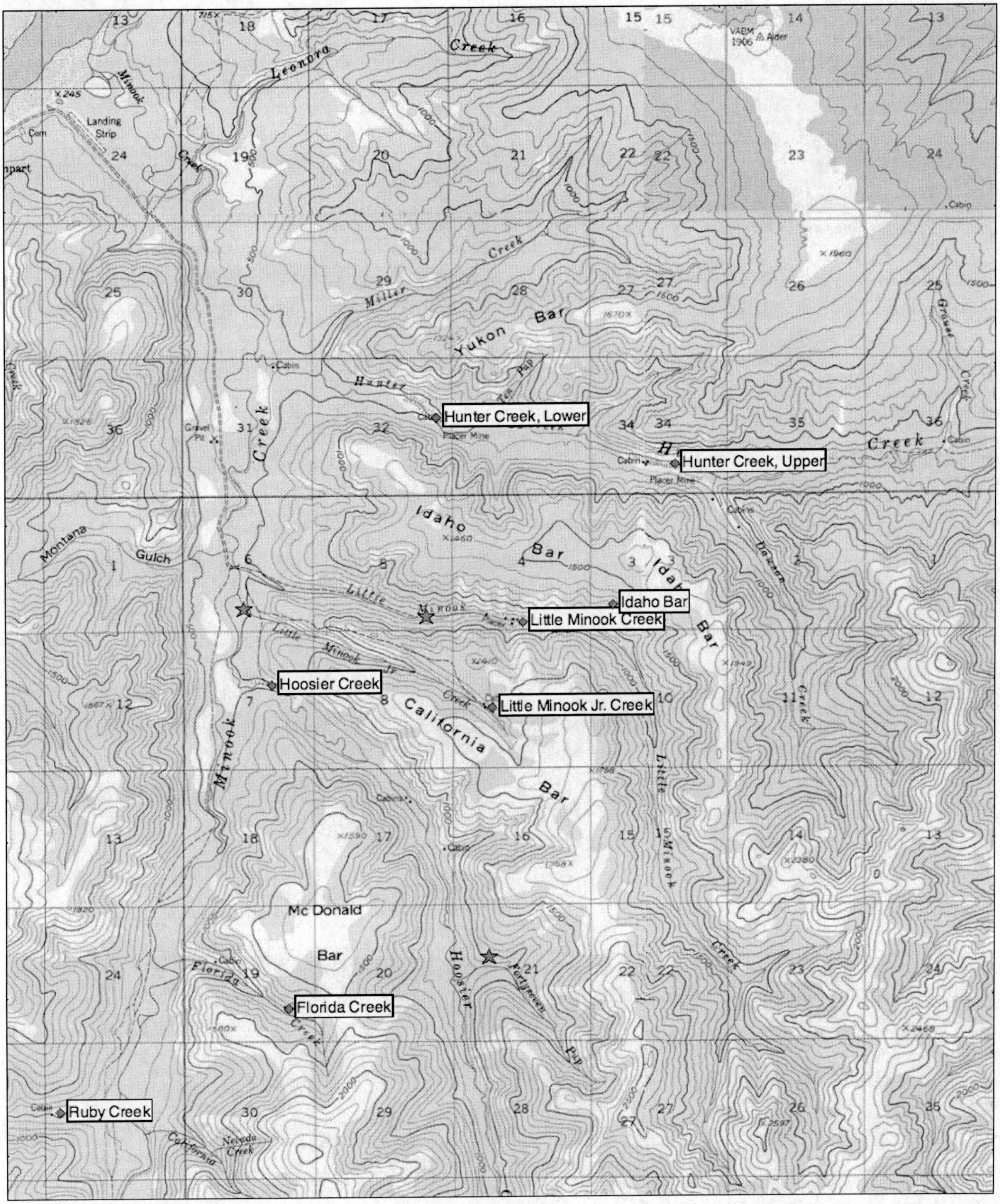
Hunter Creek, Lower
Hunter Creek, Upper
Idaho Bar
Little Minook Creek
Hoosier Creek
Little Minook Jr. Creek
Florida Creek
Ruby Creek
Leonora Creek
Miller Creek
Yukon Bar
Hunter Creek
Idaho Bar
Little Minook Creek
Little Minook Jr. Creek
California Bar
Minook Creek
Montana Gulch
Mc Donald Bar
Florida Creek
Hoosier
Forgotten Pup
Dawson Creek
Landing Strip
Gravel Pit
Placer Mine
Cabins
Nevada Creek
California

Map 4-11

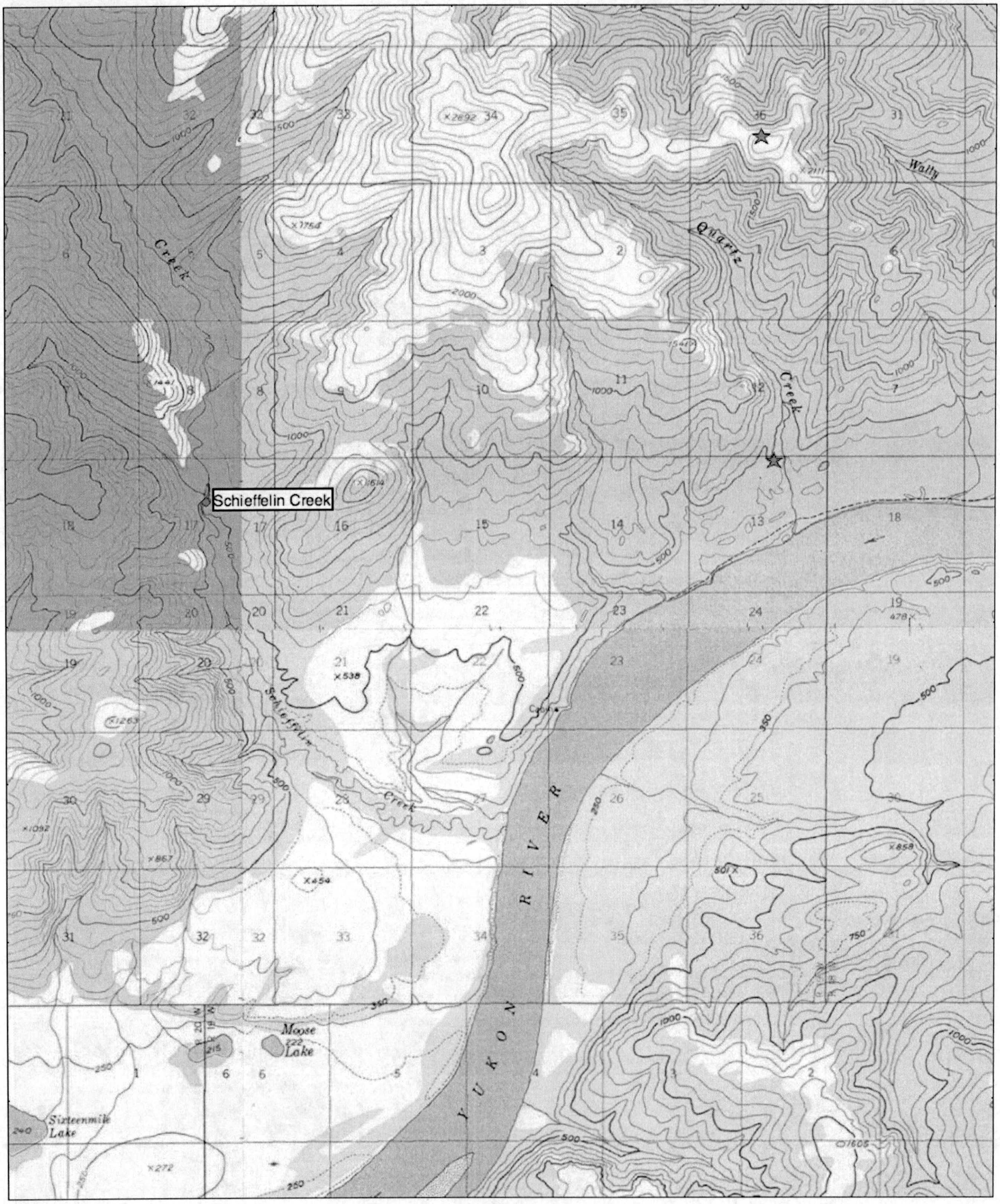
Schieffelin Creek
Creek
Quartz
Creek
Wally
Schieffelin
Creek
YUKON RIVER
Moose
222
Lake
Sixteenmile
Lake

Map 4-12

Alameda Creek

(Chromite, Gold) **Quadrangle:** Tanana (A1); **Coordinates:** 65.1677°N-150.1719°W (Mouth); **Map:** 4-1; **Worked:** 1985-; **Access:** From Elliot Highway at 65.1581°N, 150.1841°W. **Elliot Highway Right-of-Way:** ±50 of centerline; **Summary:** This placer mine is at about the midpoint of Alameda Creek. Tributary of Hutlinana Creek. **Geology:** Bedrock consists of Jurassic or Cretaceous sandstone, siltstone, and shale that are cut by thrust and high-angle faults (Chapman and others, 1982; Reifenstuhl and others, 1998). The topographic map of the Tanana A-1 quadrangle (1952, with minor revisions in 1963), shows several ditches contouring around the slopes of Alameda Creek. This development work is not mentioned in published reports. In 1989, Ed Salter worked Alameda Creek approximately 1 mile east of McCaskey Bar (Bundtzen and others, 1990). According to Bottge (1986), gold had been recovered from Alameda Creek by 1985. **Workings:** Two ditches feed into Alameda Creek, presumably to help offset the scarcity of water in the summer. According to Bottge (1986), gold had been recovered from Alameda Creek. (The Alaska Resource Data File., Tanana Quadrangle).

American Creek, Fish Lake Tributary

(Chromite, Gold) **Quadrangle:** Tanana (A3); **Coordinates:** 65.0961°N-151.1897°W (Mouth); **Map:** 4-2; **Worked:** 1911-1998; **Nugget Size:** 17 ozs; **Production:** Through 1933=34,000 ozs; **Access:** Trails from landing strip at 65.1016°N, 151.1765°W or tractor trail from Tofty Road at 65.0852°N, 150.9050°W. **Summary:** Tributary of Fish Lake. **Minerals:** magnetite, pyrite, ilmenite, barite, chromite, and hematite. **Geology:** American Creek drains the south side of Serpentine Ridge, and has been placer mined both above and below its junction with New York Gulch. The valley has an asymmetrical shape for much of its length, where the creek flows along the strike of bedding in bedrock. Serpentine Ridge consists of serpentinite and mafic intrusive rocks that apparently intrude Mesozoic, predominantly marine sedimentary rocks, through which the stream channel cuts (Chapman and others, 1982). Gold was discovered on American Creek in 1911 (Eakin, 1912). Depth to pay dirt was shallow (12 to 15 feet) in the upper stretches of the creek, and it quickly became a major producer in the Hot Springs mining district. **Workings:** Mining initially consisted of drifting and open cuts. The reports by Smith (1930) and Cobb (1977) of Ness operating a hydraulic mine in the Eureka Creek area may instead refer to Ed Ness, who operated on American Creek in 1925 and 1926 (Wimmler, 1926). Small-scale mining was reported on California Gulch (Creek) in 1932 (Smith, 1934; Cobb, 1977). In 1927, the American Creek Dredging Company began mining downstream from just below the mouth of Colorado Gulch (Smith, 1930). Parts of the placer ground consisted of 7 to 8 feet of gravels below 5 to 7 feet of muck. The lower 4 feet of the gravels was unusually coarse, and carried gold throughout, while the upper few feet of bedrock also proved rich (Mertie, 1934). The ground was frozen and had to be thawed before dredging, which continued sporadically through 1940, when the rich ground was mined out. A contiguous claim block varying from 33 to 108 claims on New York Gulch, American Creek, and tributaries of American Creek was worked by American Creek Partners from at least 1981 to 1992 (Alaska Kardex files). Placer development, using Caterpillar D-9 and D-6 bulldozers, took place over a 3 to 5 month period during those years. Gold production from this mining is not known. More recent work on American Creek included Orval McCormach's and Delima Placer's work in 1990 (Swainbank and others, 1991). In 1992 and 1994, Don Delima Placer, Inc., continued working on American Creek (Swainbank and others, 1993; Swainbank and others, 1995). In 1994, 40,000 cubic yards of material were mined. Ross Novak and Jay Hodges both recovered gold during the 1997 season (Swainbank and others, 1998). Hodges also mined in 1998 (Szumigala and Swainbank, 1999). In 1997, Stardust Mining, Inc., staked 14 state mining claims on Colorado Gulch near its mouth and planned to cut 50 small trenches across the gulch (Alaska Placer Mining Application F999579; Alaska Kardex files). Stardust Mining also planned to process a bulk sample of bench gravels from near the mouth of Colorado Gulch to determine the placer gold potential of the bench deposits. According to Mertie (1934), American Creek produced $702,000 in gold, estimated by Cobb (1977) to equal about 34,000 ounces of gold at $20.67 per ounce. This included a $600 nugget recovered in 1951 (gold at $35 per ounce) (Williams, 1951). The source of the placer gold may be quartz-carbonate veins associated with an east-trending shear zone, against which the pay streak appears to terminate (Cobb, 1981). (The Alaska Resource Data File., Tanana Quadrangle).

American Creek, Lynx Creek Tributary

(Gold) **Quadrangle:** Tanana (B6); **Coordinates:** 65.2779°N-152.9461°W; **Map:** 4-3; **Worked:** 1911-1975; **Yield:** 0.0484 oz/ft^2; **Access:** Trail from landing strip at 65.2937°N, 152.9930°W or tractor trail from the Yukon River at 65.1855°N, 153.0534°W; **Summary:** Tributary of Lynx Creek. Good placer gold prospects reported as early as 1911 but there is no report of successful mining (Cobb, 1981, OF 81-1313A p. A3). **Geology:** Bedrock in the American Creek area includes Precambrian or lower Paleozoic marine metasedimentary rocks and intermediate to mafic meta-igneous rocks; middle Paleozoic greenschist, carbonate, and chert; and Mesozoic greenstone and metasedimentary rocks (Chapman and others, 1982; WGM, Inc., 1998). The right-lateral Kaltag fault bounds these rocks to the south. Little is published about the placer deposit, and only a little work was done there in the early 1900's. Eakin (1913) reported that the gravels are 10 to 12 feet thick and are thawed placer material at the head of American Creek proved to be as rich as $1 in gold per square foot of bedrock (at $20.67 per ounce). **Workings:** The stream gradient is about 200 feet per mile. Prospect shafts were reported on American Creek, and surface mining may also have been employed. There was small-scale mining and prospecting on

American Creek in 1911 and 1912 (Eakin, 1912, 1913). No work was reported again until 1917, when some open cuts were made (Martin, 1919). The creek then appears to have been abandoned for a long time. Delima Placers and Holovics Placers operated bulldozers on American Creek in 1975 (Carnes, 1976). Some exploration for lode gold was done by Central Alaska Gold Company for Doyon, Ltd., in the American-Lynx creeks area (WGM, Inc., 1998). (The Alaska Resource Data File., Tanana Quadrangle).

Ash Creek

(Gold, Tin) **Quadrangle:** Tanana (B6); **Coordinates:** 65.3765°N, 152.8169°W; **Map:** 4-3, 4-9; **Worked:** 1900s-Present; **Yield:** 0.00697 oz/yd^3; **Access:** Trail from Yukon River at 65.2016°N, 152.7594°W **Summary:** Tributary of Tozimoran Creek. Samples from 5 drill holes within 2,400 feet of mouth of creek contained gold and tin. No record of any mining. 10 - 12 feet of sand and gravel and 1 - 4 feet of muck overlie Ordovician or Cambrian quartz-mica schist, phyllite, quartzite, and slate. See also Tozimoran Creek (Cobb, 1981, OF 81-1313A p. A3). **Geology:**The country rocks in the area of this occurrence include Paleozoic limestone, chert, quartzite, phyllite, and schist (Chapman and others, 1982). Cretaceous granite crops out north and west of Moran Dome. Ash Creek and nearby creeks were prospected following the discovery of gold in Tozimoran Creek around 1902 (Chapman and others, 1963). Placer gold was also reported in those creeks. A 12-foot-deep shaft dug by the U.S. Bureau of Mines near the mouth of Ash Creek exposed 12.4 cubic feet of frozen coarse gravel containing 4.2329 ozs of cassiterite and 0.0032 ozs of gold equals 0.00697 oz/yd^3 (Thomas and Wright, 1948 [RI 4323]). In 1952, the U.S. Bureau of Mines drilled five holes in the Ash Creek placer (Chapman and others, 1963). The work consisted of two two-hole fences and a single hole. The fences were 1,000 feet apart, and the single hole was 1,200 feet from the fences. A total of 0.1950 ozs of cassiterite were recovered from a total of 6.7144 oz of sample [representing] all of the holes. This value, compared to the results of drilling on Tozimoran Creek above Ash Creek, shows that Ash Creek contains more cassiterite. However, the bedrock source of the gold and cassiterite is unknown. (The Alaska Resource Data File., Tanana Quadrangle).

Bear Creek

(Gold) **Quadrangle:** Tanana (B3); **Coordinates:** 65.4139°N, 151.3369°W; **Map:** 4-4; **Worked:** 1900s-1990; **Access:** Trail from Yukon River at 65.3619°N, 150.9964°W **Summary:** Tributary of the Yukon River. Has been prospecting and (possibly) placer gold mining. (Cobb, 1981, OF 81-1313A p. A4). **Geology:**The upper and middle reaches of Bear Creek are underlain by Mesozoic (Rampart Group) marine metasedimentary rocks and mafic and ultramafic igneous rocks that are in thrust(?) contact with Paleozoic and/or Upper Precambrian schist, quartzite, and phyllite (Mertie, 1934; Chapman and others, 1982). The closest granitic pluton is Tertiary or Cretaceous quartz monzonite with a radiometric age of 61.8 +/- 2.5 Ma (Silberman and others, 1979), approximately 10 miles to the east. In the area around the reported placer, Bear Creek flows across Paleozoic limestone, chert, and greenstone (Chapman and others, 1982). The mining history of Bear Creek is uncertain. Mertie (1934) reported prospecting in 1932. Smith (1936) included Bear Creek in a list of gold producers, the rest of which are all on the eastern edge of theTanana quadrangle. This reference, as well as the one for the following year (1935) by Smith (1937) may refer to a different Bear Creek. According to Bottge (1986), Bear Creek had produced placer gold during or before 1985. In the 1990's, the area was explored by North Star Exploration, Inc. (North Star Exploration Inc., 2000 [DLR2000-17]), who determined that Bear Creek follows a east-southeast fault, and detected gold in stream-sediment samples. The company interpreted their geological, geophysical, and geochemical data as evidence of a potential structurally-controlled, polymetallic, tabular lode gold deposit (North Star Exploration Inc., 2000 [Highly mineralized block 20]). No follow-up work has been reported. (The Alaska Resource Data File., Tanana Quadrangle).

Big Denver Creek

(Gold) **Quadrangle:** Tanana (A2); **Coordinates:** 65.0091°N, 150.8462°W; **Map:** 4-5; **Worked:** 1981-1986; **Access:** Trail from Tofty Road at 65.0912°N, 150.8594°W **Summary:** Thirty mining claims were staked along Big Denver Creek in the early 1980s (Alaska Kardex files). The claims extended from the headwaters of Big Denver Creek near the summit of Manley Hot Springs Dome, for 4 miles downstream along the main channel of Big Denver Creek, and along two of its branches. For this record, the site is at North Rain claim number 14, where work was performed in 1983. The claim is in the southwest quarter of the southeast quarter of section 8, T. 2 N., R. 16 W., of the Fairbanks Meridian; other claims extend above and below this site. The location is accurate within about 500 feet. This site closely corresponds with the site for Big Denver, U.S. Bureau of Land Management MAS number 0020480054. **Geology:**The Big Denver Creek prospect is south of a group of cassiterite- and gold-bearing placer deposits known as the Tofty tin belt, a 12-mile-long area that trends east-northeast, between Roughtop Mountain to the north and Hot Springs Dome to the south (Thomas, 1957). Roughtop Mountain and Hot Springs Dome respectively are underlain by Cretaceous (K-Ar age date of 92 +/- 5 Ma) and Tertiary (K-Ar age date of 62 +/- 3Ma) granitic plutons (Chapman and others, 1982). The plutons intrude and contact metamorphose Mesozoicmarine sedimentary strata, which also are cut by diverse faults, including regional-scale, east-northeast striking,thrust faults (Reifenstuhl and others, 1998). Bedrock at the Big Denver Creek prospect is mainly Cretaceous sandstone, shale, and siltstone (Reifenstuhl and others, 1998). The upper end of the claim block is in the hornfels aureole of the Tertiary Hot Springs Dome granite pluton. The hornfels is black, dense rock containing un-oriented crystals or rosettes of muscovite, biotite, and andalusite. Placer mining claims were staked along Big Denver Creek in 1981 and the claims were worked from 1981 to 1986 (Alaska Kardex files). An affidavit of annual labor in 1983 lists 30 state mining claims and sample holes dug to bedrock or permafrost on North Rain claims 10-13. The material was sampled and analyzed. A trench dug to bedrock across the old dry bed of Big Denver Creek on North Rain claim 14 was also sampled. The sample analyses are not available. (The Alaska Resource Data File., Tanana Quadrangle).

Bonanza Creek

(Gold, Tin) **Quadrangle:** Tanana (B3); **Coordinates:** 65.3281°N, 151.3364°W; **Map:** 4-4; **Worked:** 1900s; **Access:** Trail from Yukon River at 65.2615°N, 151.2518°W **Summary:** Tributary of Morelock Creek. Bedrock is schist, greenstone, cherty dolomite and limestone, and barren milky quartz veins. Gold and cassiterite is basal part of 5 - 6 feet of gravel, on irregular bedrock surface, and in crevices in bedrock. Material in USBM drill holes ran 0.0443 lb tin and 0.00973 oz/yd^3 of gold. Concentrates contain gold, magnetite, cassiterite, and ilmenite. All work was prospecting rather than mining. See also Morelock Creek (Cobb, 1981, OF 81-1313A p. A4). Alluvium in valley is generally 5 - 6 feet of gravel and 2 - 3 feet of silt. Cassiterite and gold in bottom few inches of gravel, on irregular surface of bedrock, and in crevices in top few inches of weathered bedrock; weathering and erosion produced natural riffles. Tin- and gold-bearing gravels extend at least 900 feet upstream from mouth. Samples from a line of drill holes indicated averages of 0.1492 lb. Sn and 0.0328 ounces Au per square yard of bedrock (Cobb, 1977, OF 77-432, p. 12). **Geology:** Bedrock in the area of the Bonanza Creek gold-cassiterite (tin) placer deposit consists of greenstone, metachert, dolomite, marble, calcareous schist, and siliceous mica schist (Chapman and others, 1982). Although granitic rocks were reported by Eakin (1916), none were found in later mapping by Chapman and others (1963, 1982). Thomas and Wright (1948 [RI 4322]) reported that the bedrock was complexly folded and faulted, resulting in irregularities on the bedrock surface, called 'reefs.' Although this bedrock apparentlyis not the source of the gold or cassiterite, the reefs proved to be excellent for concentrating these heavy minerals.The gravels in Bonanza Creek are about 6 feet thick, with only 2.5 feet of overburden. Pebbles of cassiterite commonly contain quartz, indicating that they are derived from quartz-cassiterite veins. Both cassiterite and gold are concentrated in the bottom two feet of gravel, although in places they penetrate fractures in the upper few feet of bedrock. Chapman and others (1963) did not find other metallic minerals, although they describe limonite-stained pits (presumably after pyrite) in quartz veins along the ridges bordering Bonanza Creek. Thomas and Wright (1948 [RI 4322]) described many prospect shafts in the Bonanza Creek area that were sunk prior to a 1943 U.S. Bureau of Mines exploration project. This project included drilling on Morelock Creek and five holes at 50-foot intervals across the Bonanza Creek valley approximately 900 feet above its mouth. The average grades were 0.1492 pound of cassiterite and 0.0328 ounce of gold per square yard of bedrock on Bonanza Creek. (The Alaska Resource Data File., Tanana Quadrangle).

Boulder Creek

(Chromite, Gold, Monazite, Uraninite) **Quadrangle:** Tanana (A2, A3); **Coordinates:** 65.1553°N, 150.8839°W (Upper), 65.1438°N, 150.9723°W (Lower), 65.1440°N, 151.1394°W (Confluence with Little Boulder Creek); **Map:** 4-6; **Worked:** 1915-Present; **Purity:** 818 Au; **Access:** Trail from landing strip at 65.1501°N, 150.8784°W and trail from Tofty, Alaska; **Summary:** Tributary of Fish Lake. Asymmetrical valley with a broad bench in gentler north wall; placers are part of a Large body of low-grade gravel with a workable width of 1,200 feet that extends for several miles parallel to the stream and about 1,000 feet from it. Minerals in concentrate samples include magnetite, ilmenite, zircon, gold, rutile, garner, sphene, monazite, and aeschynite(?). Mining, 1915-17, 1930-39, and 1965-72. Total production not known. Stream heads in Cretaceous (90 m.y.) quartz monzonite of Roughtop pluton and Devonian or Permian metasedimentary rocks; most of course, however, in area underlain by Jura-Cretaceous sandstone, shale, siltstone, and argillite. Serpentinized ultramafic rock on ridge to south of creek contains disseminated grains of chromite; pieces of chromite in float are as much as 6 in. in diameter. Includes reference to Big Boulder Creek(Cobb, 1981, OF 81-1313A p. A4). Alluvial cover that contains gold is 8 - 12 feet deep. Over 200,000 square feet of bedrock cleaned in 1917. Gold assayed at 818 Au in fineness (Cobb, 1977, OF 77-432, p. 13). **Minerals:** Magnetite, ilmenite, zircon, rutile, garnet, sphene, and monazite. **Geology:** The Boulder Creek area is underlain by Jurassic or Cretaceous clastic sedimentary rocks, which have been intruded and contact metamorphosed by the Cretaceous monzodiorite and granite Roughtop Mountain pluton (Chapman and others, 1982; Reifenstuhl and others, 1998). This site is just beyond the hornfels zone ringing Roughtop Mountain, and Boulder Creek is fed by many streams that flow through the intrusion (Chapman and others, 1982). Published descriptions of the placer deposits, and of the locations of workings on Boulder Creek, are ambiguous. Waters (1934) and Moxham (1954) reported that, in addition to gold, the placer concentrates contained magnetite, ilmenite, zircon, rutile, garnet, sphene, and monazite. In 1949, pan concentrates assayed 0.21 percent equivalent uranium oxide (Uraninite), whose source was speculated to be the pluton at Roughtop Mountain (Moxham, 1954). Moxham also identified chromite float in Boulder Creek, whose source is probably serpentinite on Serpentine Ridge, south of the creek (Berg and Cobb, 1967). **Workings:** Mining on Boulder Creek was first reported in 1915 (Brooks, 1916). The creek has an asymmetrical valley (Chapin, 1919); the south side is steep with no gravels, and the north side is flat with alluvium 8 to 12 feet thick. The placer extends for several miles, with a workable width of 1,200 feet. In 1917, over 200,000 feet of bedrock were cleaned (Chapin, 1919). A hydraulic plant was operated in 1931 from May to September, and 7,000 square feet of bedrock was cleaned from a bench of Boulder Creek (Mertie, 1934). Gold 818 fine from this operation reportedly had an assay value of $16.90 per ounce (at $20.67 per ounce). There was relatively consistent small-scale production in the 1930's from Boulder Creek, although the amount of gold recovered was not made public. In 1967, two properties on Boulder Creek were being intensely prospected and another property was to be mined in 1968 (Heiner and others, 1968). Farrell & Higgins were also active on Little Boulder Creek in 1967. Boulder Creek Mining Company (Les and Dorthy Fickes) worked a cut on Boulder Creek during 1986 (Bundtzen and others, 1987). In 1992, R. Schroder and Beatrice Schafer both mined on Boulder Creek (Swainbank and others, 1993). Placer Mine Services mined along the creek in 1997 (Swainbank and others, 1998), and Richard Wilder mined in 1998 (Szumigala and Swainbank, 1999). (The Alaska Resource Data File., Tanana Quadrangle).

Cache Creek

(Chromite, Gold, Niobium, Uraninite, Silver, Tin) **Quadrangle:** Tanana (A2); **Coordinates:** 65.0999°N, 150.8157°W (Tofty Rd Xing at 65.0912°N, 150.8364°W); **Map:** 4-6, 4-10; **Tofty Road Right-of-Way:** ±50 of centerline; **Worked:** 1908-Present; **Yield:** 0.025 oz/yd^3; **Production:** Through 1956= 3,650 ozs Au, 409 ozs Ag; 1989=2,190 Au; 1990=1,074 ozs Au, 208 ozs Ag **Access:** Trail from Tofty Road at 65.0914°N, 150.8514°W; **Summary:** Tributary of Patterson Creek. Bedrock Jura-Cretaceous phyllite, slate, graywacke, quartzite, and sandstone; many essentially barren quartz veins. Gravels mainly locally derived; some cobbles of serpentinized gabbroic rock. Gravel overlain by muck and loess several tens of feet thick; entire deposit frozen. Main pay streak tin upper valley essentially mined out by 1913; very rich in places; some nuggets as much as 4 ounces. Later mining (continued until at least as recently as 1940 or 1941) farther downstream. Total production through 1956 was 3,650 ozs gold, 409 ozs silver, and 5,155 lb cassiterite concentrates. Heavy minerals in concentrates include cassiterite (mainly intergrown with tourmaline), gold, ilmenite, magnetite, barite, picotite, and aeschynite (Nb, Ti, and Ce mineral). Chromite fragments in gravel. See also Ferguson Draw and Sullivan Creek (Cobb, 1981, OF 81-1313A p. A5). Gold and cassiterite discovered 1908. Placer gravel buried by 10 - 170 feet of frozen muck and silt (loess). Gravel mainly angular pieces of country rock; some rounded quartzite boulders. Bedrock surface weathered and brecciated; gold in top foot or more. Sampling a tailing pile indicated recoverable content of 4.27 lb. concentrate, 2.26 lb Sn and 0.025 oz Au per cu. yd (Cobb, 1977, OF 77-432, p. 14-15). **Minerals**: cassiterite, magnetite, pyrite, ilmenite, barite, picotite, and aeschynite. **Geology:** The Cache Creek mine is one of a group of cassiterite- and gold-bearing placer deposits known as the Tofty tin belt, a 12-mile-long area that trends east-northeast, between Roughtop Mountain to the north and Hot Springs Dome to the south (Thomas, 1957). Roughtop Mountain and Hot Springs Dome respectively are underlain by Cretaceous (K-Ar age date of 92 +/- 5 Ma) and Tertiary (K-Ar age date of 62 +/- 3 Ma) granitic plutons (Chapman and others, 1982). The plutons intrude and contact metamorphose Mesozoic marine sedimentary strata, which also are cut by diverse faults, including regional-scale, east-northeaststriking, thrust faults (Reifenstuhl and others, 1998). A carbonatite sill(?) is in the Triassic section of these strata and there are exposures of serpentinized, Cretaceous(?) mafic and ultramafic rocks, mainly on Serpentine Ridge. The gold- and cassiterite-bearing creeks flow normal to the trend of the tin belt. They head in the plutonic, metamorphic, and mafic/ultramafic rocks of Roughtop Mountain and Serpentine Ridge, which probably are the source(s) of some of the metalliferous minerals in the placer deposits. Concentrations of gold diminish toward the south, probably due to dispersion (Thomas, 1957). The Cache Creek placer area is low-lying, with little exposed bedrock. Most of the area is covered by Quaternary loess and alluvium. The deposit is on the trace of a thrust fault that separates Triassic and Cretaceous rocks, and just south of a northwest-trending fault that offsets the thrust (Reifenstuhl and others, 1998). Farther upstream, the creek flows across carbonatite and another fault. The stream gravels in the mined area consist mainly of angular to subangular fragments of phyllite and graywacke, but also contain fragments of biotite granite, weathered monzonite, metadiorite, and serpentinized gabbro (Wayland, 1961). **Workings:** Mining in the Cache Creek area began during the winter of 1908-09 (Ellsworth, 1910), and by 1910 steam hoists were being used. The pay gravel was at an average depth of about 50 feet; it was about 10 feet thick, and rested on bedrock. The gold was concentrated at the bedrock/gravel interface. Waters (1934) reported that pan concentrates of the placer deposit contained, in addition to gold and cassiterite, magnetite, pyrite, ilmenite, barite, picotite , and aeschynite. Pebbles of cassiterite also contain tourmaline (Wayland, 1961). Little is published about mining on Cache Creek between 1917 and 1937. Mining appears to have taken place on a much smaller scale from 1937 to at least 1940. Otto Hovely drift mined on Cache Creek around 1951 (Williams, 1951). Active claims in 1967 on Cache Creek, with mining on some or all of the claims, were held by Benson, Lincoln Quartz Nos. 1 and 2, McGee, McLaughlin, and Strandberg Mines, Inc. (Heiner and others, 1968). **Production:** Through 1956, the U.S. Bureau of Mines reported that approximately 3,650 ounces of gold, 409 ounces of silver, and 5,155 pounds of cassiterite concentrate were removed from Cache Creek (Thomas, 1957). Otto Hovely drift mined on Cache Creek around 1951 (Williams, 1951). Active claims in 1967 on Cache Creek, with mining on some or all of the claims, were held by Benson, Lincoln Quartz Nos. 1 and 2, McGee, McLaughlin, and Strandberg Mines, Inc. (Heiner and others, 1968). More recent activity included mining by Shoreham Resources, Ltd., which in 1989 opened a large-scale placer operation for gold and cassiterite ('tin'), recovering 2,190 ounces of gold and 14,000 pounds of byproduct tin. In that season, the company reported that the recoverable reserves were raised from 63,000 to 74,500 ounces of gold (Bundtzen and others, 1990). The following year, it reported recovery of 1,074 ounces of gold, 208 ounces of silver and 2,600 pounds of tin from 26,000 cubic yards of processed pay (Swainbank and others, 1991). Cassiterite Placers, Inc., mined on Cache Creek during the 1990s (T.K. Bundtzen, oral commun., 2000). Quartz veins and stringers that have small amounts of cassiterite have been identified in the area, and veins containing a little gold and silver are associated with the granite of Hot Springs Dome. Such lode deposits probably are the source(s) of some of the metalliferous minerals in this placer deposit. (The Alaska Resource Data File., Tanana Quadrangle).

California Creek

(Gold) **Quadrangle:** Tanana (A1); **Coordinates:** 65.1891°N, 150.4318°W; **Map:** 4-7, 4-8; **Worked:** 1932; **Access:** Trail from Elliot Highway at 65.1775°N, 150.2459°W; **Summary:** Tributary of New York Creek. Small-scale mining in 1932 reported. Location of creek uncertain; may be tributary of the New York Creek near Eureka (Cobb, 1981, OF 81-1313A p. A5). **Geology:** Bedrock on the ridge between Allen and California creeks consists of Triassic to Permian conglomerate, argillite, sandstone, and shale (Reifenstuhl and others, 1998). The rocks form a northeast-trending syncline that is in thrust-fault contact with Mesozoic metasedimentary rocks to the south and with Paleozoic or older rocks to the north. Cretaceous monzodiorite crops out about 2 miles south and southeast of this area (Reifenstuhl and others, 1998). A sample (97SL015B) of conglomerate on the ridgeline between Allen and California creeks contained 300 parts per million (ppm) gold, 1,015 ppm copper, 13.95 percent iron, 2,810 ppm phosphorous, and 814 ppm zinc (Liss and others, 1998). The sample was not visibly altered or mineralized. There is no record of exploration or mining at this site. (The Alaska Resource Data File., Tanana Quadrangle).

Chapman Creek

(Antimony, Gold) **Quadrangle:** Tanana (B1); **Coordinates:** 65.3179°N, 150.1145°W; **Map:** 4-8; **Worked:** 1909-1912; **Access:** Trail from Rampart landing strip on the Yukon River at 65.5085°N, 150.1425°W; **Summary:** Tributary of Minook Creek. Auriferous gravel present. Small-scale mining in 1912; no data on amount of production. Lode stibnite prospect in headwaters reported; no further data. Bedrock in area is Jura-Cretaceous sedimentary rocks (Cobb, 1981, OF 81-1313A p. A5). **Geology:** Chapman Creek drains an area underlain by Triassic to Cretaceous sandstone, shale, and argillite (Chapman and others, 1982; Reifenstuhl and others, 1997 [RI 97-15a]). Granite of unknown age crops out in the headwaters of Chapman Creek (Burand and Saunders, 1966). The main channel follows a high-angle fault. **Workings:** Very little has been published about placer mining on Chapman Creek. Ellsworth (1910) reported prospecting on the creek in 1909. In 1912, four men mined on Chapman Creek (Ellsworth and Davenport, 1913). No other information about mining or production has been made public. Burand and Saunders (1966) reported a stibnite occurrence in the headwaters of Chapman Creek near the granite, and confirmed that placer mining had taken place on the creek. (The Alaska Resource Data File., Tanana Quadrangle).

Chicago Creek

(Gold) **Quadrangle:** Tanana (A1); **Coordinates:** 65.1806°N, 150.3611°W; **Map:** 4-7; **Worked:** 1904-1973; **Access:** Trail from Elliot Highway near Eureka, Alaska at 65.1768°N, 150.2214°W; **Summary:** Small tributary of Omega Creek. Gold discovered near mouth in 1904. Mining reported for years 1917, 1935-37, 1973. No reliable information on production. 5 - 6 feet of fine gravel on Cretaceous dark-gray shale. Nuggets 0.3-0.8 inches in diameter; very rough with attached quartz. A few roller-shaped nuggets. Finer fraction rough and shotty. See also Omega Creek and Thanksgiving Creek (Cobb, 1981, OF 81-1313A p. A5). **Geology:** The Chicago Creek area is marked by low, predominantly alluvium-covered topography (Chapman and others, 1982). The area is underlain by Mesozoic clastic sedimentary rocks including sandstone, siltstone, and shale, that are cut by thrust faults and locally intruded by Cretaceous monzodiorite (Reifenstuhl and others, 1998). (The Alaska Resource Data File., Tanana Quadrangle).

Chicken Creek

(Gold) **Quadrangle:** Tanana (B6); **Coordinates:** 65°22'52.0°N, 152°46'50.4°W; **Map:** 4-9; **Worked:** 1900s; **Access:** Trail from Yukon River at 65.2017°N, 152.7582°W; **Summary:** Tributary of Tozimoran Creek. Placer gold reported, but has been no systematic prospecting (Cobb, 1981, OF 81-1313A p. A6).

Colorado Gulch

(Gold, Chromite) **Quadrangle:** Tanana (A3); **Coordinates:** 65°06'00.4°N, 151°10'38.8°W; **Map:** 4-2; **Worked:** 1937-1997; **Access:** Trail from Tofty Road at 65.0891°N, 150.8784°W; **Summary:** Tributary of American Creek. Small-scale mining in 1937 reported. Location of stream unc ertain; probably the one shown on some modern maps as Colorado Gulch (Cobb, 1981, OF 81-1313A p. A6). **Minerals:** Barite, chromite, gold, hematite, ilmenite, magnetite, pyrite. **Geology:** American Creek drains the south side of Serpentine Ridge, and has been placer mined both above and below its junction with New York Gulch (TN077). The valley has an asymmetrical shape for much of its length, where the creek flows along the strike of bedding in bedrock. Serpentine Ridge consists of serpentinite and mafic intrusive rocks that apparently intrude Mesozoic, predominantly marine sedimentary rocks, through which the stream channel cuts (Chapman and others, 1982). **Workings:** In 1997, Stardust Mining, Inc., staked 14 state mining claims on Colorado Gulch near its mouth and planned to cut 50 small trenches across the gulch (Alaska Placer Mining Application F999579; Alaska Kardex files). Stardust Mining also planned to process a bulk sample of bench gravels from near the mouth of Colorado Gulch to determine the placer gold potential of the bench deposits. The source of the placer gold may be quartz-carbonate veins associated with an east-trending shear zone, against which the pay streak appears to terminate (Cobb, 1981). (The Alaska Resource Data File., Tanana Quadrangle).

Cooney Creek

(Gold) **Quadrangle:** Tanana (A2); **Coordinates:** 65.1139°N, 150.7287°W; **Map:** 4-10; **Worked:** 1912-1998; **Access:** Trail from Tofty Road at 65.0913°N, 150.8511°W; **Summary:** Tributary of Baker Creek. Large area of very low-grade placer-gold ground reported in 1912; gold irregularly distri buted. Sporadic small production through 1975. Includes reference to Coonie Creek. See also Gold Basin Creek and Killarney Creek (Cobb, 1981, OF 81-1313A p. A6). **Geology:** The Cooney Creek placer deposit is in an area of low relief, most of which is covered with a blanket of Quaternary loess and alluvium. Reifenstuhl and others (1998) mapped the underlying bedrock as Triassic sedimentary rocks. The upper reaches of Cooney Creek flow across a carbonatite sill in the Triassic rocks, and its headwaters drain an area underlain by Mesozoic granitic and metamorphic rocks (Chapman and others, 1982). **Workings:** Gold was identified in Cooney Creek in 1912, when a large area of unevenly distributed, very low grade gold placer ground was reported (Ellsworth and Davenport, 1913). Small production was reported in 1938 (Smith, 1939 [B 917-A]). Stanley Dale and J. Neubauer were mining on Cooney Creek in 1967 (Heiner and others, 1968), and Rough Top Mining Company mined with a bulldozer and sluice operation in 1975 (Carnes, 1976; Cobb, 1981). Albert Hagen mined through Claim 1 above Discovery and on about half of the Discovery claim before and during 1988, in section 25, T. 4 N., R. 16 W. (Alaska Kardex files). In 1993, Thurman Oil and Mining mined on Cooney Creek at a placer cut started in 1988 and proceeded up the valley making 20-foot-long by 90-foot-wide cuts (Bundtzen and others, 1994; Alaska Kardex files). (The Alaska Resource Data File., Tanana Quadrangle).

Dalton Gulch

(Gold, Tin) **Quadrangle:** Tanana (A2); **Coordinates:** 65.0995°N, 150.8424°W; **Map:** 4-6, 4-10; **Worked:** 1910-1956; **Purity:** NOT; **Yield:** 0.0241-0.4838 ozs/ft^2; **Production:** 1956=466 ozs; **Access:** Trail from Tofty Road at 65.0913°N, 150.8511°W; **Summary:** Tributary of Cache Creek. Mining, 1910 to World War I. Total production was 466 ounces of gold and 3,000 lb cassiterite concentrate. Source of one of the first cassiterite shipments from district. Overburden about 60 feet thick overlying thin gravels. Most pay streaks small and discontinuous localized on steeper parts of terraced surface of Jura-Creta ceous sedimentary rocks. A little cassiterite and gold in old dumps east of creek. See also Cache Creek, Ferguson Draw and Harter Gulch (Cobb, 1981, OF 81-1313A p. A6). **Minerals:** Aeschynite, columbite, monazite, and zircon. **Geology:** The Dalton Gulch mine is one of a group of cassiterite- and gold-bearing placer deposits known as the Tofty tin belt, a 12-mile-long area that trends east-northeast, between Roughtop Mountain to the north and Hot Springs Dome to the south (Thomas, 1957). Roughtop Mountain and Hot Springs Dome respectively are underlain by Cretaceous (K-Ar age date of 92 +/- 5 Ma) and Tertiary (K-Ar age date of 62 +/- 3 Ma) granitic plutons (Chapman and others, 1982). The plutons intrude and contact metamorphose Mesozoic marine sedimentary strata, which also are cut by diverse faults, including regional-scale, east-northeast striking, thrust faults (Reifenstuhl and others, 1998). A carbonatite sill(?) is in the Triassic section of these strata and there are exposures of serpentinized, Cretaceous(?) mafic and ultramafic rock, mainly on Serpentine Ridge. Bedrock in Dalton Gulch is Triassic sedimentary rocks. The gold- and cassiterite ('tin')-bearing creeks flow normal to the trend of the tin belt. They head in the plutonic, metamorphic, and mafic/ultramafic rocks of Roughtop Mountain and Serpentine Ridge, which probably are the source(s) of some of the metalliferous minerals in the placer deposits. Concentrations of gold diminish toward the south, probably due to dispersion (Thomas, 1957). **Workings:** Mining began in Dalton Gulch in 1910 and continued until World War I (Wayland,1961). One pay streak was 60 feet wide, but most of the pay was discontinuous. The gravels are thin, averaging 2 to 4 feet thick, with approximately 60 feet of overburden. The stream gradient is about 200 feet per mile. The cassiterite occurs as well-rounded to subangular particles that range in size from microscopic to several inches across. The larger sizes are commonly accompanied by quartz and tourmaline, as well as by fragments of sedimentary country rock. The smaller sizes generally are free of impurities (Thomas,1957). Dalton Gulch also includes bench placers on bedrock terraces. The richest bench placers are on the steepest slopes. They averaged $0.50 in gold per square foot, with local areas as rich as $10 per square foot (gold at $20.67 per ounce). Thomas (1957) reported that the total production of Dalton Creek through 1956 was 466 ounces of gold and 3,000 pounds of cassiterite concentrate. There was less cassite rite in Dalton Gulch than in the surrounding creeks (Wayland, 1961). Moxham (1954) reported aeschynite, columbite, monazite, and zircon in pan concentrates from Dalton Gulch. These minerals are possibly related to the carbonatite near the headwaters of the creek. The U.S. Bureau of Mines churn-drilled and channel sampled (tailings) at the Dalton Gulch mine in the 1950's. The average content of five samples was 3.90 pounds of tin and 0.014 ounce of gold per cubic yard (Thomas, 1957). (The Alaska Resource Data File., Tanana Quadrangle).

Deadwood Creek

(Chromite, Gold, Monazite, Niobium, Uraninite, Silver, Tin, Tungsten) **Quadrangle:** Tanana (A2, A3); **Coordinates:** 65.2086°N, 150.0930°W; **Map:** 4-1; **Worked:** 1902-1985; Present; **Access:** Unimproved road from Elliot Highway at 65.1767°N, 150.2209°W; **Summary:** The Deadwood Creek placer mine is about a half-mile upstream from the junction of North Fork. The site is at Deadwood claim number 1, where work was performed in 1980 (Alaska Kardex files). It is in the northeast quarter of section 1, T. 4 N., R. 13 W., of the Fairbanks Meridian. Three claims extend upstream from this site. The location is accurate within about 500 feet. This site closely corresponds with the site for Deadwood, U.S. Bureau of Land Management MAS number 0020480113. **Geology:** The Deadwood Creek area is underlain by Cretaceous sandstone, siltstone, quartzite, and shale, and by Cretaceous or Jurassic quartzite (Chapman and others, 1982; Reifenstuhl and others, 1998). Deadwood Creek follows the trace of a thrust fault between the Cretaceous and Cretaceous or Jurassic units (Reifenstuhl and others, 1998). No igneous rocks are mapped nearby. **Workings:** Placer mining in the upper Pioneer Creek-Deadwood Creek area began in 1902 (Collier, 1903). Most of the gold reportedly was mined from benches on the northwest side of Pioneer Creek or from secondary re-concentrations on the tributaries that cut across those benches (Mertie, 1934). Early work specifically on Deadwood Creek is not recorded. Placer mining claims were staked along Deadwood Creek in 1977 (Alaska Kardex files), and the claims were worked from 1977 to at least 1985. Affidavits of annual labor list digging holes up to 14 feet deep with a backhoe, and backpack drilling to 35-foot depth in 1979; sluicing ground on claim number 1 in 1980; stripping overburden and digging test holes with a backhoe in 1983 in preparation for mining the following year; stripping and ripping topsoil and frozen overburden with a D-8 Caterpillar bulldozer in 1984; and digging bedrock drains in 1984 and 1985. (The Alaska Resource Data File., Tanana Quadrangle).

Deep Creek

(Chromite, Gold, Monazite, Niobium, Uraninite, Silver, Tin, Tungsten) **Quadrangle:** Tanana (A2, A3); **Coordinates:** 65.0619°N, 150.9862°W; **Map:** 4-6; **Worked:** 1913-1950s; **Yield:** 0.194-0.290 ozs/ft^2; **Production:** Through 1956=7,684 ozs Au, 653 ozs Ag; **Access:** Unimproved road and trail from Tofty Road at 65.0888°N, 150.8784°W; **Summary:** Tributary of Woodchopper Creek. Bench north of creek is crossed by small tributaries in poorly defined valleys. In most places bedrock (mainly Jura-Cretaceous phyllite and slate) is well over 100 feet deep. Mining by drifting in basal few feet of gravel and top foot of bedrock. Placer discovered by prospect drilling in 1913; mining continued until as recently as 1955. Production was 7,684 ozs gold, 653 ozs silver, and 64,200 lbs cassiterite concentrate. Concentrates contained gold, cassiterite (some in the form of wood tin), ilmenite, picotite, zircon, pyrite, magnetite, monazite, much chromite, rutile, scheelite, columbite, aeschynite, and ellsworthite. Average recovery per yd^3 of channel samples of tailings piles was 7.08 lb concentrate containing 4 lb Sn and 0.0105 ounces gold. Includes references to Hokeley Gulch, Innesvale Gulch and Oakley Creek (Cobb, 1981, OF 81-1313A p. A6-A7). Gold found on Hokeley Gulch as the result of an extensive drilling program, 1913. Shafts sunk 130 feet to bedrock, mainly through frozen silt. Gravel 6 - 8 feet thick on bedrock. Gold on or near bedrock. Bedrock surface slopes southeastward. Tenor variable; some of ground runs 0.194 - 0.290 ounces of gold or more per bedrock foot. Cassiterite present, but not saved. Large-scale mining in 1914. Mining in 1931 was on Innesvale Gulch; ground frozen. 5 - 6 feet of gravel beneath 90 feet of muck. Gold in lower 2 feet of gravel and upper foot of weathered phyllite and slate bedrock ; bedrock very uneven. Average fineness of gold is 848 Au (Cobb, 1977, OF 77-432, p. 23-24). **Minerals:** Aeschynite, cassiterite, chromite, columbite, ellsworthite, gold, ilmenite, magnetite, monazite, picotite, pyrite, rutile, scheelite, unknown silver mineral, zircon. **Geology:** The placer mines on Deep Creek and its tributaries are part of a group of cassiterite- and gold-bearing placer deposits known as the Tofty tin belt, a 12-mile-long area that trends east-northeast, between Roughtop Mountain to the north and Hot Springs Dome to the south (Thomas, 1957). Roughtop Mountain and Hot Springs Dome respectively are underlain by Cretaceous (K-Ar age date of 92 +/- 5 Ma) and Tertiary (K-Ar age date of 62 +/- 3 Ma) granitic plutons (Chapman and others, 1982). The plutons intrude and contact metamorphose Mesozoic marine sedimentary strata, which also are cut by diverse faults, including regional- scale, east-northeast-striking, thrust faults (Reifenstuhl and others, 1998). A carbonatite sill(?) is in the Triassic section of these strata and there are exposures of serpentinized, Cretaceous(?) mafic and ultramafic rock, mainly on Serpentine Ridge. The gold- and tin-bearing creeks flow normal to the trend of the tin belt. They head in the plutonic, metamorphic, and mafic/ ultramafic rocks of Roughtop Mountain and Serpentine Ridge, which probably are the source(s) of some of the metalliferous minerals in the placer deposits. Concentrations of gold diminish toward the south, probably due to dispersion (Thomas, 1957). **Workings:** The placer deposits on Deep Creek and its tributaries stretch for more than a mile along the trace of an east-northeast-striking thrust fault. The bulk of the mining, from 1913 until at least 1941, appears to have been on benches about 1/4 of a mile south of the present channel of Deep Creek (Wayland, 1961). Although most of the gold was in gravel on the bedrock surface, gold and cassiterite ('tin') were also identified in overlying lenses of gravel. The cassiterite occurs chiefly as well-rounded to subangular particles that range in size from microscopic to several inches across. The larger sizes often are accompanied by vein quartz and tourmaline, as well as by fragments of sedimentary country rock. The smaller sizes generally are free of impurities (Thomas, 1957). A gold grain analyzed by electron microprobe contained a tiny inclusion of native nickel (Newberry and Clautice, 1997). The area of placer deposits has been divided into two sections, known locally as the Hokeley Gulch section (the western portion), and the Innesvale Gulch section (the eastern portion). The Hokeley Gulch placers were discovered in 1913 by Adolph Bock, and the first mining was reported by Eakin (1915). The shafts in the pay zone are 130 feet deep, with 6 to 8 feet of gravels lying on bedrock (Eakin, 1915). The gravels consist of angular phyllite, sandstone, and quartz boulders, along with minor quartz-tourmaline rock, chromite, and coarse-grained biotite granite. The gold is smooth and fine, suggesting that it is well traveled (Wayland, 1961). The Innesvale Gulch section was initially mined in 1918, when Bock, Handson and Albrecht did much of the work. One of the claims worked in 1931 was visited by Mertie (1934), who noted that the occurrence of the gold and cassiterite differs from that in the Hokeley Gulch section, in that the depth to bedrock is only about 65 feet, and the gravels are 5 to 6 feet thick. Irregularities in the bedrock surface trapped the cassiterite and gold. The gravels are similar in composition to those in the Hokeley Gulch section, but lack granite and include a small amount of metadiorite (Wayland, 1961). Through 1956, production from the

Deep Creek placers was 7,684 ounces of gold, 653 ounces of silver, and 64,200 pounds of cassiterite concentrate (Thomas, 1957). Barton (1962) reported that the placer tailings contained 0.1 to 5.0 percent niobium, and Kauffman and Holt (1965) published analyses of concentrates that showed 1 to 5 percent zircon. At least one sample of the concentrates below Hokeley Gulch contained 10 percent chromite (Cobb, 1973). (The Alaska Resource Data File., Tanana Quadrangle).

Dry Creek

(Gold) **Quadrangle:** Tanana (A3); **Coordinates:** 65.1671°N, 151.1423°W; **Map:** 4-2; **Worked:** 1917-1967; **Access:** Unimproved road and trail from Tofty, Alaska at 65.0933°N, 150.8820°W; **Summary:** Located in Boulder Creek basin. A branch of Trail Creek, which is a tributary of Boulder Creek. Slate, graywacke, quartzite, and schist make up bedrock and gravel. Small-scale mining in 1917 reported (Cobb, 1981, OF 81-1313A p. A7). Angular gravels 3 - 6 feet deep (Cobb, 1977, OF 77-432, p. 25). **Geology:** The Dry Creek area is underlain by Jurassic or Cretaceous clastic sedimentary rocks and by Paleozoic limestone, dolomite, greenstone, and chloritic schist, with minor amounts of phyllite, quartzite, and quartz mica schist (Chapman and others, 1982; Reifenstuhl and others, 1998). Serpentine Ridge, 1.5 miles to the south, consists of serpentinite and mafic intrusive rocks that intrude Mesozoic, predominantly marine sedimentary rocks, (Chapman and others, 1982). **Workings:** Chapin (1919) reported mining in Dry Creek in 1917. The gravels are angular, 3 to 6 feet thick, and consist of black slate, graywacke, quartzite, and schist. Water was not plentiful, but the ground was easily worked. Heiner and others (1968) reported mining in 1967, and that claims on Dry Creek were held or worked by Scotty Anderson, Farrell, Heiner and Wolff, and Higgins. (The Alaska Resource Data File., Tanana Quadrangle).

Eureka Creek

(Gold) **Quadrangle:** Tanana (A1); **Coordinates:** 65.1875°N, 150.2205°W (Lower), 65.2100°N, 150.1911°W (Upper); **Map:** 4-1, 4-8; **Eureka Road Right-of-Way:** ±50 of centerline; **Worked:** 1899-1996; **Purity:** 780-818 Au; **Production:** Through 1904=4,125 ozs; 1916=2,902 ozs; **Access:** Light duty road from Elliot Highway at 65.1627°N, 150.2183°W; **Summary:** Tributary of Baker Creek. Valley markedly asymmetrical with 2 well defined benches along gentler NW valley wall; other bedrock terraces have no surface expression. Bedrock is Lower Cretaceous quartzite, sheared grit, and argillaceous rock; small quartz veins; locally much disseminated pyrite. Gold in base of gravel and in cracks in bedrock where blocky. Gold bright and chunky; some with attached vein quartz. One small pebble of cassiterite has been found; possibly could have been planted as a prank. Mining from about 1899 to as recently as 1975; total production unknown, but large (Cobb, 1981, OF 81-1313A p. A7). Site of first gold discovery in district near mouth of Boston Creek in 1898. Stampede from Rampart in 1899. Valley marke dly asymmetrical. Steep slope southeast of creek; gentle slope to northwest. One fairly well defined old erosion level is about 250 feet above creek; another, less well defined, is 50 - 70 feet above creek. Other bedrock terraces exposed only in mines have no surface expression. Bench (called bar) deposits are on old terraces on NW side of creek from Boston Creek upstream. Upper valley follows strike of Lower Cretaceous rocks, mainly quartzite. Just below mouth of Boston Creek pay streak is 25 - 70 feet wide; gravel 10 - 18 feet thick beneath as much as 8 feet of muck. Bench and creek gravels of local origin; contains sticky clay, which makes sluicing difficult. Gold assayed at 818 Au. Low-grade placers on bench 75 feet above and East of creek. Considerable mining on sloping bench 2 miles upstream from Boston Creek; gravel 8 - 20 feet deep. Production through fall, 1904, was about 4,125 fine ounces; includes production from Dome Creek. Mining, 1916. Production worth about 2,902 fine ounces (Cobb, 1977, OF 77-432, p. 26-27). **Geology:** The bedrock of most of the upper reaches of Eureka Creek, as well as of the area that has been mined, consists of Jurassic or Cretaceous sandstone, siltstone, shale, and quartzite (Chapman and others, 1982; Reifenstuhl and others, 1998). Some of the strata are intruded by several small, northeast-trending Cretaceous monzodiorite plugs. The creek follows a southwest-trending thrust fault for four miles. This fault is manifested by the valley's asymmetry. The southeast side is steep, and consists of quartzite, and the northwest side, along which the placer benches occur, slopes gently. Two benches have been identified, one approximately 250 feet above the present channel and one 50 feet above the channel. Yeend (1990) describes four terraces at Eureka Creek, with the best-developed one 250 feet above the present channel. Yeend (1990) hypothesized that ancestral Eureka and Pioneer creeks flowed southwest toward Tofty during the Pleistocene and joined the ancestral Tanana River farther west than the present day channel. The gradient of Eureka Creek between points 1.5 miles below and 3 miles above Boston Creek is about 100 feet per mile. Above Boston Creek, water supply is small and during an average summer is barely adequate for sluicing. **Workings:** The discovery of gold on Eureka Creek in 1898 was the first one in the Hot Springs district and it sparked a stampede (Cobb, 1973). Mining on a one-mile stretch, including the discovery claim just below the confluence with Boston Creek, started by 1900 (Collier, 1903), and mining along Eureka Creek was continuous until 1940 (Cobb, 1977). Active claims or workers in 1967 included Boston Boys, Brock and Johnson, Farmer and Jones, Frank & Co., Gill, and Strandberg Mines, Inc. (Heiner and others, 1968). Both drift mining and open cuts were used. The gravels are 10 to 18 feet thick, and overlain by eight feet of muck. Pay continued for several feet into bedrock (Prindle and Hess, 1905). The width of the pay streaks ranged from 25 to 75 feet. Much of the gold occurred as chunks, commonly attached to vein quartz (Prindle and Hess, 1905; Mertie, 1934). The gold fineness was 780. Hess (1908) described small quartz veins and reported considerable disseminated pyrite in the sheared grit and argillaceous bedrock. According to Hess, the pay contained sticky clay, which made sluicing difficult. Newberry and Clautice (1997) analyzed a coarse gold nugget intergrown with quartz by electron microprobe. The gold was homogeneous with respect to silver and mercury contents, averaging 20 percent silver, 1.3 percent mercury, and a fineness of 780. The gold composition is similar to that in lode deposits on Elephant Mountain. Activity has been reported in recent years on Eureka Creek. John Cole operated a bulldozer on mining claims in 1975 (Carnes, 1976). Shimsky Mining operated along the creek in 1985 (Bundtzen and others, 1986) and reportedly completed assessment work on holdings in 1986. Mark Krenzke placer mined from 1988-90, the last year of which Orval McCormack also worked Eureka Creek (Green and others, 1989; Bundtzen and others, 1990; Swainbank and others, 1991). James Crude worked the creek in 1989, and Thurman Oil and Mining operated there in 1991 (Bundtzen and others, 1992). Kelly Mining and Ed Salter both worked Eureka Creek (Swainbank and others, 1995) in 1994, and Bob Bettisworth placer mined on the creek in 1996 (Swainbank and others, 1997). BIFS Mining and Eleven Pup Mining both explored for placer gold in 1997 (Swainbank and others, 1998). From 1968 to 1995, exploration on Boston Creek near its junction with Eureka Creek included bedrock testing, overburden stripping, building a slick plate and sluicebox, and roadwork, but no mining was reported (Alaska Kardex files). Claims were staked in 1977 on Unanimous Creek, a tributary to Eureka Creek, followed in 1979-1981 by test panning and other prospecting (Alaska Kardex files). Claims were staked in 1972 on Nugget Creek, another tributary to Eureka Creek, and on the upper part of Eureka Creek, followed in 1979-1995 by test panning and other prospecting, but no mining was reported (Alaska Kardex files). It is unclear how much gold has been recovered from Eureka Creek. Hess (1908) reported production of about $85,300 worth of gold (at $20.67 per ounce) as of the fall of 1904, but this figure included some of the production from the Pioneer Creek area. Cobb (1977) reported production in 1916 of 2,900 ounces of gold, and that total production from 1899 to 1940, although unknown, was large. (The Alaska Resource Data File., Tanana Quadrangle).

Ferguson Draw

(Gold, Tin) **Quadrangle:** Tanana (A2); **Coordinates:** 65.1006°N, 150.8133°W; **Map:** 4-6, 4-10; **Worked:** 1900s; **Access:** Unimproved road and trail from Tofty Highway at 65.0915°N, 150.8511°W; **Summary:** Small tributary of Cache Creek. Gravel thin or absent; in places 40 feet of muck rests directly on bedrock placer. Pay streak ran 0.3-0.4 oz Au and 0.1-0.4 lb cassiterite per ft^2 of bedrock. Terminated upstream against bedrock terrace. See also Cache Creek (Cobb, 1981, OF 81-1313A p. A7). Gold is fine, flaky, and well-worn (Cobb, 1977, OF 77-432, p. 29). **Geology:** The Ferguson Draw placer mine is one of a group of cassiterite- and gold-bearing placer deposits known as the Tofty tin belt, a 12-mile-long area that trends east-northeast, between Roughtop Mountain to the north and Hot Springs Dome to the south (Thomas, 1957). Roughtop Mountain and Hot Springs Dome respectively are underlain by Cretaceous (K-Ar age date of 92 +/- 5 Ma) and Tertiary (K-Ar age date of 62 +/- 3 Ma) granitic plutons (Chapman and others, 1982). The plutons intrude and contact metamorphose Mesozoic marine sedimentary strata, which also are cut by diverse faults, including regional-scale, east-northeaststriking, thrust faults (Reifenstuhl and others, 1998). A carbonatite sill(?) is in the Triassic section of these strata and there are exposures of serpentinized, Cretaceous(?) mafic and ultramafic rock, mainly on Serpentine Ridge. The gold- and tin-bearing creeks flow normal to the trend of the tin belt. They head in the plutonic, metamorphic, and mafic/ultramafic rocks of Roughtop Mountain and Serpentine Ridge, which probably are the source(s) of some of the metalliferous minerals in the placer deposits. Concentrations of gold diminish toward the south, probably due to dispersion (Thomas, 1957). Ferguson Draw is at the junction of two faults (Reifenstuhl and others, 1998). Bedrock in the creek consists of Triassic sedimentary rocks. The gravels in Ferguson Draw are thin or absent, and the gold is directly on the bedrock, often overlain by 40 feet of muck (Wayland, 1961). The gravels, when present, contain silicified phyllite, and well-traveled boulders of quartzite and quartz. Wayland reported that the ground contained 0.3-0.4 ounce of gold and 0.1 to 0.4 pound of cassiterite per square foot of bedrock. Upstream, the pay streak terminates abruptly at a bedrock bench. The gold recovered from Ferguson Draw was different from the gold at the head of Cache Creek. Gold from Ferguson Draw was characteristically fine and smooth, whereas the bulk of the gold from Cache Creek was bird-shot in size. The cassiterite mostly occurs as well-rounded to subangular particles that range in size from microscopic to several inches across. The larger sizes often are accompanied by vein quartz and tourmaline, as well as by fragments of sedimentary country rock. The smaller sizes generally are free of impurities (Thomas, 1957). (The Alaska Resource Data File., Tanana Quadrangle).

Florida Creek

(Gold) **Quadrangle:** Tanana (B1); **Coordinates:** 65.4155°N, 150.0921°W; **Map:** 4-8, 4-11; **Worked:** 1904-1951; **Nugget Size:** 1.6 ozs; **Production:** 200-300 ozs; **Access:** Trail from Rampart landing strip on the Yukon River at 65.5083°N, 150.1422°W; **Summary:** Tributary of Minook Creek. Bedrock mainly diabase. Placers narrow and as much as 15 - 20 feet deep; contain nuggets as large as 1.6 ounces. Small-scale mining in early 1900's and 1939 - 40. Total production probably no more than 200 - 300 oz of gold (Cobb, 1981, OF 81-1313A p. A8). Small stream 2 miles long flows in narrow, deep valley in bench east of Minook Creek. Florida and Little Minook Jr. are the only small eastern tributaries of Minook Creek of any economic importance (Cobb, 1977, OF 77-432, p. 30). **Geology:** Bedrock along Florida Creek consists predominantly of diabase (Hess, 1908). The general area is underlain by Paleozoic or Proterozoic siltstone and sandstone, which is in fault contact with Triassic gabbro (Reifenstuhl and others, 1997 [RI 97-15a]). These rocks in turn are overlain by Tertiary gravel terraces, such as the one on McDonald Bar (Chapman and others, 1982). **Workings:** Placer gold production was reported around 1904 on Florida Creek (Prindle and Hess, 1905). The deposits proved to be very narrow, but as deep as 15 to 20 feet. Through 1904, less than 100 ounces had been mined, and production diminished after that (Hess, 1908). Mining was reported in 1939 and 1940. Frank Dinan drift mined in 1951 on his claims on Florida Creek (Williams, 1951). (The Alaska Resource Data File., Tanana Quadrangle).

Frankie Creek

(Gold) **Quadrangle:** Tanana (B1); **Coordinates:** 65.3053°N, 152.9835°W; **Map:** 4-3; **Worked:** 1971-1981; **Access:** Trail from Yukon River at 65.1854°N, 153.0530°W; Trail from Golden Creek airstrip at 65.2937°N, 152.9928°W; **Summary:** Mining claims were staked in 1971, followed by placer mining, on Frankie Creek (Alaska Kardex files). Mining claim location notices state that the claims are about 1.25 miles from the Golden Creek airstrip. The site is at the location given in the U.S. Bureau of Land Management MAS system (MAS number 0020480091), in the northeast quarter of section 31, T. 6 N., R. 26 W., of the Fairbanks Meridian. The location is accurate within about 1000 feet. **Geology:** Bedrock in the area of the Frankie Creek placer deposit is pelitic schist with interbedded marble and greenstone (Chapman and others, 1982). The Cretaceous Melozitna Pluton, ranging in composition from quartz monzonite to granite, crops out about 5 miles north of Frankie Creek. The Kaltag fault, a major, east-northeast structure, is approximately 5 miles to the south. **Workings:** In 1971, prospecting for placer gold deposits at Frankie Creek consisted of drilling 3-inch-diameter holes through the gravels to bedrock (Alaska Kardex files). The holes were spaced 25 feet apart. Samples from the drilling were panned and good colors of gold were reported. Affidavits of annual labor from 1971 to 1985 list prospecting, drilling, and testing colors. There is no record of production. (The Alaska Resource Data File., Tanana Quadrangle).

Glenn Creek

(Gold, Tin) **Quadrangle:** Tanana (A1); **Coordinates:** 65.1733°N, 150.2510°W; **Map:** 4-1, 4-7; **Worked:** 1901-1992; **Nugget Size:** 6+ ozs; **Production:** Production to 1931=48,500 ozs; **Access:** Unimproved road and trail from Elliot Highway at 65.1627°N, 150.2183°W; **Summary:** Headwater tributary of Eureka Creek. Small stream that flows in an open, shallow valley in the bench between Eureka and Omega Creeks. Bedrock is weathered Jura-Cretaceous slate and quartzite. In some places slide-rock overlies pay gravels. Pay streak 50-100 feet. wide, 5 - 20 feet deep, and was very rich averaging more than 0.484 ounces per pan in spots. Gold reconcentrated from bench deposits of Shirle y Bar. Pay extended 2 feet into crevices in bedrock. Mined from 1901 to 1939. Production to 1931 was about 48,500 ounces. A few cassiterite pebbles that (may have been the work of pranksters) have been found. See also Shirley Bar (Cobb, 1981, OF 81-1313A p. A8). Gold discovered July 1901. Pay mainly in part of stream that cuts bench gravel. Some nuggets weighed over 6 ounces. Gold assayed at 774 Au (Cobb, 1977, OF 77-432, p. 31-32). **Geology:** The Glenn Creek area is underlain by Mesozoic clastic sedimentary strata that are cut by thrust faults and locally intruded by gabbro and diorite (Chapman and others, 1982; Reifenstuhl and others, 1998). The placer deposit is about a half-mile south of one of the thrust faults. The bedrock of Glen Creek is slate and quartzite, cut by a few quartz veins (Hess, 1908). The stream gravels locally are overlain by slide rock. The productive zones are localized where bench gravels are cut by the active creek (Hess, 1908), suggesting that the gold is reconcentrated from older bench placers (Mertie, 1934). The pay zones are up to 100 feet wide and 7 feet thick (Collier, 1903; Hess, 1908). The gold also occurs in the upper two feet of bedrock. **Workings:** Gold was discovered in this area around 1900, and mining began in Glenn Creek in 1901 (Collier, 1903). The deposit was considered to be rich, with pockets of pay dirt yielding $10 worth of gold per pan when gold was worth $20.67 per ounce. Through 1931, Glenn Creek produced more than $1,000,000 worth of gold, equivalent to about 48,380 ounces of refined gold (Mertie, 1934). Mining continued through the 1930's, when Glenn Creek was considered to be a major producer in the Eureka Creek area. A few cassiterite pebbles were found in 1939 (Wayland, 1961). Cobb's (1977) and Smith's (1930) reports of Johnson & Hensley operating a hydraulic mine in 1928 in the Eureka Creek area actually referred to an operation on Alice Bench, located at the base of the hill just east of Glenn Gulch (Wimmler, 1926). More recently, Earthmovers operated a placer mine on Glenn Creek in 1992 (Swainbank and others, 1993), and Ernest Johnson mined there in 1997 (Swainbank and others, 1997). (The Alaska Resource Data File., Tanana Quadrangle).

Gold Basin Creek

(Gold, Tin) **Quadrangle:** Tanana (A2); **Coordinates:** 65.1124°N, 150.7628°W; **Map:** 4-10; **Worked:** 1917; **Access:** Trail from Tofty Road at 65.0913°N, 150.8511°W; **Summary:** Tributary of Killarney Creek. Small amounts of rounded cassiterite and fine gold on Jura-Cretaceous phyllit e bedrock at depths of 40 - 80 feet. Considerable prospecting by shafts and drill holes, but there probably was very little gold production. See also Killarney Creek (Cobb, 1981, OF 81-1313A p. A8). Drilling and shaft prospecting beginning in about 1912 lasted a few years and was resumed in the 1930's (Cobb, 1977, OF 77-432, p. 33). **Geology:** The Gold Basin prospect is one of a group of cassiterite- and gold-bearing placer deposits known as the Tofty tin belt, a 12-mile-long area that trends east-northeast, between Roughtop Mountain to the north and Hot Springs Dome to the south (Thomas, 1957). Roughtop Mountain and Hot Springs Dome respectively are underlain by Cretaceous (K-Ar age date of 92 +/- 5 Ma) and Tertiary (K-Ar age date of 62 +/- 3 Ma) granitic plutons (Chapman and others, 1982). The plutons intrude and contact metamorphose Mesozoic marine sedimentary strata, which also are cut by diverse faults, including regional-scale, east-northeast striking, thrust faults (Reifenstuhl and others, 1998). A carbonatite sill(?) occurs in the Triassic strata and there are exposures of serpentinized, Cretaceous(?) mafic and ultramafic rocks, mainly on Serpentine Ridge. Gold Basin Creek flows along a northwest-trending fault in Triassic strata just north of one of the regional thrust faults. The gold- and tin(cassiterite)-bearing creeks flow normal to the trend of the tin belt. They head in the plutonic, metamorphic, and mafic/ultramafic rocks of Roughtop Mountain and Serpentine Ridge, which probably are the source(s) of some of the metalliferous minerals in the placer deposits. Concentrations of gold diminish toward the south, probably due to dispersion (Thomas, 1957). **Workings:** In 1917, prospectors found significant amounts of cassiterite in Gold Basin Creek, but did not find much gold. The creek was explored by drilling and prospect shafts. The gold and cassiterite were on bedrock at a depth of 40 to 80 feet (Wayland, 1961; Cobb, 1977). The exact distribution of the gold is unknown, but in the past it has proved too fine to recover. The cassiterite is similar to that in Woodchopper Creek and is well rounded (Wayland, 1961).(The Alaska Resource Data File., Tanana Quadrangle).

Gold Run Creek

(Gold) **Quadrangle:** Tanana (A1); **Coordinates:** 65.1803°N, 150.2703°W; **Map:** 4-1, 4-7; **Worked:** 1899-1967; **Purity:** 774 Au; **Yield:** 0.161 ozs/ft^2; **Production:** Total production 9,700-10,000 ozs; **Access:** Light duty road from Eureka Road at 65.1769°N, 150.2214°W; **Summary:** Tributary of Rhode Island Creek. Initially staked, 1899. Pay streak 150 feet wide and about 1,500 feet long drifted from one end to the other. Mined intermittently until at least 1973. Production through 1931 was about 9,675 ounces of gold. Fineness reportedly about 775 Au. Well-rounded gravel 16 - 18 feet thick composed mainly of slate, quartzite, and grit overlain by about 2 feet of frozen muck. Bedrock Jura-Cretaceous carbonaceous schistose grit, slate, graywacke, siltstone, and shale. Gold bright, shotty, and well worn. Some pieces show crystal faces. 4-oz nuggets reported. Gold reconcentrated from Shirley Bar. See also Glen Creek and Shirley Bar (Cobb, 1981, OF 81-1313A p. A9). Very small stream with enough water for sluicing only when snow is melting. Ground 10 - 15 feet deep and frozen; bedrock is blocky schist. Pay streak 12 - 40 feet wide. Area of 300 ft^2 mined in 1901-02 yielded more than 0.161 ounces gold per bedrock foot. Cuts western side Shirley Bench from which it has reconcentrated gold. Gold extends 3 - 4 feet into cracks in bedrock. Gold probably derived from both bench gravels and local bedrock. Gold assayed at 774 Au. Mined intermittently until 1938. Total production probably 9,700 to 10,000 fine oz. No data on other minerals in concentrates. (Cobb, 1977, OF 77-432, p. 36). **Geology:** The country rocks in the area of Gold Run Creek are chiefly Mesozoic clastic sedimentary strata that are cut by thrust faults, and intruded by Cretaceous or Tertiary monzodiorite (Chapman and others, 1982; Reifenstuhl and others, 1998). Bedrock in Gold Run Creek has been described as blocky, silicic(?) schist, and slaty to schistose grit (Collier, 1903; Hess, 1908). The placer gravels range in thickness from 16 to 18 feet, and consist of well-rounded clasts that include quartzite (Hess, 1908). The overlying muck is 2 feet thick. The ground is frozen and sluicing can only be done during snow melt (Collier, 1903), in as much as the creek flows in an open, shallow valley that is often dry (Hess, 1908). The gold occurs in a streak 150 feet wide by 1,500 feet long, both in the gravels and in the upper 3 or 4 feet of bedrock (Mertie, 1934). At least some reconcentration of gold probably took place, because Gold Run Creek cuts through the bench gravels of Shirley Bar (Prindle and Hess, 1905). Mertie reported that the gold was light colored, shotty, and somewhat worn. Some, however, exhibited crystal faces (Hess, 1908), suggesting a proximal lode source. There may be two sources of the gold: one local and one distant. Purington (1905) reported that the gold was worth $16.00 per ounce when refined gold returned $20.67 per ounce. Mining in Gold Run Creek started around 1900 and continued sporadically, probably until the 1930's. Mertie (1934) reported that the early diggings had been mined out. According to Cobb (1977), total production through 1938 was estimated at nearly 10,000 ounces of gold. Archie Pringle held active mining claims and may have mined in 1967 (Heiner and others, 1968). (The Alaska Resource Data File., Tanana Quadrangle).

Golden Creek

(Gold) **Quadrangle:** Tanana (B6); **Coordinates:** 65°18'N, 153°00'W; **Map:** 4-3; **Worked:** 1900s-1998; **Purity:** NOT; **Yield:** NOT; **Nugget Size:** NOT; **Production:** NOT ozs; **Access:** Trail from Golden Creek landing strip at 65.2936°N, 152.9933°W; Trail from Yukon River at 65.1855°N, 153.0534°W; **Summary:** Tributary of Illinois Creek. Placer gold has been mined; no cassiterite reported. See also Golden Creek, Melozitna Quadrangle (Cobb, 1981, OF 81-1313A p. A8). **Geology:** Bedrock in the area of the Golden Creek placer deposit is pelitic schist with interbedded marble and greenstone (Chapman and others, 1982). The Cretaceous Melozitna Pluton, ranging in composition from quartz-monzonite to granite, borders the schist and crops out about 5 miles north of Golden Creek. The Kaltag fault, a major, east-northeast structure, is approximately 5 miles to the south. Gold in the Golden Creek placer occurs in thin bench and shallow stream gravel deposits (Cobb, 1973). **Workings:** According to Cobb, Golden Creek was one of the most recently mined placer areas in the Melozitna mining district, and as of 1954, was the only active placer mine in the district. Chapman and others (1963) explored the placer for cassiterite, but found none. Phil Ramsted worked in the area in 1991 and was mining on Golden Creek in 1993 (Bundtzen and others, 1992, 1994). In 1993-94 and 1997-98, Wayne Gibson used a dragline to mine (Bundtzen and others, 1994; Swainbank and others, 1995, 1998; Szumigala and Swainbank, 1999). There is no public record of production, which probably has been small. (The Alaska Resource Data File., Tanana Quadrangle).

Granite Creek

(Gold, Tin) **Quadrangle:** Tanana (A6, B6); **Coordinates:** 65.2787°N, 150.2079°W; **Map:** 4-8; **Worked:** 1973-1993; **Access:** Trail from Rampart landing strip on the Yukon River at 65.5085°N, 150.1423°W; Granite Creek crosses trail at 65.3012°N, 150.1464°W; **Summary:** Mining claims were staked in 1973, followed by placer mining, along Granite Creek and on Boulder Creek at its junction with Granite Creek (Alaska Kardex files). The location given in the U.S. Bureau of

Land Management MAS system (MAS number 0020480079) is in the northwest quarter of section 10, T. 5 N., R. 13 W., of the Fairbanks Meridian. The location is accurate within about 500 feet. **Geology:** Bedrock in the Granite and Boulder creeks area consists of Triassic or Permian siliceous and locally carbonaceous argillite in thrust contact with Proterozoic(?) or Cambrian(?) cherty argillite, chert, and limestone (Reifenstuhl and others, 1997 [RI 97-15a]). No igneous rocks are mapped in the local area. The rocks are cut by northwest-trending, high-angle faults. Workings: Mining claims were staked in 1973 along Granite Creek and at the junction of Boulder Creek with Granite Creek (Alaska Kardex files). Eight to fifteen claims were worked intermittently from 1974 to at least 1993. Affidavits of annual labor list prospecting, test sluicing, and test pitting. There is no record of production.(The Alaska Resource Data File., Tanana Quadrangle).

Grant Creek

(Gold, Tin) **Quadrangle:** Tanana (A6, B6); **Coordinates:** 65.2344°N, 152.8551°W; **Map:** 4-3; **Worked:** 1909-1993; **Purity:** 866; **Yield:** 0.484-0.580 ozs/yd²; **Access:** Trail from Grant Creek, Yukon River at 65.19647°N, 152.9322°W; **Summary:** Tributary of the Yukon River. Bedrock is a complex of intensely folded Cambrian or Ordovician schist, limestone, quartzite, and greenstone. Placer gold deposits prospected or mined sporadically from 1909 to about 1975; production probably small (several hundred ounces?). Gold medium fine with some small nuggets. Fineness about 866. Depth to bedrock reported to be 5 - 30 feet. Placer tin said to have been collected in 1929; none found during mining in 1942-43. Includes references to Lynx Creek (Cobb, 1981, OF 81-1313A p. A9). Trouble operating because of live water. Some of gravel (a foo t thick) said to carry 0.484-0.580 ounces of gold a yard [does not state whether a cubic yard or a square yard of bedrock]. Report of pitchblende probably in error; material found probably was black hematite (Cobb, 1977, OF 77-432, p. 37-38)). Workings: Mining claims were staked in 1973 along Granite Creek and at the junction of Boulder Creek with Granite Creek (Alaska Kardex files). Eight to fifteen claims were worked intermittently from 1974 to at least 1993. Affidavits of annual labor list prospecting, test sluicing, and test pitting. There is no record of production. (The Alaska Resource Data File., Tanana Quadrangle).

Harter Gulch

(Gold, Tin) **Quadrangle:** Tanana (A2); **Coordinates:** 65.0928°N, 150.8516°W, 65.0913°N, 150.8537°W (Tofty Rd Xing); **Map:** 4-5, 4-6; **Tofty Road Right-of-Way:** ±50 of centerline; **Worked:** 1900s-1977; **Yield:** 0.017 oz/yd³; **Access:** Trail from Tofty Road at 65.0913°N, 150.8509°W; **Summary:** Tributary of Cache Creek. Has been placer gold mining, most recently probably in 1977. About 5,000 fine ounces of gold may have been produced. Tailings piles contain cassiterite and more gold than is usual in tailings piles in district. Recovery per yd³ from channel samples of one tailing pile averaged 0.84 lb concentrate that contained 0.27 lb Sn and 0.017 oz Au. Gravel largely angular phyllite and graywacke with some rounded sandstone clasts. Magnetite and hematite in Jura-Cretaceous crystalline limestone bed near a small Tertiary(?) mafic intrusive body. See also Cache Creek (Cobb, 1981, OF 81 -1313A p. A9). Production from a small area during one summer said to have been 4,350 fine ounces of gold (Cobb, 1977, OF 77-432, p. 39). **Geology:** The Harter Gulch mine is one of a group of cassiterite- and gold-bearing placer deposits known as the Tofty tin belt, a 12-mile-long area that trends east-northeast, between Roughtop Mountain to the north and Hot Springs Dome to the south (Thomas, 1957). Roughtop Mountain and Hot Springs Dome respectively are underlain by Cretaceous (K-Ar age date of 92 +/- 5 Ma) and Tertiary (K-Ar age date of 62 +/- 3 Ma) granitic plutons (Chapman and others, 1982). The plutons intrude and contact metamorphose Mesozoic marine sedimentary strata, which also are cut by diverse faults, including regional-scale, east-northeast striking, thrust faults (Reifenstuhl and others, 1998). A carbonatite sill(?) is in the Triassic section of these strata and there are exposures of serpentinized, Cretaceous(?) mafic and ultramafic rock, mainly on Serpentine Ridge. The gold- and tin-bearing creeks flow normal to the trend of the tin belt. They head in the plutonic, metamorphic, and mafic/ultramafic rocks of Roughtop Mountain and Serpentine Ridge, which probably are the source(s) of some of the metalliferous minerals in the placer deposits. Concentrations of gold diminish toward the south, probably due to dispersion (Thomas, 1957). The Harter Gulch mine lies just to the north of a regional thrust fault; bedrock in the gulch is Triassic strata. Little is published specifically about Harter Gulch. **Workings:** Wayland (1961) states that mining occurred there in the early days of the district, and that Richards, the principal operator, reportedly collected $90,000 worth of gold in one summer (equivalent to 4,350 ounces of refined gold at $20.67 per ounce). Wayland also states that the tailings in Harter Gulch contain significant quantities of gold and some cassiterite. Moxham (1954) tested the tailings for radioactivity, presumably with negative results. The placer gravels, which consist of angular phyllite and graywacke and some rounded sandstone, were not completely washed by the early prospectors, leaving more gold in the tailings than that found in other creeks. A cubic yard of these gravels averaged 0.84 pound of concentrate that yielded 0.27 pound of cassiterite and 0.017 ounce of gold per cubic yard (Thomas, 1957). Cobb (1981) reports that placer mining may have occurred in 1977.(The Alaska Resource Data File., Tanana Quadrangle).

Homestake Creek

(Gold) **District:** Melozitna; **Quadrangle:** Tanana (B3); **Coordinates:** 65.3213°N, 151.3417°W; **Map:** 4-4; **Worked:** 1901-1943; **Access:** Trail from Yukon River at 65.2615°N, 151.2516°W; **Summary:** Tributary of Morelock Creek. Placer gold has been mined near confluence of this creek with Morelock Creek. Little is known about this occurrence. See also

Morelock Creek (Cobb, 1981, OF 81-1313A p. A9). **Geology:** Bedrock in the area of Homestake Creek consists of greenstone, metachert, dolomite, marble, calcareous schist, and siliceous mica schist (Mertie, 1934; Chapman and others, 1982). Although granitic rocks were reported by Eakin (1916), none were found in later mapping by Chapman and others (1963; 1982). The bedrock is complexly folded, faulted, and locally cut by sulfide-bearing quartz-calcite veins (Thomas and Wright, 1948 [RI 4322]; **Workings:** North Star Exploration, Inc.,1999 [Report 99-34]). As early as 1901, gold and cassiterite ('tin') placer mining had begun in the Morelock Creek area, concentrated near the mouths of Bonanza and Homestake creeks (Thomas and Wright, 1948 [RI 4322]). Brooks and Prindle (1911) reported that placer gold had been mined near the mouth of Homestake Creek, but little else is known about the site. There are no workings known on Homestake Creek, and no other reports of mining or production. The U.S. Bureau of Mines conducted a drilling program in the area in 1943 (Thomas and Wright, 1948 [RI 4322]). At least 23 churn drill holes were tested, including two fences across Morelock Creek and one across Bonanza Creek. In addition, the Bureau used test pits, caisson shafts, and open pit mining to sample the Morelock Creek area. Drilling on line 8.5 showed 0.27 pound of tin and 0.0085 ounce of gold per square foot of bedrock. Line 15 (on Bonanza Creek) yielded weighted averages of 0.1492 pound of tin and 0.0328 ounce of gold per square foot of bedrock. Line 10 contained no tin or gold. (The Alaska Resource Data File., Tanana Quadrangle).

Hoosier Creek

(Copper, Gold, Lead, Mercury, Tungsten) **Quadrangle:** Tanana (B1); **Coordinates:** 65.4490°N, 150.0911°W; **Map:** 4-8, 4-11; **Worked:** 1898-1997; **Purity:** 941 Au, 53 Ag; **Yield:** NOT; **Nugget Size:** 12 ozs; **Production:** ~10,000 ozs; **Access:** Trail from Rampart landing strip on the Yukon River at 65.5082°N, 150.1421°W; **Summary:** Tributary of Minook Creek. Poorly defined shallow (6 - 15 foot) pay streak 100 - 150 feet wide. Some of Gold is coarse, a nugget weighing 12 ounces was recovered. Gold is high grade; one assay indicated 941 Au and 53 Ag. **Minerals:** Hematite, ilmenite barite, magnetite, pyrite, gold, garnet, picotite, scheelite, zircon, native copper, galena, and a few grains of cinnabar. Bedrock mainly greenstone containing quartz veins as much as 18 inches thick. Gold was discovered in 1898 or soon thereafter and was mined in most years until as recently as 1952. Total production not known, but may have been about 10,000 ounces (Cobb, 1981, OF 81-1313A p. A10). Live water makes prospecting difficult. Gold mostly on bedrock. Upper part of valley has not been productive. Preparations for hydraulicking 2 miles above mouth in 1904. (Cobb, 1977, OF 77-432, p. 41-42). **Geology:** Hoosier Creek drains an area underlain upstream by Cambrian or late Proterozoic quartzite, feldspathic quartzite, grit, calcareous siltstone, limestone, and chert (Reifenstuhl and others, 1997 [RI 97-15a]). The placer operations are underlain by Triassic(?) Rampart Group mafic and intermediate volcanic and intrusive rocks, chert, and tuff. Gold was discovered before 1902 (Mertie, 1934), and by 1904 preparations were made for hydraulicking 2 miles above the mouth of the creek (Prindle and Hess, 1905). The dimensions of the pay zone were never well defined, but averaged between 100 and 150 feet wide (Mertie, 1934). By 1928, Hoosier Creek was being drift mined, mostly during the winter, because there was not enough water or enough gradient in the stream to run the sluiceboxes during the summer (Mertie, 1934). **Workings:** U.S. Geological Survey reports describe mining on the stream fairly continuously through 1938 (Smith, 1939 [B 917- A]; Cobb, 1977). In 1957, Saunders (1957 [MR 194-17]) reported that the Weisner Trading Company intended to mine on Hoosier Creek, and implied that mining had occurred on the creek in the recent past. Mining or mining claims were active in 1967 (Heiner and others, 1968). Eakin and others (1985) and Bundtzen and others (1987) reported that the Hoosier Creek Mining Company was successful. In 1988, Willford Mining cleared ground, stripped overburden and constructed settling ponds in preparation for mining, and mined into 1990 (Green and others, 1989; Bundtzen and others, 1990; Swainbank and others, 1991). In 1991, Lucas Mining, and in 1992, Jimmy Dale, operated on Hoosier Creek (Bundtzen and others, 1992; Swainbank and others, 1993). Mining was reported on Hoosier and Little Hoosier creeks in 1996 (Swainbank and others, 1997), and in 1997, Frank Willford mined a tributary to Hoosier Creek (Swainbank and others, 1998). Some of the gold mined was coarse, including a $250 nugget (gold at $20.67? per ounce). The gold was about 941 fine (Mertie, 1934). The heavy minerals collected with the gold included hematite, ilmenite, barite, magnetite, pyrite, garnet, picotite, scheelite, zircon, native copper and cinnabar (Mertie, 1934; Waters, 1934). The stream gradient is 100 feet per mile. Eight gold grains, including several small nuggets and one large nugget from Hoosier Creek were analyzed by electron microprobe (Newberry and Clautice, 1997). Five of the grains had nearly identical compositions characterized by cores with fineness of 800 to 810, Cu and Hg below detection, and Bi and Te at detection levels. Two other grains had cores of higher fineness, and rims with finenesses up to 990. Most of the grain compositions are similar to those of the gold recovered from the nearby Elephant Mountain deposit (Newberry and Clautice, 1997). Diabase bedrock in Hoosier Creek contains auriferous quartz-pyrite veins that possibly are the source of some of the placer gold (Hess, 1908). The diabase is covered by 6 to 15 feet of alluvium. White mica from one of the veins yielded an Ar40/Ar39 age of 72.2 +/- 0.3 Ma (Reifenstuhl and others, 1997 [PDF 97-29h]). (The Alaska Resource Data File., Tanana Quadrangle).

Hunter Creek

(Gold) **Quadrangle:** Tanana (B1); **Coordinates:** 65.4779°N, 150.0549°W (Lower); 65.4729°N, 149.9947°W (Upper); **Map:** 4-8, 4-11; **Worked:** 1896-1997; **Purity:** 919 Au; **Nugget Size:** 10 ozs; **Access:** Trail from Rampart landing strip on the Yukon River at 65.5082°N, 150.1421°W; **Summary:** Tributary of Minook Creek. Stream cuts through benches east of Minook Creek and into fractured greenstone of Permian to Triassic Rampart Group in a narrow V-shaped valley with low gravel benches. Both creek and bench deposits have been mined; in places 16 inches of bedrock had to be taken up with the gravel. Much of gold is "pumpkin seed," but some is nuggets as large as 10 ounces. One assay indicated fineness of 919. Mining began by 1896 and continued until at least as recently as 1940. Most concentrate samples and much of mining in part of creek in Livengood quad. 'Total production not known, but probably was a few tens of thousands of ounces. See also Hunter Creek, Livengood quad (Cobb, 1981, OF 81-1313A p. A10). **Minerals:** Barite, cassiterite, magnetite, ilmenite, cinnabar, pyrite, galena, silver, hematite, and native copper also occur in the placer concentrates (Prindle and Hess, 1905; Hess, 1908; Waters, 1934). Three nuggets of mercurian to nearly pure silver were examined by the Alaska Division of Geological & Geophysical Surveys (Forbes and Cannon, 1991). One nugget was compositionally similar to silver from a Bonanza-type silver deposit (such as the Chanarcillo deposit in Chile; Guilbert and Park, 1986), which suggests the possibility of a silver lode nearby. Eleven gold grains, including several small nuggets and one large nugget from Hoosier Creek were analyzed by electron microprobe (Newberry and Clautice, 1997). Seven of the grains had Ag-Hg depleted rims indicative of significant residence in the near-surface environment; the other 4 grains were essentially unzoned (Newberry and Clautice, 1997). One of the gold grains contained up to 2 percent copper and inclusions of bornite-chalcocite and galena. **Geology:** Hunter Creek is a deeply incised tributary of Minook Creek, which it joins about 3 miles above its junction with the Yukon River. It was named for William Hunter, who discovered gold there in 1896. The discovery claim is approximately 8,000 feet upstream from the confluence with Minook Creek (Waters, 1934). The original mining claims apparently stretched at least several miles up the creek into the feeder creeks, and into the adjacent Livengood quadrangle (Freeman and Schaefer, 1999; ARDF site LG185). Reifenstuhl and others (1997 [RI 97-15a]) mapped the underlying bedrock as undivided Triassic Rampart Group gabbro and intermediate plutonic and volcanic rocks, tuff and chert. These rocks locally are weakly foliated and metamorphosed. Tertiary volcanic rocks lie nearby to the north and west. **Workings:** Mining on Hunter Creek began in 1896 (Spurr, 1898). Spurr reported that gold was found throughout the stream but that mining initially took place 1.25 miles above the mouth. Mining quickly spread upstream into the Livengood quadrangle (Freeman and Schaefer, 1999). Hunter Creek also contains the only bench placer deposit known in the Rampart district (Mertie, 1934). The discovery claim was 16 feet above the creek and the gravels were 5 to 12 feet thick with up to 40 feet of muck overburden (Mertie, 1934). The gold was concentrated in the upper 18 inches of fractured diabase and quartzite bedrock. The gold was mostly smooth and fine-grained, although some nuggets weighed as much as 10 ounces. A bench south of the valley mined in 1957 produced coarser gold than the middle and north parts of the valley, where the stream now flows (Saunders, 1957 [MR 194-17]). From about 1900 to 1940, Hunter Creek was mined by surface methods (Cobb, 1977). Hydraulic mining on Hunter Creek in the late 1920s was the largest operation in the Rampart district. Williams (1951) reported that Emil and Albert Swanson were mining on lower Hunter Creek in the early 1950's. They mined six cuts covering about 80,000 square feet of bedrock. Bill Thomas (T and T Mining Company) mined the left-limit bench on lower Hunter Creek in 1957, using hydraulic giants (Saunders, 1957 [MR 194-17]). In 1962, the Idaho Bar Mining Company was preparing to mine Idaho Bar, a ridge between Hunter and Little Minook Creeks (Saunders, 1962). Mining claims were held or worked by Hunter and Swenson in 1967 (Heiner and others, 1968). William Carlo operated a hydraulic giant on Hunter Creek in 1975 (Carnes, 1976). More recent mining on Hunter Creek includes work in 1989 by Bill Carlo and Steve Losonsky, and in 1991 by Bob Bettisworth (Bundtzen and others, 1990, 1992). In 1996, Green Mining and Exploration operated upstream from Losonsky's smaller operation. Losonsky continued mining in 1997 (Swainbank and others, 1998). (The Alaska Resource Data File., Tanana Quadrangle).

Hutlinana Creek

(Gold) **Quadrangle:** Tanana (A1); **Coordinates:** 65.1852°N, 150.0118°W, 65.1595°N, 150.1542°W (Elliott Hwy Xing, location 53); **Map:** 4-8; **Eureka Road Right-of-Way:** ±50 of centerline; **Worked:** 1902-???; **Access:** Trail from Elliot Highwat at 65.1588°N, 150.1428°W; **Summary:** Tributary of the Hutlitakwa Creek. Placer gold in upper part of basin, in northern tributaries, but there has been no successful mining. The ground is deep and unfrozen, so small-scale operations are not practicable. Some (or possibly all) of the reported gold may be in the part of the basin in the Livengood quadrangle. Includes references to Hootlenana Creek and Hutlina Creek (Cobb, 1981, OF 81-1313A p. A11). Live water in gravels prevented small-scale mining (Cobb, 1977, OF 77-432, p. 47). **Geology:** Bedrock in the area of the Hutlinana Creek placer gold prospects consists of Jurassic or Cretaceous clastic sedimentary strata that are cut by thrust faults and locally intruded by one or more dikes of unknown composition (Chapman and others, 1982; Reifenstuhl and others, 1998). Reifenstuhl and others (1998) mapped thrust faults along both sides of this part of Hutlinana Creek. A stampede to the Hutlinana Creek area occurred in 1902 and colors and occasional good prospects were reported (Collier, 1903; Hess, 1908). Mertie (1934) reported that some gold had been found on most of the northwest tributaries, but no good pay streak had been located. (The Alaska Resource Data File., Tanana Quadrangle).

Idaho Bar

(Gold)) **Quadrangle:** Tanana (A2); **Coordinates:** 65.4580°N, 150.0104°W; **Map:** 4-8, 4-11; **Worked:** 1913-1989; **Access:** Trail from Rampart landing strip on the Yukon River at 65.5082°N, 150.1421°W; **Summary:** This site represents an area of placer mining on Idaho Bar, the upland between Hunter and Little Minook creeks. Placer prospecting and mining have been described on the south slopes of Idaho Bar near Little Minook Creek, extending into the adjacent Livengood quadrangle (Freeman and Schaefer, 1999 [LG186]). For this record, the site is on a small tributary of Little Minook Creek, known locally as Pup 8, in the southwest quarter of section 3, T. 7 N., R. 12 W., of the Fairbanks Meridian.

The location is accurate within 1000 feet. This site roughly corresponds with the site for Idaho Bar, U.S. Bureau of Land Management MAS number 0020480074. **Minerals:** Ilmenite, hematite, and magnetite (Mertie, 1934). **Geology:** Bedrock, poorly exposed in the Idaho Bar area, consists of volcanic and sedimentary rocks of the Triassic to Mississippian, Rampart Group (Reifenstuhl and others, 1997 [RI 97-15a]). The sedimentary rocks include argillite, phyllitic argillite, volcaniclastic rocks, chert, and impure limestone. Tertiary rhyolite flows or domes are on the western slopes of Idaho Bar and Minook Creek, and on the northern slopes of Hunter Creek. Coarse and shotty gold occurs in Pliocene(?) gravels about 1,000 feet higher in elevation than the mouth of Little Minook Creek (Mertie, 1934). The coarse gravels contain quartzite, quartzite breccia, vein quartz, and chert (Prindle and Hess, 1905). Boulders 2 to 3 feet in diameter are common. Although there was not enough water for large-scale mining, prospecting occurred in 1913, the late 1920's, and the 1930's (Wimmler, 1926; Mertie, 1934; Smith, 1934; Smith, 1936; Smith, 1937). Sporadic mining at Idaho Bar took place from 1953 through 1989 (Alaska Kardex files). Idaho Bar Mining Co., operated by Ace Parker and Kosta Melinkoff, began a new operation in 1962 on Idaho Bar (Saunders, 1962). Their preparatory work included planning to pump water for mining from the upper part of Little Minook Creek. Stripping overburden and mining with a bulldozer occurred in 1972 and 1973 (Alaska Kardex files). Resource Associates of Alaska, Inc. (RAA) trenched and sampled in 1981, and stripped, mined, and tested the gold content of the gravels in 1984. (The Alaska Resource Data File., Tanana Quadrangle).

Idaho Gulch

(Gold, Monazite, Niobium, Uraninite, Silver, Tin) **Quadrangle:** Tanana (A2); **Coordinates:** 65.0785°N, 150.9188°W; **Map:** 4-5, 4-6; **Worked:** 1912-1956; **Yield:** 0.02 oz/yd^3; **Production:** 1907-1956=61 ozs; **Access:** Light duty road from Tofty at 65.0796°N, 150.9232°W (gulch Xing); **Summary:** Tributary of Sullivan Creek. Placer deposit, small, discontinuous, not particularly rich concentrations in as much as 40 feet of gravel; total depth to bedrock is 24 - 85 feet. Total recorded production, 1907-56, was 61 ounces gold and 300 lbs cassiterite concentrate; probably total production was much more and mainly included with that from Sullivan Creek. **Minerals:** Gold, ca ssiterite, aeschynite, columbite, monazite, and zircon. Tailing piles contain cobbles of brecciated quartz with brown tourmaline and cassiterite. Recovery from samples of tailings averaged 2.67 lb concentrate per yd^3 containing 1.00 lb Sn and 0.02 oz Au. See also Sullivan Creek (Cobb, 1981, OF 81-1313A p. A11). Drift mining in 1919-21 recovered 0.77 lb cassiterite per bedrock foot. Gold found farther upstream than cassiterite (Cobb, 1977, OF 77-432, p. 48). **Geology:** The Idaho Gulch mine is one of a group of cassiterite- and gold-bearing placer deposits known as the Tofty tin belt, a 12-mile-long area that trends east-northeast, between Roughtop Mountain to the north and Hot Springs Dome to the south (Thomas, 1957). Roughtop Mountain and Hot Springs Dome respectively are underlain by Cretaceous (K-Ar age date of 92 +/- 5 Ma) and Tertiary (K-Ar age date of 62 +/- 3 Ma) granitic plutons (Chapman and others, 1982). The plutons intrude and contact metamorphose Mesozoic marine sedimentary strata, which also are cut by diverse faults, including regional-scale, east-northeast striking, thrust faults (Reifenstuhl and others, 1998). A carbonatite sill(?) is in the Triassic section of these strata and there are exposures of serpentinized, Cretaceous(?) mafic and ultramafic rock, mainly on Serpentine Ridge. The gold- and tin-bearing creeks flow normal to the trend of the tin belt. They head in the plutonic, metamorphic, and mafic/ultramafic rocks of Roughtop Mountain and Serpentine Ridge, which probably are the source(s) of some of the metalliferous minerals in the placer deposits. Concentrations of gold diminish toward the south, probably due to dispersion (Thomas, 1957). The Idaho Gulch deposit lies on or just south of the trace of an east-northeast-striking thrust fault. The bedrock surface of Idaho Gulch is marked by bold scarps that do not have any surface expression (Eakin, 1915). Well-rounded cobbles of brecciated quartz with tourmaline and cassiterite are common in the boulder piles (Wayland, 1961). Cassiterite placer deposits have been found as far as 1,500 feet upstream from the access road, and as much as 2,500 feet downstream from it. The cassiterite tends to occur as well-rounded to subangular particles that range in size from microscopic to several inches across. The larger sizes often are accompanied by vein quartz and tourmaline, as well as by fragments of sedimentary country rock. The smaller sizes generally are free of impurities (Thomas,1957). The pay streaks in Idaho Gulch are discontinuous. **Workings:** Extensive drilling in 1912 led to the discovery of gold and the start of drift mining in 1912-1913 (Eakin, 1915). Since then, some drilling was done by the Alaska Gold Dredging (1929) and Cleary Hill Mines (1940-1941) companies (Wayland, 1961). The results of the latter work led to drift mining of a small pay streak. The tailings from this mining contained a higher proportion of light-colored gravels and boulders. McGee and Strandberg Mines, Inc., were active on Idaho Gulch in 1967 (Heiner and others, 1968). The U.S. Bureau of Mines studied the placer tin deposits around Tofty in 1954-1956 (Thomas, 1957). Their channel samples of tailings in Idaho Gulch yielded an average of 1.00 pound of 'tin' (cassiterite) and 0.02 ounce of gold per cubic yard. They also reported sporadic tin in drill holes on both Idaho and Tofty gulches. By 1956, production from Idaho Gulch was 61 ounces of gold and 300 pounds of cassiterite (Thomas, 1957). In Idaho Gulch and Miller Gulch, pan concentrates contained 0.2-7.0 percent niobium-bearing minerals, and concentrates from churn-drill cuttings contained up to 0.6 percent CeO2 and 1.8 percent Nb2O5 (Warner and others, 1986). These minerals are often associated with carbonatite, and may be the result of weathering of the carbonatite sill(?) in the Triassic country rocks. Moxham (1954) reported that the eU content of the concentrates ranged from 0.015 percent to 0.035 percent, and that one sample contained 2.3 percent eU. Cobb (1977) speculated that these values implied the presence of aeschynite, columbite, monazite, and zircon. (The Alaska Resource Data File., Tanana Quadrangle).

Killarney Creek

(Gold, Tin) **Quadrangle:** Tanana (A1); **Coordinates:** 65.1143°N, 150.7404°W; **Map:** 4-10; **Worked:** 1912-1997; **Access:** Trail from Tofty Highway at 65.0913°N, 150.8511°W; **Summary:** Tributary of Cooney Creek. Prospected for several years after 1912 by churn drilling and shafts. Small rounded pebbles of cassiterite and very fine gold on Jura-Cretaceous phyllite bedrock at depths of 40 - 80 feet in belt 1/2 mile wide north of junction of Gold Basin and Killarney Creeks. May mark northeast limit of the Tofty "tin belt." Creek heads in Roughtop Mountain quartz monzonite intrusive (Cretaceous; 90 m.y.). No record of mining. See also Gold Basin Creek and Coonie Creek (Cobb, 1981, OF 81-1313A p. A12). **Geology:** The Killarney Creek prospect is one of a group of cassiterite- and gold-bearing placer deposits known as the Tofty tin belt, a 12-mile-long area that trends east-northeast, between Roughtop Mountain to the north and Hot Springs Dome to the south (Thomas, 1957). Roughtop Mountain and Hot Springs Dome respectively are underlain by Cretaceous (K-Ar age date of 92 +/- 5 Ma) and Tertiary (K-Ar age date of 62 +/- 3 Ma) granitic plutons (Chapman and others, 1982). The plutons intrude and contact metamorphose Mesozoic marine sedimentary strata, which also are cut by diverse faults, including regional-scale, east-northeast striking, thrust faults (Reifenstuhl and others, 1998). A carbonatite sill(?) occurs in the Triassic section of these strata and there are exposures of serpentinized, Cretaceous(?) mafic and ultramafic rock in the area, mainly on Serpentine Ridge. Bedrock at the Killarney Creek prospect is Triassic or Permian siliciclastic rocks (argillite, sandstone, and shale), characterized by subphyllitic to phyllitic textures (Reifenstuhl and others, 1998). The gold- and tin-bearing creeks flow normal to the trend of the tin belt. They head in the plutonic, metamorphic, and mafic/ultramafic rocks of Roughtop Mountain and Serpentine Ridge, which probably are the source(s) of some of the metalliferous minerals in the placer deposits. Concentrations of gold diminish toward the south, probably due to dispersion (Thomas, 1957). Killarney Creek delineates the northeastern boundary of the tin belt (Thomas, 1957). The creek gravels contain cobbles of a coarse-grained biotite granite probably from Roughtop Mountain (Wayland, 1961). Bedrock in Killarney Creek consists of Triassic sedimentary rocks (Reifenstuhl and others, 1998). **Workings:** Prospecting began on Killarney Creek in 1912 (Wayland, 1961). Potential gold-bearing gravel deposits were drilled the following years by Hanson and Bock, and by Howell and Bargery (Brooks, 1916). The cassiterite and gold occur on phyllite bedrock for approximately 1,000 feet along Killarney Creek, at a depth of 40 to 80 feet. Unlike the other deposits in the tin belt, the cassiterite is fine; the largest piece recovered from one drilling program was only one inch in diameter (Wayland, 1961). More recent work on Killarney Creek has included testing the gravels for gold and tin by Bruce Savage in 1989 (Bundtzen and others, 1990). In 1997, Bed Rock Enterprises conducted placer exploration on Killarney Creek, as well as on Roughtop Mountain (Swainbank and others, 1998). (The Alaska Resource Data File., Tanana Quadrangle).

Lancaster Creek

(Gold) **Quadrangle:** Tanana (A6); **Coordinates:** 65.2051°N, 152.7485°W; **Map:** 4-3; **Worked:** 1917-; **Access:** Trail from Yukon River at 65.2016°N, 152.7584°W; **Summary:** Tributary of the Yukon River. Open cut, 1917. No production reported (Cobb, 1981, OF 81-1313A p. A12). **Geology:** The country rocks in the Lancaster Creek area are upper Precambrian or lower Paleozoic schist, quartzite, phyllite, and slate, and Cretaceous granite (Chapman and others, 1982). Cobb (1977) does not describe any other work at this site. (The Alaska Resource Data File., Tanana Quadrangle).

Little Boulder Creek

(Gold, Tungsten) **Quadrangle:** Tanana (A2, A3); **Coordinates:** 65.1540°N, 151.0589°W; **Map:** 4-2; **Worked:** 1918-2003; **Access:** Light duty road and Trail from Tofty, Alaska at 65.0935°N, 150.8815°W; **Summary:** Tributary of Boulder Creek. Prospecting, 1918-19, of low-grade placer-gold ground. Layers and lenses of angular slate fragments in 6 - 12 feet of silt (Cobb, 1981, OF 81-1313A p. A12). This site represents an area of placer mining on Little Boulder Creek and its northern tributaries (Queen Twin Creek and 4 unnamed tributaries) that extends approximately 5 miles in an east-to-northeast direction. The site is in the southwest quarter of section 20, T. 4 N., R. 17 W., of the Fairbanks Meridian. The location is accurate within several hundred feet. This site roughly corresponds with the site for Little Boulder Creek, U.S. Bureau of Land Management MAS number 0020480072. **Geology:** The Little Boulder Creek area is underlain mainly by Jurassic or Cretaceous clastic sedimentary rocks, which are intruded and contact metamorphosed by the Cretaceous monzodiorite to granite Roughtop Mountain pluton (Chapman and others, 1982; Reifenstuhl and others, 1998). Locally, Triassic or Permian phyllitic argillite, sandstone, and shale are in thrust contact with the Jurassic or Cretaceous sedimentary rocks (Reifenstuhl and others, 1998). North-trending, high-angle faults cut all of the rocks. **Workings:** Early references to Little Boulder Creek reported that the placer ground was thin, with low gold grades (Brooks, 1918). Chapin (1919) described 6 to 12 feet of alluvium consisting of silt with layers and lenses of angular slate fragments. Farrell & Higgins were active on Little Boulder Creek in 1967 (Heiner and others, 1968). Other placer mining-related activities from 1967 to 1983 included overburden stripping, trenching, test hole digging, and portable sluicebox sampling (Alaska Kardex files). Sample cuts were made with a bulldozer in 1989 and gold and scheelite were recovered from the samples. In 1990, J.L. Wood and Vern Petefish worked on Little Boulder Creek. In 1991, Wood continued to work the creek, while Windy Hill Mining prospected some areas (Bundtzen and others, 1992). Placer mining continued through the 1990's, including sluicing 9,000 cubic yards of material with a bulldozer-backhoe operation in 1992 (Alaska Kardex files). Wood also mined on Little Boulder Creek in 1997-98 (Swainbank and others, 1998; Szumigala and Swainbank, 1999) and planned mining through

2002 (Swainbank and others, 2000; Szumigala and others, 2001; Swainbank and others, 2002; Szumigala and others, 2003). (The Alaska Resource Data File., Tanana Quadrangle).

Little Minook Creek

(Barium, Chromite, Copper, Gold, Lead, Manganese, Mercury(?), Silver, Tungsten) **Quadrangle:** Tanana (B1); **Coordinates:** 65.4562°N, 150.0332°W; **Map:** 4-11; **Worked:** 1893-1991; **Purity:** 930 Au, 75 Ag; **Nugget Size:** 17.5 ozs; **Production:** 65,000 ozs; **Access:** Trail from Rampart landing strip on the Yukon River at 65.5083°N, 150.1425°W; **Summary:** Tributary of Minook Creek. Creek cuts auriferous gravel benches east of Minook Creek. Gold in placers largely reconcentrated from bench gravels. Some of gold coarse. One 17.5 ounce nugget reported. Assays show fineness of about 920 Au and 75 Ag. Gold in base of 5 - 12 feet of locally derived stream gravels beneath as much as 30 feet of frozen silt and muck and in top 1 - 2 feet of bedrock. Bedrock is Permian-Triassic slate, shale, limestone, chert, and greenstone; rocks jointed and sheared; mineralized with pyrite and chalcopyrite; many quartz-carbonate and quartz veins, at least one of which is auriferous. Lode occurrence of rhodonite or rhodochrosite reported. **Minerals:** Gold, native copper, native silver, hematite, barite, pyrite, galena, chromite, ilmenite, magnetite, argentite, tetradymite, picotite, scheelite, cinnabar(?), garnet, zircon, and sphene. Gold discovered in 1893; mining began in 1896 and continued to as recently as 1975; nearly all by drifting, groundsluicing, and shoveling in. Total production probably about 65,000 fine ounces of gold; over half of total for district. Included reference to Little Mynook Creek (Cobb, 1981, OF 81-1313A p. A12). Samples when panned yielded small quantities of fine gold. All mining in lower 3 miles of course which flows through a bench east of Minook Creek. Gold in placers largely reconcentrated from bench gravels. Mammalian bones reported. Most of gold smooth and worn. Nuggets weigh 12+ ounces. No mineable placers upstream from bench gravels, the source of most of the gold. All productive placers are in the present creek gravels. Gold assayed at 930 Au (Cobb, 1977, OF 77-432, p. 55-57). **Geology:** Little Minook Creek drains a relatively small area of about 6 square miles. It is straight, V-shaped, and is incised to a depth of 500-700 feet (Hess, 1908). The bedrock consists of Triassic Rampart Group mafic intrusive and volcanic rocks and sedimentary rocks (Reifenstuhl and others, 1997 [RI 97-15a]). California and Idaho Bars, respectively to the south and north, were mapped by Chapman and others (1982) as covered by Tertiary gravels. **Workings:** Gold was discovered on Little Minook Creek in 1893 by John Minook, and was mined continuously from 1896 (Hess, 1908) to 1940 (Smith, 1942). The gold in the stream placers may be reconcentration of gold in bench placers (Hess, 1908). Spurr (1898) reported that crushed quartz veins in shear zones contained a small amount of gold. Native copper and silver were found in concentrates (Spurr, 1898). The stream placer deposits ranged from 100 to 150 feet wide and contained coarse, flattened gold (Mertie, 1934). At least one report described brown material sticking to the gold, making it difficult to clean (Williams, 1951). Williams hypothesized that this material was a natural cement. One placer gold sample had a fineness of 920 (Mertie, 1934). As of 1904, Prindle and Hess (1905) reported that approximately $475,000 worth of gold (about 23,000 ounces) had been mined (Cobb, 1977), all from the lowest 3 miles. Claim 8 above Discovery (approximately three miles from the mouth of Little Minook Creek) reportedly was the richest claim, producing $1,000,000 worth of gold (about 48,400 ounces) by 1931 (Mertie, 1934). Claims 1 and 2 below Discovery contained coarse cobbles and 5 to 6 feet of gravels covered by only a few feet of muck. Native copper occurred on many of the claims, as well as some native sulfur, hematite, barite, pyrite, galena, and chromite (Mertie, 1934; Waters, 1934). Following World War II, public reports on mining activity were sporadic. In 1951, the Little Minook Mining Company worked a pay streak 100 feet wide and reportedly had been moving downstream approximately 1,000-1,200 feet per year, with enough ground for several more years of mining (Williams, 1951). In 1957, Saunders (1957 [MR 194-17]) reported intent of the Weisner Trading Company to mine on Little Minook Creek during that year and implied that recent mining had occurred. Cunningham & Associates, Langford, and/or Minook Ltd. were active in 1967 (Heiner and others, 1968). Weisner Trading Company was operating a bulldozer on Little Minook Creek in 1975 (Carnes, 1976). In 1991 Munsell Mining, a new operator, started working on Little Minook Creek (Bundtzen and others, 1992). (The Alaska Resource Data File., Tanana Quadrangle).

Little Minook Jr. Creek

(Gold, Lead) **Quadrangle:** Tanana (B1); **Coordinates:** 65.4472°N, 150.0409°W; **Map:** 4-11; **Worked:** 1893-1997; **Purity:** 915-920 Au, 75-80 Ag; **Nugget Size:** 10 ozs; **Production:** Through 1904=7,250 ozs; **Access:** Trail from Rampart landing strip on the Yukon River at 65.5083°N, 150.1425°W; **Summary:** Tributary of Minook Creek. Placer gold discovered in 1898 or soon thereafter; intermittent small-scale production until at least World War II. Total production not known, but more than 14,325 ounces of gold. Short frequently dry creek with headwaters in Pliocene(?) high bench east of Minook Creek. Lower part of valley cut in Triassic-Permian diabase of Rampart Group; father upstream cut in chert and iron-stained tuff. Most of placers in gravels of present stream. Gravels 4 - 5 feet thick beneath 18 - 25 feet of muck and colluvium. Gold mainly in upper 1.5 feet of bedrock. Gold flat and flaky; largest reported nugget about 10 ounces. Fineness about 915-920 Au and 75-80 Ag. **Minerals:** Barite, galena, gold, hematite, ilmenite, magnetite, pyrite (Cobb, 1981, OF 81-1313A p. A13). Short, 2 mile long creek, frequently dry creek with headwaters in high gravel bench East of Minook Creek. Lower part of course sharply cut into greenstone bedrock. Narrow lower valley with a more open depression in bench near head. Width of pay about 60 feet. Said to be worked out as of 1904. Few bedrock benches with no surface expression. Some of gold very coarse. One nugget weighed 9.7 ounces. Most of gold smooth and probably reconcentrated from bench gravels; some rough and probably of local derivation. Many vertebrate remains in overburden (Cobb, 1977, OF

77-432, p. 58-59). **Geology:** Bedrock in the area of Little Minook Jr. Creek consists of gabbro, mafic volcanic and volcaniclastic rocks, argillite, chert, and limestone of the Triassic Rampart Group (Chapman and others, 1982; Reifenstuhl and others, 1997 [RI 97-15a]). The drainage area of the creek includes the Tertiary, gravel-covered California and Idaho bars, and the placer deposit is along the contact between Rampart Group rocks and the Tertiary gravels (Reifenstuhl and others, 1997 [RI 97-15a]). The gold is mainly in the present stream channel, but some is also in local benches. On one claim, the upper 1.5 feet of irregular greenstone bedrock surface contained most of the gold. Overlying this pay streak was 4 to 5 feet of frozen gravel, overlain in turn by 18 to 25 feet of frozen muck and slide rock (Mertie, 1934). The stream gradient is 385 feet per mile. The gold is flattened, suggesting that it probably was reconcentrated from bench placers (Hess, 1908). According to Mertie (1934), the gold is of comparable fineness to that of Little Minook Creek. The largest nugget was worth $200 (gold at $18 per ounce). **Workings:** Both drifting and surface mining techniques have been used in Little Minook Jr. Creek. Excess water hampered the drift mining, and a meager summer water supply hampered washing (Mertie, 1934). Mining reportedly was continuous on Little Minook Jr. Creek until 1935 (Smith, 1937; Cobb, 1977). In 1957, Saunders (1957 [MR 194-17]) reported that the Weisner Trading Company intended to mine on Little Minook Jr. Creek during that year and implied that mining had occurred on the creek in the recent past. Reports of more recent work are incomplete. Fairbanks Mining Company mined for 90 days in 1997 using a JD450 tractor and backhoe (Alaska Kardex files). As of 1904, Little Minook Jr. Creek had produced a total of $150,000 worth of gold (Hess, 1908), equivalent to approximately 7,250 ounces of gold (at $20.67 per ounce). Production since then has not been made public. Six placer gold grains from Little Minook Jr. Creek were analyzed by electron microprobe (Newberry and Clautice, 1997). They found three different composition populations, but all grains examined exhibited evidence for extensive leaching. Several grains contained native nickel inclusions. (The Alaska Resource Data File., Tanana Quadrangle).

McCaskey Bar

(Gold, Mercury) **Quadrangle:** Tanana (A1); **Coordinates:** 65.1675°N, 150.2131°W; **Map:** 4-1; **Worked:** 1920-1975; **Purity:** 802 Au, 191 Ag; **Nugget Size:** 0.387 oz; **Access:** Unimproved road from Elliot Highway at 65.1664°N, 150.2174°W; **Summary:** Low-grade placer at least 100 feet wide in bench gravels about 250 feet above level of present stream. D eposit covered by 15 - 18 feet of frozen overburden. Bedrock is altered phyllite and argillite cut by quartz veins. Most of gold within 1 foot of bedrock. One assay indicated fineness of 802 Au and 191 Ag. **Minerals:** ilmenite, picotite, gold, cinnabar, magnetite, zircon, and tourmaline. Mining, 1920-39 and probably since World War II (Cobb, 1981, OF 81-1313A p. A13). Low-grade placer on bench east of Eureka Creek at end of spur between Pioneer and Kentucky Creeks. Ground opened up in 1924 and mined continuously through 1931. Mining reported from 1924-39 (Cobb, 1977, OF 77-432, p. 60). **Geology:** Bedrock in the area of the McCaskey Bar placer deposit consists of Jurassic or Cretaceous clastic sedimentary strata that are cut by thrust faults and locally intruded by one or more dikes of unknown composition (Chapman and others, 1982; Reifenstuhl and others, 1998). Mertie (1934) described the bedrock in the mined area as altered phyllite and argillite, often so decomposed that no hard rock could be found. A dike that cuts the country rock is just as altered, and quartz veins in the bedrock are also decomposed and iron stained. Mertie (1934) described McCaskey Bar as an old terrace placer east of Eureka Creek, approximately 250 feet above the active stream channel. The gold appears to be concentrated in the upper foot or so of the bedrock, but minor amounts also occur throughout the gravels in streaks at least 100 feet wide. Mertie reported the fineness of gold in one sample as 802.75. The largest nugget reported as of 1931 was worth $8.00 (gold at $20.67 per ounce). Mining has been mainly by hydraulicking the frozen overburden and allowing the ground to thaw naturally. Water is scarce in summer and mining was intermittent (Mertie, 1934). Cobb's (1977) and Smith's (1933) reports of Farmer & Jones operating a hydraulic mine in 1928 in the Eureka Creek area actually refer to an operation on McCaskey Bar (Wimmler, 1926). Ray C. Pittman operated a bulldozer just south of McCaskey Bar on Eureka Creek in 1975 (Carnes, 1976). (The Alaska Resource Data File., Tanana Quadrangle).

McKinley Creek

(Gold) **Quadrangle:** Tanana (A1); **Coordinates:** 65.1662°N, 150.3184°W; **Map:** 4-7; **Worked:** 1902-2003; **Access:** Unimproved road and trail from Eureka, Alaska (Elliot Highway) at 65.1777°N, 150.2466°W; **Summary:** Tributary of Rhode Island Creek. Prospects reported in 1902 near Glenn Creek. No further mention of this creek in the literature (Cobb, 1981, OF 81-1313A p. A13). The location of placer prospect(s) along McKinley Creek is poorly documented, and references to McKinley Creek could pertain to any portion of the creek. For this record, the site is on the creek at the boundary between sections 13 and 24, T. 4 N., R. 14 W., of the Fairbanks Meridian, where mining claims were active in 1984. This site roughly corresponds with the site for McKinley Creek, U.S. Bureau of Land Management MAS number 0020480025. **Geology:** The McKinley Creek area is underlain by Mesozoic clastic sedimentary strata that are cut by thrust faults (Chapman and others, 1982; Reifenstuhl and others, 1998). A 100-foot-thick carbonatite sill or dike in Triassic or Permian argillite, sandstone, and shale crosses Kinley Creek and is cut along strike by thrust faults. Cretaceous or Jurassic rocks, mostly quartzite, sandstone, and shale north of Omega Creek are locally intruded by Cretaceous monzodiorite (Reifenstuhl and others, 1998). **Workings:** Good placer prospects on McKinley Creek were reported in 1902, but their exact locations were not known (Collier, 1903). Claims were staked by McIntyre and Youngstrom in 1966 and the property was active during 1967 (Heiner and others, 1968). Mining claims were maintained in the late 1990's through 2003 by Richard Ott, who mined on Omega Creek (Swainbank and others, 1998; Szumigala and Swainbank, 1999, Szumigala and others, 2001). (The Alaska Resource Data File., Tanana Quadrangle).

Melozimoran Creek

(Gold, Tin) **Quadrangle:** Tanana (A2); **Coordinates:** 65.4276°W , 152.9226°W; **Map:** 4-3; **Worked:** 1913-1943; **Access:** Trail from Yukon River at 65.2016°N, 152.7594°W to Ash Creek then via foot over the ridge at the head of Ash Creek to Melozimoran Creek; **Summary:** This record represents approximately located placer gold prospects and occurrences in Melozimoran Creek. The site is on the upper part of the creek, between the junctions of Grimm and Webories creeks, in the south half of section 16, T. 7 N., R. 26 W., of the Fairbanks Meridian. The prospects are referred to in Cobb (1977), and the site corresponds to location 26 of Eberlein and others (1977). This site roughly corresponds to the U.S. Bureau of Land Management location for Melozimoran Creek (MAS number 0020480130). **Geology:** Bedrock in the area of Melozimoran Creek consists of Cretaceous granitic rocks, and of lower Paleozoic quartzite and minor mafic meta-igneous rocks cut by barren, lenticular quartz veins (Eberlein and others, 1977; Chapman and others, 1982). Original bedding is obliterated in most places, but appears to be parallel to schistosity, which dips north more steeply than it does on the south flank of Moran Dome (Chapman and others, 1982). Coarsely crystalline to porphyritic granite locally underlies the ridges on the north side of Webories and Grimm creeks. Alluvium is largely confined to the floor of the modern valley. Prospect pits dug from 1913 to 1918 reportedly yielded a little gold and cassiterite (Martin, 1920). Claims were staked in the Grimm and Webories creeks area during 1943, but there was no production or significant development. Thirteen test pits dug by the U.S. Geological Survey in 1943 near the junction of Melozimoran and Webories creeks failed to find cassiterite or gold in either the bench or the modern stream gravels, but only one pit reached bedrock (at a depth of 6.5 feet) (Chapman and others, 1963). Twelve feet of gravel overlies bedrock in a prospect pit on Webories Creek a short distance above its mouth. (The Alaska Resource Data File., Tanana Quadrangle).

Miller Gulch

(Gold, Monazite(?), Niobium, Uraninite(?), Silver, Tin) **Quadrangle:** Tanana (A2); **Coordinates:** 65.0693°W, 150.9314°W; **Map:** 4-5, 4-6; **Worked:** 1912-1967; **Yield:** 0.17 oz/yd^3; **Production:** Through 1956=17,576 ozs Au, 2,668 ozs Ag; **Access:** Unimproved road from Tofty, Alaska. Road crosses Miller Gulch at 65.0722°N, 150.9363°W; **Summary:** Tributary of Sullivan Creek. Discovered by drilling program in 1912. Sporadic drift mining and re-sluicing of old tailing piles until 1940. Old channel cut across a series of terraces in bedrock with local enrichment at each channel-terrace intersection. Pay gravel in fairly continuous narrow pay streak beneath 35 - 120 feet of muck. Bedrock surface slopes gently to south except for some prominent terraces. Much of ground was worked out by 1931. **Minerals:** Gold, cassiterite, columbite, and probably aeschynite, monazite, and zircon (reference, Moxham, 1954 (C 317), p. 5) is not clear). (Cobb, 1977, OF 77-432, p. 63-64). **Geology:** The Miller Gulch mine is one of a group of cassiterite- and gold-bearing placer deposits known as the Tofty tin belt, a 12-mile-long area that trends east-northeast, between Roughtop Mountain to the north and Hot Springs Dome to the south (Thomas, 1957). Roughtop Mountain and Hot Springs Dome respectively are underlain by Cretaceous (K-Ar age date of 92 +/- 5 Ma) and Tertiary (K-Ar age date of 62 +/- 3 Ma) granitic plutons (Chapman and others, 1982). The plutons intrude and contact metamorphose Mesozoic marine sedimentary strata, which also are cut by diverse faults, including regional-scale, east-northeast striking, thrust faults (Reifenstuhl and others, 1998). A carbonatite sill(?) is in the Triassic section of these strata and there are exposures of serpentinized, Cretaceous(?) mafic and ultramafic rock, mainly on Serpentine Ridge. The gold- and tin-bearing creeks flow normal to the trend of the tin belt. They head in the plutonic, metamorphic, and mafic/ultramafic rocks of Roughtop Mountain and Serpentine Ridge, which probably are the source(s) of some of the metalliferous minerals in the placer deposits. Concentrations of gold diminish toward the south, probably due to dispersion (Thomas, 1957). The Miller Gulch placer deposit lies on or just south of the trace of an east-northeast-striking thrust fault. **Workings:** High-grade placers were discovered on the United States Association claim in lower Miller Gulch in 1912. By 1914, there were three different mines operating in the area, employing 112 men. The depths to bedrock over the area of interest ranged from 85 to 120 feet over a lateral distance of only 500 feet, and to as little as 45 feet upstream. The upper portions of the creek appear to have multiple horizons of continuous pay streak (Eakin, 1915). The pay streaks reportedly were long and narrow compared to other Tofty area placers. The stream gravels consisted of angular cobbles of dark phyllite, graywacke, and metadiorite. Cobbles of brecciated quartz with tourmaline and cassiterite were also reported. Cobb (1973) reported that fine brown tourmaline replaced all micaceous layers in one piece of subangular phyllite, and dense tourmaline filled crosscutting fractures. Cassiterite tends to occur as well-rounded to subangular particles that range in size from microscopic to several inches across. The larger sizes often are accompanied by vein quartz and tourmaline, as well as by fragments of sedimentary country rock. The smaller sizes generally are free of impurities (Thomas,1957). A considerable amount of gold and silver, and cassiterite were mined from Miller Gulch. Through 1956, 17,576 ounces of gold, 2,668 ounces of silver, and 101,875 pounds of cassiterite concentrate had been recovered (Thomas, 1957). The average grade was about 0.17 ounce of gold per cubic yard. Channel sampling of tailings by the U.S. Bureau of Mines in 1954-56 showed that the efficiency of tin recovery from different workings varied dramatically. One tailings pile yielded 1.26 pounds of tin per cubic yard, while another yielded 0.29 pound of tin per cubic yard. Claims were staked at this site in 1953 and 1958 and there was mining in 1967 (Heiner and others, 1968). In addition to gold and cassiterite, other economic minerals were reported in the Miller Gulch-Idaho Gulch area . Thomas (1957) and Barton (1962) reported that 0.2-7 percent niobium-bearing minerals were collected in pan concentrates, and Warner and others (1986) reported up to 0.6 percent CeO_2 and 1.8 percent Nb_2O_5 in concentrates from churn drill cuttings. A calculation of possible reserves of niobium, based on this sampling and on earlier placer reserve estimates by Thomas and others (1948) and Wayland (1961), indicated that there may be 100,000 pounds of Nb_2O_5 reserves in the Tofty placers (Southworth, 1984). Moxham (1954) reported that the eU content of 2 [concentrate?] samples was 0.026 percent and 0.017 percent, and Cobb (1977) speculated that these values implied the presence of aeschynite, columbite, monazite, and zircon. (The Alaska Resource Data File., Tanana Quadrangle).

Minook Creek

(Gold, Silver) **Quadrangle:** Tanana (A1, B1, C1); **Coordinates:** 65.3839°N, 150.1255°W ; **Map:** 4-8; **Worked:** 1882-1967; **Yield:** 0.145-0.193 oz/yd^3; **Nugget Size:** 1 to 4 ozs; **Production:** Through 1904=435 ozs; **Access:** Trail from Rampart landing strip on the Yukon River at 65.5085°N, 150.1425°W; **Summary:** Tributary of the Yukon River. Gold discovered in 1890's. Valley narrow; incised into closely folded middle Paleozoic slate, limestone, quartzite, and schist and Permian to Triassic diabase, shale, and tuff; Tertiary continental rocks near mouth. Valley has several terraces on east side at elevations above creek of 10 to about 1,000 feet near mouth; some terraces extend up tributary valleys; terraces approach creek gradient upstream. Top terrace capped by 100 feet of Pliocene(?) gravel. Terrace gravels auriferous; gold reconcentrated into rich placers on some of tributaries, but placers on main stream too low grade to support small-scale mining. Much of mining was bar sniping and small operations in richer spots. In places most of gold (probably of local derivation rather than from the benches) was in nuggets weighing 1 - 4 ounces. Total production certainly was no more than a few thousand ounces; may have been less than 1,000 ounces; most of reported mining probably was on tributaries. Nuggets of native silver accompany gold. Includes references to Big Minook Creek and Mynook Creek (Cobb, 1981, OF 81-1313A p. A14). Prospected as early as 1882; coarse gold found in 1893. Small areas said to have run 0.145-0.194 ounces per square yard of bedrock. High bench probably represents old valley floor of Minook Creek. Large hydraulic plant installed in 1905. Mining in 1931 was on west bank 1.5 miles below Slate Creek. Bedrock chert and phyllite with gravel 3 - 5 feet thick. Several large-scale operations were planned, but none materialized (Cobb, 1977, OF 77-432, p. 65-66). **Geology:** Minook Creek is in a steeply incised, V-shaped valley, and has several gold-bearing tributaries. On the east, they include Hunter, Little Minook, Hoosier, Chapman, Little Minook Jr., and Florida creeks. On the west, they include Ruby and Slate creeks. It is possible that much of the reported mining on Minook Creek was on tributaries. Minook Creek flows along a high-angle fault, with indicated west-side uplift (Reifenstuhl and others, 1997 [RI 97-15a]). In the vicinity of the placer mining, the creek flows across Lower Paleozoic or Proterozoic calcareous siltstone, sandstone, chert, and quartzite (Chapman and others, 1982). Mining on Minook Creek started around 1902 (Collier, 1903). The gold is not in the present stream bed, but in multiple, variously developed terraces ranging from several feet to approximately 800 feet above the present channel (Eakin, 1913). The lowest bench is 25 feet above the present stream bed and has proved to be the most productive. Hess (1908) reported that most of the gold was taken from tributaries, where small sections contained as much as $3-$4 per cubic yard (gold at $20.67 per ounce). The gradient on Minook Creek is approximately 40 feet per mile in the lower stretches where gold has been reported (Waters, 1934). Native silver nuggets also occur in the placers (Hess, 1908). **Workings:** Around 1931, after the early period of small-scale mining, larger scale mining was considered (Mertie, 1934), but none was ever undertaken, because the distance to water for sluicing is too great. Mining has remained small-scale, and the last recorded activity was in 1967 (Smith, 1942; Saunders, 1962; Heiner and others, 1968). The only production specific to Minook Creek was $10,000 through the fall of 1904 (Hess, 1908), equivalent to approximately 485 ounces of gold (Cobb, 1977). White mica from a quartz vein one mile west of Minook Creek, and due west of McDonald Bar has an Ar40/Ar39 plateau age of 73.1 +/- 0.3 Ma (Reifenstuhl and others, 1997 [PDF 97-29h]). The sample site is downstream from the Minook Creek placer, but similar quartz veins upstream may be the source of some of the placer gold. (The Alaska Resource Data File., Tanana Quadrangle).

Moose Creek

(Gold, Tin) **Quadrangle:** Tanana (B3, B4); **Coordinates:** 65.1377°N, 151.2503°W; **Map:** 4-2; **Worked:** 1963-1993; **Access:** Trail from Tofty, Alaska at 65.0933°N, 150.8818°W; **Summary:** Mining claims were staked in the 1960's, followed by placer mining, along Moose Creek, lower Boulder Creek, and Value Creek (Alaska Kardex files). The discovery claim notice for Moose Creek states that one of the corner claim posts is 150 feet from L.E. Anderson's old cabin. This cabin is presumed to be the one at the junction of Moose and Boulder creeks. Mining claim location notices show 18 claims in this claim block. The site is in the southeast quarter of section 29, T. 4 N., R. 18 W., of the Fairbanks Meridian. The location is accurate within about 500 feet. This site closely corresponds with the site for Moose Creek, U.S. Bureau of Land Management MAS number 0020480115. This record also includes information about Value Creek, MAS number 0020480116. **Geology:** The Moose Creek area is underlain mainly by Jurassic or Cretaceous clastic sedimentary rocks, and by Paleozoic limestone, dolomite, greenstone, and chlorite schist (Chapman and others, 1982). Serpentine Ridge, 1.5 miles to the south, consists of serpentinite and of mafic intrusive rocks that cut Mesozoic, predominantly marine sedimentary rocks. **Workings:** Eighteen or nineteen mining claims were staked between 1963 and 1966 by J & M Mining Co. along Moose Creek, Value Creek, and the part of Boulder Creek between Moose and Value creeks (Alaska Kardex files). The location notices state that gold was found on the discovery claim. Affidavits of annual labor from 1964 to 1993 list sampling, prospect holes, drilling, stripping, and sluicing. Recorded placer mining included 150 days in 1980, 40 days sluicing in 1981,

50 days sluicing in 1982, 20,000 feet of open cut in 1984, 22,000 feet of open-cut in 1985, and mining in 1990 and 1991. There is no record of production. (The Alaska Resource Data File., Tanana Quadrangle).

Morelock Creek

(Gold, Tin) **Quadrangle:** Tanana (B3, B4); **Coordinates:** 65°15'51.6°N, 151°18'07.8°W; **Map:** 4-4; **Worked:** 1901-1998; **Purity:** 922.2 Au; **Yield:** 0.00378-0.0274 oz/yd^3; **Nugget Size:** 1.25 ozs; **Production:** Less then 300 ozs Au; **Access:** Trail from Yukon River at 65.2615°N, 151.2516°W; **Summary:** Tributary of the Yukon River. Bedrock is Paleozoic (possibly some Precambrian and Mesozoic) metamorphosed sedimentary and mafic igneous rocks; many quartz veins barren except for minute pyrite crystals. Bedrock surface irregular; average thickness of gravel 5 - 6 feet beneath 2 - 3 feet of silt. Heavy minerals in basal few inches of gravel, on bedrock, and in crevices in bedrock. **Minerals:** Coarse gold, cassiterite, magnetite, ilmenite, hematite, and garnet. No lode source for gold or cassiterite has been found. Exploration by prospectors, U.S. Bureau of Mines, and Geological Survey. Sporadic prospecting and small-scale mining from about 1901 into 1940's. Total production probably no more than a few hundred ounces of gold (Cobb, 1981, OF 81-1313A p. A14-A15). Gold discovered, 1901. Sluicing on bench claim at confluence of Bonanza and Morelock Creeks in 1908. Two lines of churn-drill holes indicated averages of 0.1227 lb. Sn and 0.00378 ounces Au per yd^3. Small cuts mined from "reefs" in 1943 yielded 6.054 lb. Sn concentrate and 1.951 ounces Au from 71.3 yd^3 of material excavated. Gold is coarse with one nugget weighing 1.25 ounces and assayed at 922.2 fine. Gold- and tin-bearing gravels probably extend for 1.5 miles along creek and are 400 - 500 feet wide (Cobb, 1977, OF 77-432, p. 67-68). **Geology:** Bedrock in the area of the Morelock Creek gold-cassiterite (tin) placer deposit consists of greenstone, metachert, dolomite, marble, calcareous schist, and siliceous mica schist (Mertie, 1934; Chapman and others, 1982). Although granitic rocks were reported by Eakin (1916), none were found in later mapping by Chapman and others (1963, 1982). The bedrock is complexly folded, faulted, and locally cut by sulfidebearing quartz-calcite veins (Thomas and Wright, 1948 [RI 4322]; North Star Exploration, Inc., 1999 [Report 99-34]). **Workings:** As early as 1901, gold and tin mining had begun on Morelock Creek, mainly near the mouths of Bonanza and Homestake creeks (Thomas and Wright, 1948 [RI 4322]). The placer deposits consist of approximately 6 feet of thawed gravels that lie on an irregular bedrock surface. The gravels, which contain boulders as large as 24 inches, are overlain by 2.5 feet of organic matter. Although the local bedrock apparently is not the source of gold or cassiterite, the irregularities in the bedrock surface are excellent sites for collecting these heavy minerals. The gold and cassiterite are concentrated in the lower two feet of the gravels and occasionally in the upper few inches of the bedrock, where they have worked into joints (Thomas and Wright, 1948 [RI 4322]). Early reports describe mining on Morelock Creek in the years 1937 and 1938 (Smith, 1939 [B 910-A, B 917-A]). The U.S. Bureau of Mines conducted a drilling program in the area in 1943 (Thomas and Wright, 1948 [RI 4322]). At least 23 churn drill holes were tested, including two fences across Morelock Creek and one across Bonanza Creek. In addition, test pits, caisson shafts, and open-pit mining were used to sample the Morelock Creek area. Drilling on line 8.5 showed 0.27 pound of cassiterite and 0.0085 ounce of gold per square foot of bedrock. Line 15 (on Bonanza Creek) yielded weighted averages of 0.1492 pound of cassiterite and 0.0328 ounce of gold per square foot of bedrock. Line 10 contained no cassiterite or gold. North Star Exploration, Inc. (1999 [Report 99-34]) worked Doyon block 20 in 1998, and reported that Morelock Creek (along with other prospects) had potential for stockwork and vein gold deposits. Trenching of some of the placer deposits exposed mineralized quartz-calcite veins. (The Alaska Resource Data File., Tanana Quadrangle).

New York Gulch

(Gold) **Quadrangle:** Tanana (A3); **Coordinates:** 65.1031°N, 151.1578°W; **Map:** 4-2; **Worked:** 1921-1992; **Access:** Trail from Tofty, Alaska at 65.0930°N, 150.8815°W; **Summary:** Prospecting or mining reported in 1921 and 1926. **Minerals:** Magnetite, barite, ilmenite, picotite, and gold. See also American Creek (Cobb, 1981, OF 81-1313A p. A15). **Geology:** New York Gulch is in the Hot Springs mining district, approximately 15 miles west of the old town of Tofty. It drains the south side of Serpentine Ridge, and flows into American Creek. Serpentine Ridge consists of serpentinite and mafic intrusive rocks that apparently intrude Mesozoic, predominantly marine, sedimentary rocks, through which the stream channel cuts (Chapman and others, 1982). There is little in the public record about the placer deposit in New York Gulch. **Workings:** Prospecting or mining occurred in 1921 (Cobb, 1977); Smith (1939 [B 910-A]) reported mining there in 1926; and Wimmler (1929) reported open-cut mining during 1929. Waters (1934) stated that the mineral content of the placer deposit is similar to that of samples from American Creek. A contiguous claim block varying from 33 to 108 claims on New York Gulch, American Creek, and tributaries of American Creek was worked by American Creek Partners from at least 1981 to 1992 (Alaska Kardex files). Placer development, using Caterpillar D-9 and D-6 bulldozers, took place over a 3 to 5 month period during those years. Placer gold production from this mining is not known. Newberry and Clautice (1997) analyzed three small, irregularly-shaped nuggets from New York Gulch by electron microprobe. Each of the nuggets had a different composition, probably indicating different sources and/or degrees of exposure to erosion. One of the nuggets contained electrum and another nugget contained a tiny inclusion of native nickel. (The Alaska Resource Data File., Tanana Quadrangle).

Omega Creek

(Gold, Mercury, Tungsten) **Quadrangle:** Tanana (A1); **Coordinates:** 65.1031°N, 151.1579°W; **Map:** 4-7; **Worked:** 1899-1998; **Production:** Through 1904=880 ozs; **Access:** Unimproved road from Eureka, Alaska (Elliot Highway) at 65.1775°N, 150.2464°W; **Summary:** Tributary of Baker Creek. Bedrock mainly Lower Cretaceous black slate; quartzite in valley wall to south. Upper part of creek course along strike. North wall gentler than south wall. Gravel (containing a few boulders of quartzite and vein quartz) 5 - 10 feet thick beneath variable thin layer of muck. **Minerals:** Gold (some exhibiting crystal faces), many pyrite crystals, ilmenite, zircon, picotite, garnet, scheelite, and cinnabar. Gold discovered in 1899; mining from 1901 to as recently as 1975. No data on total production; probably was many thousands of ounces of gold. (Cobb, 1981, OF 81-1313A p. A15). One of principal producing creeks in Baker Creek area. Much of gold is rough. Nuggets contain quartz (Cobb, 1977, OF 77-432, p. 72-73). **Geology:** The Omega Creek area is underlain by Mesozoic clastic sedimentary strata that are cut by thrust faults and locally intruded by Cretaceous monzodiorite (Chapman and others, 1982; Reifenstuhl and others, 1998). In part, the upper valley of the creek is parallel to the N70E strike of the country rocks and of the thrust faults. The local bedrock is fissile slate and sandy phyllite (Mertie, 1934). Gravel deposits range in thickness from 5 to 10 feet under very thin overburden, and contain cobbles of slate, vein quartz, and quartzite. According to Prindle (1906), the gold-bearing bench gravels stretched over a couple of miles. Newberry and Clautice (1997) analyzed four gold grains by electron microprobe. Three have almost identical compositions, lack zoning, and contain anomalous tellurium. The other grain has a core with abnormally low (less than 0.2 percent) mercury content and low mercury-silver rims. Gold compositions are similar to those of placer gold at Rhode Island Creek and Eureka Creek. **Workings:** According to Prindle and Hess (1905), mining began in 1903, and by 1904 Omega Creek was established as economically important, having produced, in conjunction with Thanksgiving Creek, $18,200 in gold, which Cobb (1977) estimated to equal about 880 ounces. There are reports of fairly consistent mining on Omega Creek after 1904. In the late 1930's Omega Creek was considered to be a major producer in the Eureka area, and in 1938, it was the largest producer (Smith, 1939 [B 917-A]). Tony Lanning mined on Omega Creek In 1951 (Williams, 1951); Trumpey mined there in 1967 (Heiner and others, 1968); and Lloyd Hubbard ran a bulldozer on the creek in 1975 (Carnes, 1976). The only recent reference to mining was in 1998, when Richard Ott mined on Omega Creek (Szumigala and Swainbank, 1999). No official production figures have been made public, but Cobb (1977) estimated that several tens of thousands of ounces of gold were probably mined from Omega Creek. (The Alaska Resource Data File., Tanana Quadrangle).

Patterson Creek

(Gold, Silver, Tin) **Quadrangle:** Tanana (A2, A3); **Coordinates:** 65.0432°N, 150.9341W; **Map:** 4-5; **Worked:** 1907-Present; **Production:** Through 1956= 2,599 ozs Au, 385 Ag ; **Access:** From Tofty, Alaska, trail crosses Patterson Creek at 65.03964°N, 150.9716°W; **Summary:** Tributary of Williams Slough. Gold discovered in 1907 at mouth of Easy Money and Sullivan Creek. Most (if not all) of mining was on tributaries and benches between tributaries. According to Thomas (1957 (RI 5373), p. 7) production through 1956 was 2,599 ounces gold, 385 ounces silver, and 20,282 lb cassiterite concentrate. He also lists production from many of tributaries, so the source of the reported production is not clear. See also Cache Creek, Deep Creek, Sullivan Creek and Woodchopper Creek (Cobb, 1981, OF 81-1313A p. A15-A16). Gold discovered, 1907, at mouth of Easy Money and Sullivan Creek. Slate bedrock; gravels 35 - 70 feet deep (Cobb, 1977, OF 77-432, p. 74). **Geology:** The Patterson Creek placer mine is one of a group of cassiterite- and gold-bearing placer deposits known as the Tofty tin belt, a 12-mile-long area that trends east-northeast, between Roughtop Mountain to the north and Hot Springs Dome to the south (Thomas, 1957). Roughtop Mountain and Hot Springs Dome respectively are underlain by Cretaceous (K-Ar age date of 92 +/- 5 Ma) and Tertiary (K-Ar age date of 62 +/- 3 Ma) granitic plutons (Chapman and others, 1982). The plutons intrude and contact metamorphose Mesozoic marine sedimentary strata, which also are cut by diverse faults, including regional-scale, east-northeast striking, thrust faults (Reifenstuhl and others, 1998). A carbonatite sill(?) is in the Triassic section of these strata and there are exposures of serpentinized, Cretaceous(?) mafic and ultramafic rock, mainly on Serpentine Ridge. The gold- and tin-bearing creeks flow normal to the trend of the tin belt. They head in the plutonic, metamorphic, and mafic/ultramafic rocks of Roughtop Mountain and Serpentine Ridge, which probably are the source(s) of some of the metalliferous minerals in the placer deposits. Concentrations of gold diminish toward the south, probably due to dispersion (Thomas, 1957). Exactly how much mining was done on Patterson Creek is unknown. There are no remains of workings, although Brooks (1908, 1909) reported that gold was discovered in the creek in 1907. References were made to mining in 1913 and 1915 (Brooks, 1914, 1916) as well as in 1919 (Brooks and Martin, 1921). No specifics are given, however, and as Cobb (1977) and Eberlein and others (1977) point out, the references may actually have been to tributaries of Patterson Creek, like Sullivan Creek, Cache Creek, or Woodchopper Creek. (The Alaska Resource Data File., Tanana Quadrangle).

Pioneer Creek

(Gold, Mercury, Tungsten) **Quadrangle:** Tanana (A1); **Coordinates:** 65.1835°N, 150.2111°W (Lower), 65.2056°N, 150.12792°W (Middle), 65.2106°N, 150.1022°W (Upper); **Map:** 4-1; **Worked:** 1902-1998; **Production:** Through 1904=4,125 ozs; **Access:** Unimproved road from Eureka, Alaska at 65.1830°N, 150.2202°W; **Summary:** Tributary of Eureka Creek. Asymmetrical valley (steeper on south side) parallel to upper Eureka Creek. Several gravel-covered terraces on north valley wall. Richest is about 250 feet above and 2,000 feet north of creek; on lower benches placers are lower grade, but mineable in places. Bedrock is Jura-Cretaceous(?) pyritized sheared sandstone and slate, phyllite, and quartzite; cut by a few quartz veins. Gravels (locally derived) are 3 - 12 feet thick. Gold reconcentrated into rich placers where benches are crossed by small tributary streams. **Minerals:** Picotite, ilmenite, pyrite zircon, gold, sphene, barite, magnetite, garnet, scheelite, cinnabar, and tourmaline. Mining from 1902 to as recently as 1975. No data on total production; probably was many tens of thousands of ounces of gold. Includes references to Boothby Creek, Doric Creek, Jordan Bar, Last Bench, Seattle Bar, Seattle Jr. Creek, Skookum Creek and What Cheer Bar (Cobb, 1981, OF 81-1313A p. A16). All tributaries enter from NW cross several gravel-covered terraces. The most prominent terrace is 250 feet above creek (What Cheer Bar) SW of Doric Creek. All mining has been on benches or on tributaries that cut them. Lower bench at upstream end of productive terrace gravels (Jordan Bar) is of lower grade than deposits on higher benches. Creek gravels of Pioneer Creek generally too low grade to mine. Pay discovered on What Cheer Bar in 1902. Further discoveries extend limits of possible productivity from What Cheer Bar toward head of creek in 1905. Nuggets from What Cheer Bar contain slate. Gold reconcentrated into rich placers where benches are crossed by small tributary streams. Gold in basal gravel and top 1 - 2 feet of bedrock. Mining from 1902 to as recently as 1940. Production from Eureka and Doric Creeks and bench bars to fall of 1904 was about 4,125 fine ounces (Cobb, 1977, OF 77-432, p. 75-76). **Geology:** The area of Pioneer Creek (including its tributaries) is underlain by Cretaceous or Jurassic sandstone, siltstone, and shale (Chapman and others, 1982; Reifenstuhl and others, 1998). Pioneer Creek follows the trace of a fault, and all of the placer gold deposits are north of this fault. No intrusive rocks are mapped nearby. Yeend (1990) described four terraces at Eureka Creek, with the best-developed one at 250 feet above the present channel of Eureka and Pioneer creeks. Yeend (1990) also hypothesized that ancestral Eureka and Pioneer creeks flowed southwest toward Tofty during the Pleistocene and joined the ancestral Tanana River farther west than the present channel. Workings: Mining on Pioneer Creek and its tributaries began in 1902 (Collier, 1903). Most of the gold reportedly was mined, not from Pioneer Creek itself, but from benches on the northwest side of the creek or from secondary reconcentrations on the tributary creeks that cut across those benches (Mertie, 1934). Gold was discovered in 1902, just southwest of the mouth of Doric Creek. According to Eakin (1913), this site, locally named What Cheer Bar, was typical of the bench deposits, and consisted of quartzite and conglomerate gravels 3 to 10 feet thick. This, the richest bench, is approximately 250 feet above Pioneer Creek and 2,000 feet to the north. Mertie's (1934) description of the bedrock included pyritized sandstone, slate, and phyllite. Hess (1908) described the bedrock at the Doric Creek mine as slate, grit, and quartzite that strike N70E and dip steeply north. Brooks (1907) reported mining on Boothby and Skookum creeks in 1906. Chapin (1914) described mining on Seattle Jr. Creek. Hydraulic mining by J.R. Frank Co. on Doric Creek and What Cheer Bar in the late 1920s was one of the principal operations in the Eureka area. Mining on some or all of these creeks continued until 1940, after which there is a hiatus in published reports of activity. In 1967, Strandberg Mines, Inc., held active mining claims on Boothby, Doric, Pioneer, and Skookum creeks; Frank & Co. were active on Doric, Seattle Jr., and Pioneer creeks; and claims were active on North Fork and Skookum creeks (Heiner and others, 1968). in 1975, John Cole ran a bulldozer on Skookum Creek (plotted location looks closer to Eureka Creek), and Mary Sue Mines ran a bulldozer on placer claims in Skookum Creek during (Carnes, 1976). In 1991, Ross Novak finished mining on the Pioneer bench (Bundtzen and others, 1992). From 1990-92, Rick Swenson operated on Doric Creek (Swainbank and others, 1991; Bundtzen and others, 1992; Swainbank and others, 1993). In 1992, Rick Swenson and partner Ross Novak open-cut mined on Doric Creek during a 110-day season. In 1997, Rick Swenson mined again, as did Salter and Associates (Swainbank and others, 1998). In 1990-91 and 1993, Salter and Associates mined gold on Joe Bush Creek (Bundtzen and others, 1992, 1994), where the biggest obstacle to mining was 6 feet of blue clay just above the pay gravels. The clay was removed carefully in order to preserve the gold. In 1995, Salter and Associates trenched on Joe Bush Creek, and Kelley Mining Company mined North Fork Creek (Bundtzen and others, 1996). In 1998, Salter and Associates mined on Joe Bush Creek (Alaska Kardex files). Official production figures are not available for Pioneer Creek. As of the fall of 1904, Hess (1908) reported that production from Doric Creek , Eureka Creek, and associated bench bars was around $85,300, estimated by Cobb (1977) to have equaled 4,125 ounces of gold. Cobb also estimated that many tens of thousands of ounces of gold were mined from the Pioneer Creek area. (The Alaska Resource Data File., Tanana Quadrangle).

Quartz Creek

(Gold, Tin, Yttrium) **Quadrangle:** Tanana (A2); **Coordinates:** 65.1208°N, 150.8840°W; **Map:** 4-6; **Worked:** 1908-1998; **Access:** From Tofty; **Summary:** Tributary of Sullivan Creek. Bench deposit, about 2 miles north of Tofty tin belt, consists largely of creep and slope-wash material (3 - 4 feet thick) with a thin (3-foot) cover of silt. Gold placer not rich; small-scale mining in 1908-14 and 1930 reported. Extensive prospect drilling, 1940. A concentrate sample contained a little xenotime and a few grains of clear brown, angular cassiterite. Both gold, which has a greenish cast, and cassiterite are different from those in samples from within the tin belt (Cobb, 1981, OF 81-1313A p. A16). Bench deposit consists largely of creep and slope-wash material with a thin cover of silt. Homestake Bar is on slightly sloping hillside 1/4 mile from creek. Bedrock surface slopes toward creek more gently than ground surface. Bedrock surface steepens at uphill margin and marks limit of richer deposit. Water supply inadequate for continuous sluicing. Coarse gold reported 1908-09. **Minerals:** Ilmenite, pyrite, zircon, gold, xenotime [yttrium phosphate], tourmaline, and a few grains of cassiterite (Cobb, 1977, OF 77-432, p. 78). **Geology:** The country rocks in the area of this site include Mesozoic marine clastic sedimentary strata, and Cretaceous granitic plutons, mainly on Roughtop Mountain, that intrude and contact metamorphose the sedimentary rocks (Mertie, 1934; Chapman and others, 1982; Reifenstuhl and others, 1998). The bedded rocks are also cut by diverse faults, including regional-scale, east-northeast-striking thrust faults (Reifenstuhl and others, 1998). A carbonate sill(?) is in the Triassic section of these strata, and there are exposures of serpentinized Cretaceous(?) mafic and ultramafic rocks, mainly on Serpentine Ridge. Quartz Creek drains Serpentine Ridge and Roughtop Mountain, and crosses a north-dipping regional thrust fault. The gradient of the creek is about 100 feet per mile. **Workings:** The mined area on Quartz Creek was Homestake Bar, a bench approximately a quarter mile from the creek. Eakin (1913) reported that the bedrock surface dips towards the creek at a lesser angle than the ground surface. The uphill extent of the richer portion of the deposit is marked by a steepening of the bedrock surface. The overlying sediments consist of three to four feet of gravels, with only the basal foot well worn. These gravels in turn are overlain by 3 feet of silt. Waters (1934) reported that the heavy minerals in his samples of the placer included ilmenite, pyrite, zircon, gold, xenotime, tourmaline, and a few grains of cassiterite. In 1940, Cleary Hill Mines Co. ran a 185-hole drill program in the Quartz Creek area. No mention of cassiterite was made in any of the drill logs. Delima Placers mined Quartz Creek in 1998 (Szumigala and Swainbank, 1999). (The Alaska Resource Data File., Tanana Quadrangle).

Rhode Island Creek

(Gold) **Quadrangle:** Tanana (A1); **Coordinates:** 65.1755°N, 150.2608°W (Lower), 65.1837°N, 150.2758°W (Upper); **Map:** 4-1, 4-7; **Worked:** 1900s-1990s; **Purity:** 750-840 Au; **Nugget Size:** 0.87 ozs; **Access:** Rhode Island Creek crosses unimproved road from Eureka, Alaska at 65.1760°N, 150.2671°W; **Summary:** Tributary of Omega Creek. Creek flows across strike of Lower Cretaceous phyllite with quartz stringers. Both stream and bench deposits mined. Many vertebrate remains in silt overburden on benches. Gold on or in cracks in top few feet of bedrock. Largest nugget reported was about 0.87 ounces. No data on amount of production; probably in the tens of thousands of ounces of gold (Cobb, 1981, OF 81-1313A p. A17). Auriferous terrace deposits extend from Thanksgiving Creek to and beyond Rhode Island Creek nearly to Eureka Creek (Cobb, 1977, OF 77-432, p. 78). **Geologic description:** The country rocks in the area of Rhode Island Creek consist of thrust-fault-juxtaposed Cretaceous sandstone, siltstone, and shale, Triassic clastic sedimentary rocks, and Jurassic or Cretaceous quartzite (Chapman and others, 1982; Reifenstuhl and others, 1998). The area also contains small, elongate intrusions of Cretaceous or Tertiary monzodiorite, which have a northeast trend. Rhode Island Creek flows across the strike of the bedrock. Mining concentrated both on bench deposits and the modern stream channel. The bench deposits on both sides of the creek are 300 feet above the present channel. The easternmost bench deposit is known as Shirley Bar (Mertie, 1934). Collier (1903) reported that the local bedrock is schist without quartz veins, but bedrock in the area being hydraulically mined in 1931 consisted of phyllite with quartz stringers (Mertie, 1934). The mined gravels consisted of schist and mafic igneous rocks (Collier, 1903), and the overburden of silt contained mammoth, bison, and other vertebrate remains (Mertie, 1934). The gold occurs on bedrock and in places has worked into cracks in the bedrock to a depth of several feet (Mertie, 1934). Large amounts of bedrock were often moved in order to recover the gold. The pay streak above the mouth of Gold Run Creek was 70 feet wide. **Workings:** Hydraulicking, open cuts, and drift mining were employed, and Rhode Island Creek was economically significant by 1904 (Prindle and Hess, 1905). Most of the mining took place between 1930 and 1940 (Cobb, 1977). Much of it was done by Archie W. Pringle, who, as of 1951, had been mining there since 1933 (Williams, 1951). His company, Rhode Island Creek Mines, mined the east bench below the mouth of Gold Run Creek (Saunders, 1962). Rhode Island Creek Mines was mining with two bulldozers and a sluice in 1975 (Carnes, 1976). In 1993, Thurman Oil and Mining ran a large placer operation on Rhode Island and Cooney creeks (Bundtzen and others, 1994). Newberry and Clautice (1997) analyzed gold nuggets from Rhode Island Creek by electron microprobe. The gold in the nugget cores is homogeneous with respect to silver content, averaging 20 percent silver. Mercury content in the cores varies from 0.8 to 5 percent, and the fineness of the cores varies from 750 (high Hg) to 840 (low Hg). The rims of most of the gold grains are depleted in mercury, and two of the nuggets also have silver-depleted rims. (The Alaska Resource Data File., Tanana Quadrangle).

Rosa Creek

(Gold) **Quadrangle:** Tanana (B3); **Coordinates:** 65.3047°N, 151.2850°W; **Map:** 4-4; **Worked:** 1934-1998; **Access:** Trail from Yukon River at 65.2615°N, 151.2518°W; **Summary:** Tributary of Morelock Creek on which there was a little placer mining in about 1950-52. See also Morelock Creek (Cobb, 1981, OF 81-1313A p. A17). **Geology:** The bedrock across which Rosa Creek flows consists of Paleozoic limestone-dolomite, greenstone, chlorite schist, phyllite, quartz-mica schist, and chert (Chapman and others, 1982). The closest granitic pluton, a Tertiary quartz monzonite stock (radiometric age of 61.8 +/- 2.5 Ma), is about 1 mile to the north (Silberman and others, 1979). **Workings:** The first report of placer gold production from Rosa Creek was in 1934 (Alaska minerals locations database; MAS no. 0020480033), and there was a little placer mining on the creek in about 1950-52 (Cobb, 1981). Production figures from that period have not been published, and it is likely that the production from the Rosa Creek mine was included in the totals for placer mining on Morelock Creek. Information about more recent mining on Rosa Creek has not been made public. According to North Star Exploration, Inc. (1999 [Report 99-34]), this and other similar placer deposits seem to occur near major faults associated with quartz vein systems and stockwork veins. (The Alaska Resource Data File., Tanana Quadrangle).

Ruby Creek

(Bismuth, Gold, Silver, Copper, Lead) **Quadrangle:** Tanana (B1); **Coordinates:** 65.4042°N, 150.1484°W; **Map:** 4-8,

4-11; **Worked:** 1901-1991; **Nugget Size:** 8.5 ozs; **Production:** Through 1904=650 ozs; **Access:** Trail from Rampart landing strip on the Yukon River at 65.5085°N, 150.1427°W; **Summary:** Tributary of Minook Creek. Placer gold mining carried on sporadically from 1901 to at least as recently as 1971. Production data incomplete; about 650 fine ounces through 1904 and 11 ounces in clean up in 1970 from 2,000 ft^2 of open cut. Stream drains area underlain by Paleozoic schists and phyllite. Alluvium in lower valley 300 - 400 feet wide consists of locally derived medium-sized gravel and 1 - 4 feet of muck. Gold, probably also locally derived, on bedrock across entire width. Nuggets of gold as large as 8.5 ounces and of native silver as large as 2 ounces. Concentrates also contain native bismuth and abundant barite and garnet (Cobb, 1981, OF 81-1313A p. A17). All pay within 1.5 miles of mouth on bedrock. Gravel about 10 feet thick. Much garnet and some native silver with the placer gold. Small-scale mining intermittently from 1901 to 1916. Hydraulic plant being installed in 1905. Operated in 1906. Pyrite and garnets in concentrates (Cobb, 1977, OF 77-432, p. 81). **Minerals:** Galena, garnet, gold, native bismuth, native copper, native silver. **Geology:** Ruby Creek drains an area underlain by Triassic (Rampart Group) argillite and chert, Triassic gabbro, and diverse schists (Mertie, 1934; Chapman and others, 1982; Reifenstuhl and others, 1997 [RI 97-15a]). The placer mine is at the easternmost point of contact between Ruby Creek and the gabbro. The lower portion of the creek is deeply incised, and contains alluvial deposits 300 to 500 feet wide and 5 to 7 feet thick (Mertie, 1934). Mining on Ruby Creek was first reported by Collier (1903). The gold is concentrated on the bedrock surface and is evenly distributed across the entire width of the gravel deposits (Mertie, 1934). The deposits extend a mile and a half upstream from the mouth of Ruby Creek (Hess, 1908). The gold from Ruby Creek is iron-stained and rougher than the gold from Minook and Hunter creeks (Mertie, 1934). Other heavy minerals in the placer concentrates included garnet, barite, and silver nuggets up to 2 ounces (Cobb, 1973). Burand and Saunders (1966) described native bismuth in a concentrate and also referred to reports of native copper and galena. Ruby Creek contains unfractured, euhedral garnets of possible gem quality (T.K. Bundtzen, oral commun., 2001). **Workings:** Mining in the area was intermittent from 1902 to 1916. Production through the fall of 1904 was worth about $13,500 in gold (Hess, 1908), equivalent to about 650 ounces of gold at $20.67 per ounce. Harry Havrilack mined on Ruby Creek 1/4 mile from the mouth in 1957 and 1962 (Saunders, 1957 [MR 194-17]; Saunders, 1962), and mining was underway in 1967 (Heiner and others, 1968). Harry Havrilack operated a bulldozer on Ruby Creek in 1975 (Carnes, 1976). There is no record of further activity until Bundtzen and others (1992) reported that Williams Mining operated on Ruby Creek in 1991. An auriferous, copper-bearing quartz vein 3.5 miles upstream from the Ruby Creek placer deposit yielded an Ar40/Ar39 isotopic age of 61.2 +/- 0.3 Ma (Reifenstuhl and others, 1997 [PDF 97-29h]). Such veins may be the source of some of the placer gold. Hess (1908) speculated that the gold was locally derived from carbonaceous slate containing disseminated pyrite. (The Alaska Resource Data File., Tanana Quadrangle).

Schieffelin Creek

(Gold) **Quadrangle:** Tanana (A3, B4); **Coordinates:** 65.2632°N, 151.5083°W; **Map:** 4-12; **Worked:** 1907-1908; **Purity:** NOT; **Yield:** NOT; **Nugget Size:** NOT; **Production:** NOT ozs; **Access:** Yukon river at 65.228°N, 151.4361°W; **Summary:** Tributary of the Yukon River. Gold-bearing gravels prospected 1907-08. No further data on this creek. Includes reference to Shevlin Creek (Cobb, 1981, OF 81-1313A p. A17). This record represents the approximate location of placer gold prospects in Schieffelin Creek. The site is approximately 3.5 miles above its junction with the Yukon River, in the north half of section 17, T. 5 N., R. 19 W., of the Fairbanks Meridian, but the prospects could be almost anywhere along this approximately 12- mile long creek. The site corresponds to location 33 of Eberlein and others (1977). It is approximately 2 miles upstream from the U.S. Bureau of Land Management location for Schieffelin Creek (MAS number 0020480098). **Geology:** Bedrock in the area of the Schieffelin Creek placer gold prospect consists of greenstone, metachert, dolomite, marble, calcareous schist, and siliceous mica schist (Chapman and others, 1982). The bedrock is complexly folded and faulted. Brooks (1909) reported that there was prospecting for placer gold during 1907 and 1908, with discovery of gravels containing gold on Shevlin Creek. Cobb (1977) reported that there is no additional information about minerals or mining on Schieffelin Creek. (The Alaska Resource Data File., Tanana Quadrangle).

Seattle Creek

(Gold) **Quadrangle:** Tanana (A1); **Coordinates:** 65.1860°N, 150.2798°W; **Map:** 4-1, 4-7; **Worked:** 1904-1997; **Purity:** 780; **Access:** Unimproved road from Eureka, Alaska at 65.1768°N, 150.2212°W; **Summary:** Headwater tributary of Rhode Island Creek. Bedrock is Jurassic or Lower Cretaceous slaty graphitic arkose and schistose calcareous grit. Overlain by 8 - 30 feet of gravel and 1 - 3 feet of frozen muck. Gravel is fragments of local bedrock, quartzite, and vein quartz. Gold fine, bright, and shotty; probably derived from older bench gravels. A little mining in 1904 and 1931. See also Gold Run, Shirley Bar, Rhode Island Creek and Glen Creek (Cobb, 1981, OF 81-1313A p. A18). Bedrock is slaty graphitic arkose. Overlain by 8-30 feet of gravel and 1-3 feet of frozen muck. Gravel is fragments of local bedrock, quartzite, and vein quartz. Gold recovered to fall of 1904 was less than 5 fine ounces (Cobb, 1977, OF 77-432, p. 83). **Geology:** The country rocks in the area of Seattle Creek are Mesozoic sandstone, siltstone, and shale that are cut by thrust faults and intruded by Cretaceous or Tertiary monzodiorite (Chapman and others, 1982; Reifenstuhl and others, 1998). The local bedrock is graphitic, slaty arkose, overlain by 8 to 30 feet of fine gravel consisting of quartzite, vein quartz, and slate, overlain in turn by 1 to 3 feet of muck (Hess, 1908). Hess postulated that the gold was a placer reconcentration from older gravels. The gold's fineness is approximately 780 (Ed Salter, oral commun., 2000). Very little information about mining on Seattle Creek has been made public. Through the fall of 1904, only $100 worth of gold had been recovered from the creek, equal to less than 5 ounces at $20.67 per ounce. Mertie (1934) reported only sporadic mining at a small open cut in 1931. Strandberg Mines, Inc., and Pringle had active claims on Seattle Creek in 1967 (Heiner and others, 1968). Salter and Associates mined in the area for 28 years (Green and others, 1989), but stopped mining in 1997 due to depressed gold prices (Ed Salter, oral commun., 2000). Although production figures have not been made public, significant amounts of gold probably were recovered in the last few decades of the 20th Century. (The Alaska Resource Data File., Tanana Quadrangle).

Shirley Bar

(Gold, Lead, Mercury) **Quadrangle:** Tanana (A1); **Coordinates:** 65.1713°N, 150.2572°W; **Map:** 4-1, 4-7; **Worked:** 1901-1967; **Purity:** 792 Au, 204 Ag; **Nugget Size:** 2.75 ozs; **Access:** Unimproved road from Eureka, Alaska at 65.1768°N, 150.2212°W; **Summary:** Bench between Glenn Creek and Gold Run. Gravel 2 - 10 feet thick; gold distributed throughout, with largest nuggets about 2.75 ounces near surface. Most of gold fine and shotty; fineness about 792 Au and 204 Ag. Bar contributed gold to gulches to form high-grade placers. Bench placers semi-residual; angular gravel derived mainly from local Jura-Cretaceous sedimentary rocks. Sporadic mining from 1901 until as recently as 1938; no reliable data on amount of production. **Minerals:** Gold, pyrite, cinnabar, picotite, barite, galena, ilmenite, garnet, and sphene (Cobb, 1981, OF 81-1313A p. A18). Contributed gold to gulches to form high-grade placers. Some gold is coarse-and rough. Pay streak continues downhill as a body of subangular gravel on muck (Cobb, 1977, OF 77-432, p. 84). **Geology:** Shirley Bar is a bench placer deposit between Glenn Creek and Rhode Island Creek. It is on a broad, gently sloping bedrock spur that is covered with [Pleistocene?] alluvial deposits. The bedrock in the area (and at the workings) is predominantly Cretaceous sandstone, shale, and siltstone (Chapman and others, 1982; Reifenstuhl and others, 1998). A thrust-emplaced lenticular block of Triassic sedimentary rocks and Jurassic or Cretaceous quartzite are mapped to the north and west. No intrusive rocks are exposed in the area. The bench was originally mined in 1901. Collier (1903) reported that the pay was in the gravel on bedrock and in a clay layer that often lies on the bedrock. Later reports state that the gold is distributed evenly throughout the 2- to 9-foot thickness of gravel in the bench (Prindle and Hess, 1905; Hess,1908). Mertie (1934) described the deposit as a semi-residual body of auriferous, angular gravel, and that the pay streak continues down the hillside as a body of subangular gravel on muck. Collier (1903) described the gold as coarse and rough. An assay of the gold yielded a value of $16.45 an ounce (at $20.67 per ounce). Mertie (1934) reported a gold fineness of 792. Published reports indicate that by 1904 there had been considerable production from Shirley Bar (Prindle and Hess, 1905). After 1904, mining was recorded in 1931, 1937, and 1938 (Smith, 1933; Smith, 1939 [B 910-A; B 917-A]; Cobb, 1977), and again in 1951 (Williams, 1951). Tony Landing mined in 1962 (Saunders, 1962), and Johnson & Toftaker were active in 1967 (Heiner and others, 1968). (The Alaska Resource Data File., Tanana Quadrangle).

Slate Creek, Minook Creek Tributary

(Copper, Gold, Silver) **Quadrangle:** Tanana (B1); **Coordinates:** 65.3629°N, 150.1683W; **Map:** 4-8; **Worked:** 1902-1998; **Purity:** 915; **Nugget Size:** 5 ozs; **Production:** Through 1904=725 ozs; **Access:** Trail from Rampart landing strip on the Yukon River at 65.5082°N, 150.1425°W; **Summary:** Tributary of Minook Creek. Narrow valley in lower Paleozoic schistose chert, slate, and phyllite; limestone near mouth; many quartz veins, the probable source of the gold in the placers. Creek placer about 26 feet deep; gold in basal 3 feet of gravel and in crevices in top 1.5 feet of bedrock over a 50 foot width. Bench placer 15 - 20 feet above creek on NW side. Gold rough and coarse with nuggets of as much as 5 ounces; fineness about 915 Au. Concentrates contain nuggets of native copper and native silver and abundant barite. Mining 1902-16, 1926-39, and as recently as 1975. No data on amount of production (Cobb, 1981, OF 81-1313A p. A18). Production to fall of 1904 was about 725 fine ounces. No data on amount of production (Cobb, 1977, OF 77-432, p. 85-86). **Geology:** Bedrock in the Slate Creek area consists chiefly of northeast-trending, lower Paleozoic or Proterozoic cherty argillite, calcareous siltstone and sandstone, chert, metavolcanic rocks, and limestone (Mertie, 1934; Chapman and others, 1982; Reifenstuhl and others, 1997 [RI 97-15a]). The rocks are cut and stacked by a series of thrust faults. The creek follows a fault for much of its length. Slate Creek heads in lower Paleozoic chert, cherty argillite, and limestone. As the creek takes a northeasterly course and follows a fault, the north bank is calcareous siltstone and sandstone and the south bank is mainly chert. The placer deposit is centered at the approximate intersection of the northeasterly fault with a north-trending fault. Slate Creek has a V-shaped valley near its mouth and an open valley farther upstream. According to Hess (1908), there are numerous quartz veins that he believed were the source of the gold. The gravels consist mainly of various sedimentary and metasedimentary rocks, along with some vein quartz and greenstone. **Workings:** Mining began in 1902 (Collier, 1903) and was concentrated about 2 miles above the mouth of the creek. Drifting techniques were used initially (Hess, 1908). Prindle and Hess (1905) reported that the gravels were up to 26 feet thick with the gold concentrated in the bottom 3 feet of gravel and top 1.5 feet of bedrock. The gold-bearing gravels are both in bench and stream placers. The bench placers are only on the northwest side of the creek, 15 to 20 feet above the current channel, and consist of 5 feet of worn gravel overlain by 10-12 feet of muck. Mertie (1934) reported that the weathered bedrock contains clayey material that is yellowish brown and may contain gold. He hypothesized that much of the gold appears to have been derived from this bedrock. The gold recovered from Slate Creek included nuggets that were several ounces in size (Mertie, 1934), and had a fineness of 915. Sluicing is aided by the usually good supply of water (one measurement equaled 560 miners inches) (Avnet, 1948). The placer also reportedly contains cobbles of barite, and nuggets of native silver and native copper (Prindle and Hess, 1905). One property, mined on Slate Creek from 1912 continuously until 1941, consisted of 13 placer claims and stretched from the mouth of the creek upstream for 3.25 miles (Avnet, 1948). In the approximately 30 years that the property was mined, about 46,000 square feet of placer ground was mined by sluicing to wash away gravels to bedrock. Exposed nuggets were picked up by hand. This property subsequently was sold to Hilliard Avnet. By 1948, approximately 450,000 square feet of placer ground remained to be mined below the old workings. Two small deposits of gravel, one 205 square feet in surface area and the other a 17-cubic yard body in a drift mine, respectively produced $141.69 and $121.00 in gold (gold at $35 per ounce) (Avnet, 1948). Assessment work by Hilliard Avnet was performed in 1957. There were active mining claims on Slate Creek in 1967 (Heiner and others, 1968), and William Anderson ran a bulldozer on the creek in 1975 (Carnes, 1976). In 1984, Thanksgiving Mining Company used a backhoe to trench placer ground on Slate and Thanksgiving creeks (Eakins and others, 1985). John Shilling sluiced on Slate Creek in 1990 (Swainbank and others, 1991). In 1997, Slate Creek Mining explored for placer gold and Don Harris mined (Swainbank and others, 1998). Don Harris stripped 5,000 cubic yards of overburden with bulldozers and a backhoe, and sluiced 6,500 cubic yards of gravel (Alaska Kardex files). Szumigala and Swainbank (1999) reported active mining on Slate Creek by Slate Creek Mining in 1998. The only published estimate of production from Slate Creek was $15,000 worth of gold (about 725 ounces at $20.67 per ounce), mined through the fall of 1904 (Hess, 1908). Newberry and Clautice (1997) analyzed four gold grains from Slate Creek by electron microprobe. Two of the grains are not zoned, contain bismuth-tellurium inclusions, and have respective silver and mercury compositions of 20 percent and 1 percent. They concluded that the grains were derived from a plutonic gold source like that at Elephant Mountain. (The Alaska Resource Data File., Tanana Quadrangle).

Slate Creek, Tozimoran Creek Tributary

(Gold) **Quadrangle:** Tanana (B6); **Coordinates:** 65.3826°N, 152.7089W; **Map:** 4-9; **Worked:** 1942-; **Access:** Trail from Yukon River at 65.2016°N, 152.7608°W; **Summary:** Tributary of Tozimoran Creek. Placer gold reported; no systematic prospecting. See also Tozimoran Creek (Cobb, 1981, OF 81-1313A p. A18). This placer occurrence is at the junction of Slate Creek and Tozimoran Creek. The site is in the northeast quarter of section 4, T. 6 N., R. 25 W., of the Fairbanks Meridian. The occurrence includes part of Tozimoran Creek just below the mouth of Slate Creek. The location is accurate within several hundred feet. The site corresponds to location 3 of Eberlein and others (1977). **Geology:** Bedrock in the Tozimoran Creek area consists of metamorphosed sedimentary and mafic igneous rocks (Chapman and others, 1963). The local bedrock is mainly quartz-mica schist. Coarsely crystalline to porphyritic Cretaceous granite, cut by smaller granitic dikes, crops out about 7 miles north and west of this site (Chapman and others, 1982). Placer deposits in the Moran Dome area are alluvial deposits formed by the modern streams. Older alluvial deposits are limited in extent and occur in several places along Tozimoran Creek above its confluence with Slate Creek. The average width of alluvium in the present valley bottom of Tozimoran Creek near the mouth of Slate Creek is 250 feet (Chapman and others, 1963). **Workings:** In 1942, the U.S. Geological Survey examined the Moran Dome area for cassiterite-bearing placer deposits (Chapman and others, 1963). In 1952, areas of Tozimoran Creek were drilled and trenched under a Defense Minerals Exploration Administration contract. A shaft dug in 1942 near the mouth of Slate Creek reportedly reached bedrock at a depth of 25 feet. Material from the shaft yielded $1.65 in gold (gold at $35 per ounce) and 7 pounds of 'tinstone' (cassiterite and other minerals in pebbles and cobbles). Microscope examination of the tinstone showed cassiterite, associated with pale-brown tourmaline, in narrow veinlets in micaceous quartzite (Chapman and others, 1963). (The Alaska Resource Data File., Tanana Quadrangle).

Sullivan Creek

(Chromite, Copper, Gold, Lead, Monazite, Niobium(?), Uraninite, Tin); **Quadrangle:** Tanana (A2); **Coordinates:** 65.0938°N, 150.8786°W, 65.0943°N, 150.8818°W (Tofty Rd Xing); **Map:** 4-5, 4-6; **Tofty Road Right-of-Way:** ±50 of centerline; **Worked:** 1907-1995; **Yield:** 0.169-0.968 ozs/ft^2; **Nugget Size:** 0.532 ozs; **Production:** Through 1956= 58,156 ozs Au, 5,463 ozs Ag ozs; **Access:** Unimproved road from Tofty, Alaska. Creek crosses road at 65.0943°N, 150.8818°W; **Summary:** Tributary of Patterson Creek. Mining on benches, 1907 to 1940's; 2 - 14 feet of gravel overlain by 40 - 70 feet of frozen silt and muck containing many vertebrate remains. Heavy minerals mainly in basal 1 - 2 feet of gravel and in crevices in top few inches of bedrock (mainly crumpled Jura-Cretaceous soft graphitic phyllite and graywacke). Some very rich "spots" in benches; in places selected pans ran 0.483-0.726 ounces of gold and as much as 1/2 lb. cassiterite. Total production through 1956 (including that from reworked old tailings) was 58,156 ounces gold, 5,463 ounces silver, and 215,445 lb cassiterite concentrate. **Minerals:** Cassiterite, pyrite, ilmenite, picotite, magnetite, zircon, monazite, aeschynite, xenotime, gold, native copper, galena, arsenopyrite, chromite, apatite, brookite, and anatase; columbite may also be present. In upper basin (65°07'N, 150°53'W) north of the tin belt the gold is greenish and more angular than in tin belt. See also Idaho Gulch, Miller Gulch, Quartz Creek, trib. Sullivan Creek and Tofty Gulch (Cobb, 1981, OF 81-1313A p. A19). Gold discovered on January 1, 1907. Ground averaged about 0.169 ounces per bedrock foot, but some ran as high as 0.968 ounces of gold a bedrock foot. Largest nugget weighed 0.532 ounces. Overall samples for tailing piles averaged 0.241 lb. Sn and 0.0015 oz. Au per yd^3; averages of selected parts were 0.777 lb. Sn and 0.0028 Au (Cobb, 1977, OF 77-432, p. 88-90). **Geology:** The Sullivan Creek mine is one of a group of cassiterite- and gold-bearing placer deposits known as the Tofty tin belt, a 12-mile-long area that trends east-northeast, between Roughtop Mountain to the north and Hot Springs Dome to the south (Thomas, 1957). Roughtop Mountain and Hot Springs Dome respectively are underlain by Cretaceous (K-Ar age date of 92 +/- 5 Ma) and Tertiary (K-Ar age date of 62 +/- 3 Ma) granitic plutons (Chapman and others, 1982). The plutons intrude and contact metamorphose Mesozoic marine sedimentary strata, which also are cut by diverse faults, including regional-scale, east-northeast striking, thrust faults (Reifenstuhl and others, 1998). A carbonatite sill(?) is in the Triassic section of these strata and there are exposures of serpentinized, Cretaceous(?) mafic and ultramafic rock, mainly on Serpentine Ridge. The gold- and tin-bearing creeks flow normal to the trend of the tin belt. They head in the plutonic, metamorphic, and mafic/ultramafic rocks of Roughtop Mountain and Serpentine Ridge, which probably are the source(s) of some of the metalliferous minerals in the placer deposits. Concentrations of gold diminish toward the south, probably due to dispersion (Thomas, 1957). The Sullivan Creek placer mine is on the trace of a northwest-trending fault with small relative right lateral displacement, and approximately 0.5 mile north of a regional thrust fault (Reifenstuhl and others, 1998). Bedrock in Sullivan Creek comprises Triassic clastic sedimentary strata. The principal pay streak was the Sullivan bench placer, approximately 2,000 feet east of the present course of Sullivan Creek (Wayland, 1961). The bedrock of the placer operations has been described as graphitic phyllite containing oxidized pyrite (Wayland, 1961), and phyllite containing quartz stringers and pyrite (Mertie, 1934). The thickness of overburden ranges from 30 to 70 feet and the gravels are 10 to 35 feet thick (Eakin,1913). The gold was concentrated in the lower 2 to 3 feet of the gravels and in fractures in shattered bedrock. The overburden contains the remains of Quaternary mammoth, bison, and horses (Heiner and others, 1968). Eakin (1915) described the bench placer deposits as elliptical in plan view. The centers of these deposits were very rich and graded outward to leaner ground. For example, one mine produced $200,000 worth of gold from 5,000 square feet (gold at $20.67 per ounce) (Eakin, 1915). Cassiterite occurs with the gold throughout the area. Hess (1912) reported that the pebbles are very smooth. This was confirmed by Eakin (1913), who described pans worth $10-$15 in gold (at $20.67 per ounce), and containing as much as a half-pound of cassiterite. The sources of the chromite, picotite, magnetite, and ilmenite probably are the serpentinized mafic and ultramafic rocks. **Workings:** Cobb (1977) summarized mining activity on Sullivan Creek to 1975. Mining began on the benches in 1907 and continued into the 1940's. Initially, only the gold was kept, but by about 1918 the cassiterite became an important byproduct. In 1951, L. McGee bought the Cleary Hill Mines, Inc., mining claims and equipment and began mining (Williams, 1951). By 1956, Sullivan Creek had produced 58,136 ounces of gold, 5,463 ounces of silver, and 215,445 pounds of cassiterite concentrate, which was between 1/3 and 1/2 of the total production of placer deposits in the Tofty tin belt (Thomas, 1957). It appears that far more cassiterite might have been recovered if different sluicing techniques had been employed (Wayland, 1961). Placer mining was underway in 1967, with over a dozen claim blocks or working groups on the creek (Heiner and others, 1968). More recent mining has included drifting on Sullivan bench by Burgess Mining in 1985 (Bundtzen and others, 1986). As of 1988, Shoreham Resources was preparing to mine on the bench (Green and others, 1989), and they mined it in 1991. Their operation recovered 1,746 ounces of gold, 324 ounces of silver, and 6,800 lbs of tin metal as cassiterite (Bundtzen and others, 1992). Harold Bergman spent part of 1990 working a cut (Swainbank and others, 1991). Bundtzen and others (1996) reported that Cassiterite Placers, Inc., worked Sullivan Bench in 1995. (The Alaska Resource Data File., Tanana Quadrangle).

Thanksgiving Creek

(Gold) **Quadrangle:** Tanana (A1); **Coordinates:** 65.1774°N, 150.3693°W; **Map:** 4-7; **Worked:** 1903-1994; **Purity:** 734 Au; **Access:** Unimproved road from Eureka, Alaska at 65.1774°N, 150.2496°W; **Summary:** Tributary of Omega Creek. Creek runs in broad bench north of Baker Creek. Bedrock is Jura-Cretaceous schistose grit, shale, and siltstone cut by white quartz veins. Gravel mainly subangular pieces of quartzite, grit, vein quartz, slate, and monzonitic rock. Magnetite in concentrates. Pay streak 25 - 45 feet wide, 1.5 - 9 feet thick; some gold in overlying 10 - 12 feet of muck. Sporadic mining, 1903-75; no data on production; probably no more than a few thousand ounces (Cobb, 1981, OF 81-1313A p. A19). Gold discovered in 1903. Gold generally rough and somewhat iron stained; some bright and "shotty." Gold assayed at 734 Au (Cobb, 1977, OF 77-432, p. 91). **Geology:** Thanksgiving Creek flows in a shallow, open valley about 5 miles long. The area is underlain by Mesozoic clastic sedimentary rocks that are cut by thrust faults and locally intruded by small Cretaceous monzodiorite bodies (Chapman and others, 1982; Reifenstuhl and others, 1998). The upper part of Thanksgiving Creek flows along the trace of one of the thrust faults. According to Hess (1908) and Mertie (1934), the local bedrock is yellowish, sheared arkosic sandstone. It is overlain by 6 to 18 feet of gravels, overlain in turn by 1 to 4 feet of muck. The gravels contain subangular cobbles of quartzite, schistose arkose, vein quartz, slate, and small amounts of monzonite, along with intermixed sticky yellowish clay (Hess, 1908). Hess reported that the pay streak was 25 to 45 feet wide and 1.5 to 9 feet thick. The gold was rough and shotty, sometimes bright, and sometimes iron-stained. The source of the gold is controversial. Prindle and Hess (1905) thought that it was derived directly from small quartz veins in the schistose bedrock. Mertie (1934), however, believed that the gold in Thanksgiving Creek was reconcentrated from adjacent bench or terrace gravels. **Workings:** Gold was discovered on Thanksgiving Creek in 1903 (Prindle and Hess, 1905), and there was sporadic mining from 1903-1936. After a long hiatus in reported mining, Anthony Lanning and two men mined on Thanksgiving Creek in 1967 and 1975 with a bulldozer and sluice (Heiner and others, 1968; Carnes, 1976). Thanksgiving Mining Company was active in 1984, using a backhoe to trench placer ground on both Thanksgiving and Slate creeks (Eakin and others, 1985). In 1989, John Shilling worked Thanksgiving and Slate creeks (Bundtzen and others, 1990). In 1992, Thanksgiving Mining (John Shilling) worked Thanksgiving Creek, and collected a bulk sample (Swainbank and others, 1993). Mining continued in 1993 (Bundtzen and others, 1994), and a small amount of sluicing was done in 1994 (Swainbank and others, 1995). Separate production figures have not been published for Thanksgiving Creek. According to Hess (1908), combined mining of Thanksgiving and Omega creeks had produced $18,200 worth of gold by the fall of

1904. Cobb (1977) estimated this to equal approximately 880 ounces of gold, and speculated that no more than a few thousand ounces of gold had been produced through 1977 from Thanksgiving Creek. (The Alaska Resource Data File., Tanana Quadrangle).

Tofty Gulch

(Chromite, Gold, Monazite(?), Niobium(?), Uraninite(?), Silver, Tin) **Quadrangle:** Tanana (A2); **Coordinates:** 65.0845°N, 150.8993°W; **Map:** 4-5, 4-6; **Worked:** 1906-1998; **Yield:** 0.000867 oz/yd^3; **Production:** Through 1956=8,855 ozs Au, 1,376 ozs Ag; **Access:** At Tofty, Alaska; **Summary:** Tributary of Sullivan Creek. Open-cut mining; gravel as much as 7 feet thick beneath 10 - 25 feet of frozen silt and muck that contained many remnants of trees. Heavy minerals in gravel and upper 2 - 3 feet of Jura-Cretaceous phyllite and interbedded graywacke with many quartz veins that carry pyrite but no gold or cassiterite. Minerals: Gold, cassiterite, and chromite, columbite, aeschynite, monazite, and zircon. Gold discovered in winter of 1906-07; mining 1909-12, 1917, 1929, 1941, 1975, and probably in other years as well. Total production through 1956 was 8,855 ounces of gold, 1,376 ounces of silver, and 19,600 lbs of cassiterite concentrate. Includes reference to Tufty Gulch. See also Sullivan Creek (Cobb, 1981, OF 81-1313A p. A19). **Geology:** The Tofty Gulch mine is one of a group of cassiterite- and gold-bearing placer deposits known as the Tofty tin belt, a 12-mile-long area that trends east-northeast, between Roughtop Mountain to the north and Hot Springs Dome to the south (Thomas, 1957). Roughtop Mountain and Hot Springs Dome respectively are underlain by Cretaceous (K-Ar age date of 92 +/- 5 Ma) and Tertiary (K-Ar age date of 62 +/- 3 Ma) granitic plutons (Chapman and others, 1982). The plutons intrude and contact metamorphose Mesozoic marine sedimentary strata, which also are cut by diverse faults, including regional-scale, east-northeast striking, thrust faults (Reifenstuhl and others, 1998). A carbonatite sill(?) is in the Triassic section of these strata and there are exposures of serpentinized, Cretaceous(?) mafic and ultramafic rock, mainly on Serpentine Ridge. The gold- and tin-bearing creeks flow normal to the trend of the tin belt. They head in the plutonic, metamorphic, and mafic/ ultramafic rocks of Roughtop Mountain and Serpentine Ridge, which probably are the source(s) of some of the metalliferous minerals in the placer deposits. Concentrations of gold diminish toward the south, probably due to dispersion (Thomas, 1957). The Tofty Gulch placer deposit is at the intersection of an east-northeast-striking thrust fault that dips north, and a high-angle fault that dextrally displaces the thrust fault by about 1/4 mile. The local bedrock consists of phyllite and minor interbedded graywacke that strikes N85E and dips steeply to the north (Wayland, 1961). The rocks are sheared and contain barren, milky quartz-calcite-pyrite veinlets. According to Mertie (1934) and Ellsworth (1910), gold was discovered in Tofty Gulch in the winter of 1906-07. There is no well-defined pay zone. Some of the gold is in a bench placer approximately 1,000 feet from Sullivan Creek (Eakin, 1913). The gravels, which are about 4 to 6 feet thick (Eakin, 1913), consist chiefly of angular graywacke, phyllite, and quartz. Other cobbles include metadiorite, quartzite, and altered phyllite containing quartz-tourmaline veins (Wayland, 1961). Cassiterite particles in the goldbearing gravels range from well-rounded (Hess, 1912) to subangular, and are up to several inches across. The larger sizes often are accompanied by vein quartz and tourmaline, as well as by fragments of sedimentary country rock. The smaller sizes generally are free of impurities (Thomas,1957). In addition to gold and cassiterite, Wayland (1961) reported chromite. Moxham (1954) reported elevated radioactivity measurements, and suggested that columbite, aeschynite, monazite, and zircon may also be present. Newberry and Clautice (1997) analyzed a gold grain by electron microprobe that contained a tiny inclusion of native nickel. Five other gold grains had cores with silver contents varying from 10 to 17 percent, mercury contents of 1 percent, and strongly silver- and mercury-depleted rims. **Workings:** Gold was discovered in Tofty Gulch in the winter of 1907, and presumably mined by drifting or shaft sinking. There was open-cut mining in 1909-12, 1917, 1929 and 1941 (Cobb, 1977). L. McGee mined the property in 1951 (Williams, 1951). The total production reported by Thomas (1957) was 8,855 ounces of gold, 1,376 ounces of silver, and 19,600 pounds of cassiterite concentrate. Drilling took place west of the gulch in 1941 (Wayland, 1961). Man Mining Company had a washing plant, three bulldozers, two draglines, and five men working a placer claim in 1975 (Carnes, 1976). In 1992, GHD Resources reacquired the Tofty (Gulch?) property and began to prepare the ground for the 1993 season. In September, the company sluiced 150,000 cubic yards of pay gravels, which yielded 130 ounces of gold and 1,500 pounds of cassiterite (Swainbank and others, 1993). In 1993, the company sluiced 89,000 cubic yards of pay, producing gold, silver and byproduct tin (Bundtzen and others, 1994). In January, 1994, the property was sold back to Cassiterite Placers, Inc., who mined that year and again in 1998 (Swainbank and others, 1995; Szumigala and Swainbank, 1999). (The Alaska Resource Data File., Tanana Quadrangle).

Tonawanda Creek

(Gold, Lead, Silver, Tin) **Quadrangle:** Tanana (B5, B6); **Coordinates:** 65.1219°N, 150.6661°W; **Map:** 4-10; **Worked:** 1980-1994; **Access:** From Tofty Road; **Summary:** Mining claims were staked in 1980, followed by placer mining along Tonawanda Creek (Alaska Kardex files). The discovery claim was on the creek at the northeast corner of section 6, T. 3 N., R. 15 W., of the Fairbanks Meridian, and claims extended above and below the discovery claim. The site is at the discovery claim. The location is accurate within about 500 feet. This site roughly corresponds with the site for Tonawanda Creek, U.S. Bureau of Land Management MAS number 0020480128. **Geology:** The Tonawanda Creek mine is one of a group of cassiterite- and gold-bearing placer deposits known as the Tofty tin belt, a 12-mile-long area that trends east-northeast, between Roughtop Mountain to the north and Hot Springs Dome to the south (Thomas, 1957). Roughtop Mountain and Hot Springs Dome respectively are underlain by Cretaceous (K-Ar age date of 92 +/- 5 Ma) and Tertiary (K-Ar age date of

62 +/- 3 Ma) granitic plutons (Chapman and others, 1982). The plutons intrude and contact metamorphose Mesozoic marine sedimentary strata, which also are cut by diverse faults, including regional-scale, east-northeast striking, thrust faults (Reifenstuhl and others, 1998). A carbonatite sill(?) is in the Triassic section of these strata and there are exposures of serpentinized, Cretaceous(?) mafic and ultramafic rock, mainly on Serpentine Ridge. The gold- and cassiterite ('tin')-bearing creeks flow normal to the trend of the tin belt. They head in the plutonic, metamorphic, and mafic/ultramafic rocks of Roughtop Mountain and Serpentine Ridge, which probably are the source(s) of some of the metalliferous minerals in the placer deposits. Concentrations of gold diminish toward the south, probably due to dispersion (Thomas, 1957). Bedrock at the Tonawanda Creek mine is Triassic or Permian, subphyllitic to phyllitic, argillite, sandstone, and shale, and a steeply-dipping Jurassic carbonatite sill (Reifenstuhl and others, 1998). Mining claims were staked along Tonawanda Creek in 1980 (Alaska Kardex files), and worked semi-continuously from 1980 to 1994. An affidavit of annual labor from 1986 listed 22 state mining claims and stripping on No. 1 below Discovery. (The Alaska Resource Data File., Tanana Quadrangle).

Tozimoran Creek

(Gold, Lead, Silver, Tin) **Quadrangle:** Tanana (B5, B6); **Coordinates:** 65.3792°N, 152.7892°W; **Map:** 4-9; **Worked:** 1902-1980s; **Purity:** 835-895 Au; **Yield:** 0.0228 ozs/yd^3; **Nugget Size:** NOT; **Production:** NOT ozs; **Access:** Trail from Yukon River at 65.2016°N, 152.7589°W; **Summary:** Tributary of the Tozitna River. Gold discovered in 1902. Considerable prospecting and a little hand mining on a bench 2000 feet downstream from Ash Creek. Alluvium on bench consists of 4 feet of gravel beneath 3 feet of muck; a little gold and cassiterite in bottom 2 feet of gravel; gold, but not cassiterite, in cracks in upper part of bedrock. A little gold and cassiterite in unfrozen creek alluvium. Deposits explored and sampled by U.S. Bureau of Mines, Geological Survey, and owner under DMEA contract. Combined samples of gold assayed 835 fine. Total production was probably no more than a few ounces of gold and a few hundred pounds of cassiterite concentrate (which may not have been sold). Bedrock is metamorphosed lower Paleozoic sedimentary and mafic igneous rock. Quartz-calcite veins in quartz-mica schist contain some argentiferous galena and cerussite. Includes references to Columbe Creek, Moraine Creek, and to Moran Creek unless specifically to Melozimoran Creek. See also Ash Creek (Cobb, 1981, OF 81-1313A p. A20). Most of gold and cassiterite on bench; a little in valley-floor gravels below bench. Bedrock is metamorphosed sedimentary and mafic igneous rock; large granite pluton cut by younger granite dikes. Argentiferous galena and cerussite in quartz-calcite veins in quartz-mica schist are only un-weathered sulfides found. Cassiterite (1/16 to 1-1/2 inches in diameter) and gold in bottom 2 feet. of gravel accompanied by considerable clay. Heavy minerals do not penetrate bedrock more than 6 inches. Valley-floor deposits similar, but gravel not frozen and heavy minerals and clay distributed through entire thickness. Two ages of alluvium. Older in isolated remnants of benches one of which carries gold and cassiterite. Younger is deposits of present stream and tributaries and tin area tested is about 15 feet thick. Origins of gold and cassiterite not known; tinstone pebbles are micaceous quartzite with veinlets of cassiterite and light-brown tourmaline. Cassiterite in older alluvium is angular and subangular fragments mainly in coarse (1/4 in. or greater) grains or fragments as heavy as 3 lb. Gold is clean and angular; fineness determinations of 895 and 835 have been reported. Both near or on bedrock. Gold penetrates cracks for a few inches; cassiterite does not. Gold and cassiterite in younger alluvium also, but few data on these deposits. Channel samples from bench indicated averages of 0.731 lb. Sn and 0.0228 ounces Au per yd^3 in block 650 feet long and 80 feet wide. Silver-lead ore has been found in headwaters area; galena in quartz and calcite stringers and veins cutting quartz-mica schist (Cobb, 1977, OF 77-432, p. 94-95). **Geology:** Bedrock in the Tozimoran Creek area consists of metamorphosed sedimentary rocks and mafic igneous rocks (Chapman and others, 1963). The local bedrock is mainly quartz-mica schist. Coarsely crystalline to porphyritic Cretaceous granite, cut by small granitic dikes, crops out about 5 miles north and west of the Tozimoran Creek mine (Chapman and others, 1982). The deposit consists of a bench placer on the north side of Tozimoran Creek, and of alluvium in the present stream channel, 2,000 feet downstream from the mouth of Ash Creek (Thomas and Wright, 1948 [RI 4323]). Both bench and stream gravels are approximately 4 feet thick, and the bench also contains 3 feet of muck, frozen in many places. Gold, along with cassiterite, is concentrated in the bottom 2 feet of the bench gravels and is evenly distributed throughout the stream gravels. The gold and cassiterite do not appear to penetrate bedrock beyond 6 inches (Thomas and Wright, 1948 [RI 4323]). The cassiterite occurs as subangular to angular grains, pebbles, and cobbles up to three pounds; it is dark brown to black and translucent to opaque. **Workings:** Little mining and prospecting has actually been done on Tozimoran Creek, where gold was discovered in 1902 (Thomas and Wright, 1948 [RI 4323]). Early reports of mining in Columbe Creek, Moraine Creek, and Moran Creek (Brooks, 1908; Smith, 1932, 1933, 1936, 1937) likely refer to Tozimoran Creek (Cobb, 1977). There are no records of large-scale mining in the 1930s, only of sporadic prospecting. I.W. Purkeypile and Martin Webories hand-mined the upper portion of Tozimoran Creek from 1938 until at least 1944 (Thomas and Wright, 1948 [RI 4323]). Wright (1940) examined the property and concluded that the gold and cassiterite pay streaks were not continuous or rich enough to warrant mining. Purkeypile and Webories maintained active claims through at least 1967 (Heiner and others, 1968). The only reported recent activity was by Wayne Gibson, who mined placer cassiterite in the 1980's (Bundtzen and others, 1986). Wedow and others (1952) reported silver-lead ore in the headwaters of Tozimoran Creek, where galena occurs in quartz-calcite stringers and veins that cut the quartz-mica schist. There is no cassiterite in the gravels near these veins. (The Alaska Resource Data File., Tanana Quadrangle).

Trail Creek

(Gold) **Quadrangle:** Tanana (B6); **Coordinates:** 65.1482°N, 151.1584°W; **Map:** 4-2; **Worked:** 1917-1993; **Access:** Trail from Tofty, Alaska at 65.0934°N, 150.8820°W; **Summary:** This record represents placer gold prospects in the Boulder Creek drainage, chiefly its tributaries Trail Creek, Salt Creek, and Dry Creek, an unnamed branch of Trail Creek (Chapin, 1919). The site is at the junction of Salt Creek and Trail Creek, in the northwest quarter of section 26, T. 4 N., R. 18 W., of the Fairbanks Meridian. The location is accurate, but it was not plotted by Cobb (1972). The site roughly corresponds with the U.S. Bureau of Land Management location for Salt Creek (MAS number 0020480073). **Geology:** The Salt Creek-Dry Creek area is underlain by Jurassic or Cretaceous clastic sedimentary rocks and by Paleozoic limestone, dolomite, greenstone, and chloritic schist, with minor amounts of phyllite, quartzite, and quartz mica schist (Chapman and others, 1982; Reifenstuhl and others, 1998). Serpentine Ridge, 1.5 miles to the south, consists of serpentinite and mafic intrusive rocks that intrude Mesozoic, predominantly marine sedimentary rocks, (Chapman and others, 1982). **Workings:** Chapin (1919) reported mining in Dry Creek in 1917. The gravels are angular, 3 to 6 feet thick, and consist of black slate, graywacke, quartzite, and schist. Water was not plentiful, but the ground was easily worked. Heiner and others (1968) reported mining in 1967, and that claims on Dry Creek were held or worked by Scotty Anderson, Farrell, Heiner and Wolff, and Higgins. Salt Creek and the immediate area of the junction of Salt Creek with Trail Creek was under claim and mined sporadically from 1973 to 1993 (Alaska Kardex files). Twelve claims were staked in the area in 1973, with 10 placer claims on Salt Creek and 2 placer claims downstream from the junction of Salt and Trail creeks. Mining was a bulldozer and loader sluicebox operation. (The Alaska Resource Data File., Tanana Quadrangle).

Unnamed, Baker Creek Tributary

(Gold) **Quadrangle:** Tanana (B6); **Coordinates:** 65.0834°N, 150.6864°W; **Map:** 4-8, 4-10; **Worked:** 1967-1972; **Access:** From Tofty Road near location of deposit; **Summary:** Mining claims were staked in 1967, followed by placer mining, along a tributary of Baker Creek (Alaska Kardex files). The location given in the U.S. BLM MAS system (MAS number 0020480059), is in the southwest quarter of section 18, T. 3 N., R. 15 W., of the Fairbanks Meridian. The location is accurate within about 1000 feet. **Geology:** This placer prospect is about 2 miles south of the east edge of the Tofty tin belt, a group of cassiterite- and gold-bearing placer deposits centered on the town of Tofty and extending for approximately 12 miles in an east-northeast trend (Thomas, 1957). The gold- and tin-bearing creeks flow normal to the trend of the tin belt. Concentrations of gold diminish toward the south, probably due to dispersion (Thomas, 1957). Bedrock at the prospect is Lower Cretaceous sandstone, shale, and siltstone (Reifenstuhl and others, 1998). A square-mile outcrop area of the Tertiary (58 Ma) Hot Springs granite is approximately one mile to the southeast. Numerous east-trending and northwest-trending high-angle faults cut through the area. **Workings:** Placer mining claims were staked on this tributary of Baker Creek in 1967, and were worked from 1967 to at least 1972 (Alaska Kardex files). Affidavits of annual labor list trenching in 1971 and drilling in 1972. There is no record of mining or production. (The Alaska Resource Data File., Tanana Quadrangle).

Unnamed, Boulder Creek Tributary

(Gold, Chromite, Uraninite) **Quadrangle:** Tanana (B6); **Coordinates:** 65.1476°N, 151.0945°W; **Map:** 4-2; **Worked:** 1915-1991; **Purity:** 818 Au; **Access:** Trail from Tofty, Alaska at 65.0853°N, 150.9050°W; **Summary:** This site represents an area of placer mining on Boulder Creek near its junction with Little Boulder Creek. The site is at a (placer) mine symbol in the northwest quarter of section 30, T. 4 N., R. 17 W., of the Fairbanks Meridian. The location is accurate within several hundred feet. The site corresponds to location 22 of Cobb (1972). **Geology:** The Boulder Creek area is underlain by Jurassic or Cretaceous clastic sedimentary rocks, which have been intruded and contact metamorphosed by the Cretaceous monzodiorite to granite Roughtop Mountain pluton (Chapman and others, 1982; Reifenstuhl and others, 1998). Locally, the area is covered by unconsolidated Quaternary deposits (Chapman and others, 1982). Published descriptions of the placer deposits, and of the locations of workings, on Boulder and Little Boulder creeks are ambiguous. The following description herein is assumed to apply to the placer deposits in the area of this site. **Workings:** Mining on Boulder Creek was first reported in 1915 (Brooks, 1916). The creek has an asymmetrical valley (Chapin, 1919); the south side is steep with no gravels, and the north side is flat with alluvium 8 to 12 feet thick. The placer extends for several miles, with a workable width of 1,200 feet. In 1917, over 200,000 feet of bedrock were cleaned (Chapin, 1919). A hydraulic plant was operated in 1931 from May to September, and 7,000 square feet of bedrock was cleaned from a bench of Boulder Creek (Mertie, 1934). Gold 818 fine from this operation reportedly had an assay value of $16.90 per ounce (at $20.67 per ounce). There was relatively consistent small-scale production in the 1930's from Boulder Creek, although the amount of gold recovered was not made public. Early references to Little Boulder Creek reported that the placer ground was thin, with low gold grades (Brooks, 1918). Chapin (1919) reported prospecting the following year in 6 to 12 feet of alluvium consisting of silt with layers and lenses of angular slate fragments. Waters (1934). **Minerals:** Moxham (1954) reported that, in addition to gold, the placer concentrates contained magnetite, ilmenite, zircon, rutile, garnet, sphene, and monazite. In 1949, pan concentrates assayed 0.21 percent equivalent uranium oxide, whose source was speculated to be the pluton at Roughtop Mountain (Moxham, 1954). Moxham also identified chromite float in Boulder Creek, whose source is probably serpentinite on Serpentine Ridge, south of the creek (Berg and Cobb, 1967). In 1967, two properties on Boulder Creek were being intensely prospected and another property was to be mined in 1968 (Heiner and others, 1968). Farrell & Higgins were also active on Little Boulder Creek in 1967. Boulder Creek Mining Company (Les and Dorthy Fickes) worked a cut on Boulder Creek during 1986 (Bundtzen and others, 1987). In 1992, R. Schroder and Beatrice Schafer both mined on Boulder Creek (Swainbank and others, 1993). Placer Mine Services mined along the creek in 1997 (Swainbank and others, 1998), and Richard Wilder mined in 1998 (Szumigala and Swainbank, 1999). In 1990, J.L. Wood and Vern Petefish worked on Little Boulder Creek. In 1991, Wood continued to work in Little Boulder Creek, while Windy Hill Mining prospected some areas (Bundtzen and others, 1992). Wood also mined on Little Boulder Creek in 1997-98 (Swainbank and others, 1998; Szumigala and Swainbank, 1999).(The Alaska Resource Data File., Tanana Quadrangle).

Utah Creek

(Gold) **Quadrangle:** Tanana (B6); **Coordinates:** 65.1260°N, 150.6023°W; **Map:** 4-8; **Worked:** 1970s-1982; **Purity:** NOT; **Yield:** NOT; **Nugget Size:** NOT; **Production:** NOT ozs; **Summary:** Mining claims were staked in the 1970's, followed by placer mining, on Utah Creek (Alaska Kardex files). Mining claim location notices show 3 claims along Utah Creek about 2 miles upstream from its junction with Baker Creek. The site is in the southeast quarter of section 33, T. 4 N., R. 15 W., of the Fairbanks Meridian. The location is accurate within about 1,000 feet. This site closely corresponds with the site for Utah Creek, U.S. Bureau of Land Management MAS number 0020480110. **Geology:** The Utah Creek placer prospect is one of a group of cassiterite- and gold-bearing placer deposits known as the Tofty tin belt, a 12-mile-long area that trends east-northeast, between Roughtop Mountain to the north and Hot Springs Dome to the south (Thomas, 1957). Roughtop Mountain and Hot Springs Dome respectively are underlain by Cretaceous (K-Ar age date of 92 +/- 5 Ma) and Tertiary (K-Ar age date of 62 +/- 3 Ma) granitic plutons (Chapman and others, 1982). The plutons intrude and contact metamorphose Mesozoic marine sedimentary strata, which also are cut by diverse faults, including regional-scale, east-northeast striking, thrust faults (Reifenstuhl and others, 1998). A carbonatite sill(?) is in the Triassic section of these strata and there are exposures of serpentinized, Cretaceous(?) mafic and ultramafic rock, mainly on Serpentine Ridge. The gold- and cassiterite-bearing creeks flow normal to the trend of the tin belt. They head in the plutonic, metamorphic, and mafic/ultramafic rocks of Roughtop Mountain and Serpentine Ridge, which probably are the source(s) of some of the metalliferous minerals in the placer deposits. Concentrations of gold diminish toward the south, probably due to dispersion (Thomas, 1957). Bedrock at the Utah Creek prospect is Triassic or Permian, phyllitic argillite, sandstone, and shale, and a steeply-dipping Jurassic carbonatite sill (Reifenstuhl and others, 1998). **Workings:** Three mining claims were staked along Utah Creek in the 1970's (Alaska Kardex files). Affidavits of annual labor from 1979 to 1982 list excavation, an adit 18 feet high and 15 feet deep, and drilling. There is no record of production from Utah Creek, but if placer gold has been recovered, the creek would be the easternmost placer gold producer in the Tofty tin belt. (The Alaska Resource Data File., Tanana Quadrangle).

Wells Creek

(Gold) **Quadrangle:** Tanana (B6); **Coordinates:** 65.3961°N, 152.6572°W; **Map:** 4-3, 4-9; **Summary:** Tributary of Tozimoran Creek; placer gold reported; no known systematic prospecting (Cobb, 1981, OF 81-1313A p. A20).

Woodchopper Creek

(Gold, Silver, Tin) **Quadrangle:** Tanana (A3); **Coordinates:** 65.0516°N, 151.0122°W; **Map:** 4-6; **Worked:** 1913-1997; **Production:** Through 1956=28,501 ozs; **Access:** Unimproved road and trails from Tofty, Alaska; **Summary:** Tributary of Patterson Creek. Gold and cassiterite discovered in deep channel and on buried benches 100 - 200 feet below surface in 1913. Large-scale drift mining preceded by prospect drilling with few interruptions until 1956 and probably later. Total production through 1956 was 28,501 ounces gold, 3,402 ounces silver, and 40,300 lbs cassiterite concentrate. Average per cubic yard channel samples of tailing pile was 1.54 lb concentrate containing 0.72 lb Sn and 0.045 oz Au. Gravel contains quartz-tourmaline cobbles. **Minerals:** Gold, cassiterite, ilmenite, picotite, pyrite, and magnetite (Cobb, 1981, OF 81-1313A p. A20). Mine on bench just east of Woodchopper Creek (visited in 1922) showed 20 feet of gravel on bedrock overlain by 20 feet of muck, 40 feet of gravel, and another 40 feet of muck (Cobb, 1977, OF 77-432, p. 97-98). **Geology:** The Woodchopper Creek placer mine marks the western end of the Tofty tin belt, a 12-mile long area of cassiterite- and gold-bearing placer deposits that trends east-northeast, between Roughtop Mountain to the north and Hot Springs Dome to the south (Thomas, 1957). Roughtop Mountain and Hot Springs Dome respectively are underlain by Cretaceous (K-Ar age date of 92 +/- 5 Ma) and Tertiary (K-Ar age date of 62 +/- 3 Ma) granitic plutons (Chapman and others, 1982). The plutons intrude and contact metamorphose Mesozoic marine sedimentary strata, which also are cut by diverse faults, including regional-scale, east-northeast striking, thrust faults (Reifenstuhl and others, 1998). A carbonatite sill(?) is in the Triassic section of these strata and there are exposures of serpentinized, Cretaceous(?) mafic and ultramafic rock, mainly on Serpentine Ridge. The gold- and tin-bearing creeks flow normal to the trend of the tin belt. They head in the plutonic, metamorphic, and mafic/ultramafic rocks of Roughtop Mountain and Serpentine Ridge, which probably are the source(s) of some of the metalliferous minerals in the placer deposits. Concentrations of gold diminish toward the south, probably due to dispersion (Thomas, 1957). The pay gravels on Woodchopper Creek occupy a deep channel and lie as deep as 200 feet below the surface (Brooks, 1918). Mertie (1934) described one of the mines on the east side of Woodchopper Creek as containing 20 feet of gravel on bedrock, overlain by 20 feet of muck, then 40 feet more of gravel, capped in turn by 40 feet of muck. These gravels consist predominantly of light phyllite, dark phyllite, and quartzite, with sandstone increasing downstream (Wayland, 1961). Tourmaline-bearing quartz boulders occur in the tailings (Thomas, 1957). The gold and cassiterite generally occur together. However, Wayland (1961) reported that drilling between 1926 and 1941 by Adolph Brock yielded 5 holes with cassiterite and no gold. The cassiterite generally occurs as rounded pebbles, and at least some of the gold is coarse (Wayland, 1961). **Workings:** Cobb (1977) has summarized the mining history on Woodchopper Creek. Mining on the creek started in 1913 with the discovery of gold gravels near the mouth of the creek (Chapin, 1914). There was large-scale drift mining from 1915 to 1916, employing more than 100 men (Brooks, 1918). This project disbanded, but mining on different claims continued through 1941, and many prospecting holes were dug by different people (Wayland, 1961). Woodchopper Creek was the most productive creek in the Hot Springs district in 1926 (Smith, 1929). The Woodchopper Mining Company drift mined from a 167-foot-deep shaft on the Loraine Claim in 1951 (Williams, 1951). Seven groups held active mining claims on Woodchopper Creek in 1967 (Heiner and others, 1968). Jack Neubauer mined in the area in 1997 (Swainbank and others, 1998).(The Alaska Resource Data File., Tanana Quadrangle).

Wiseman Quadrangle

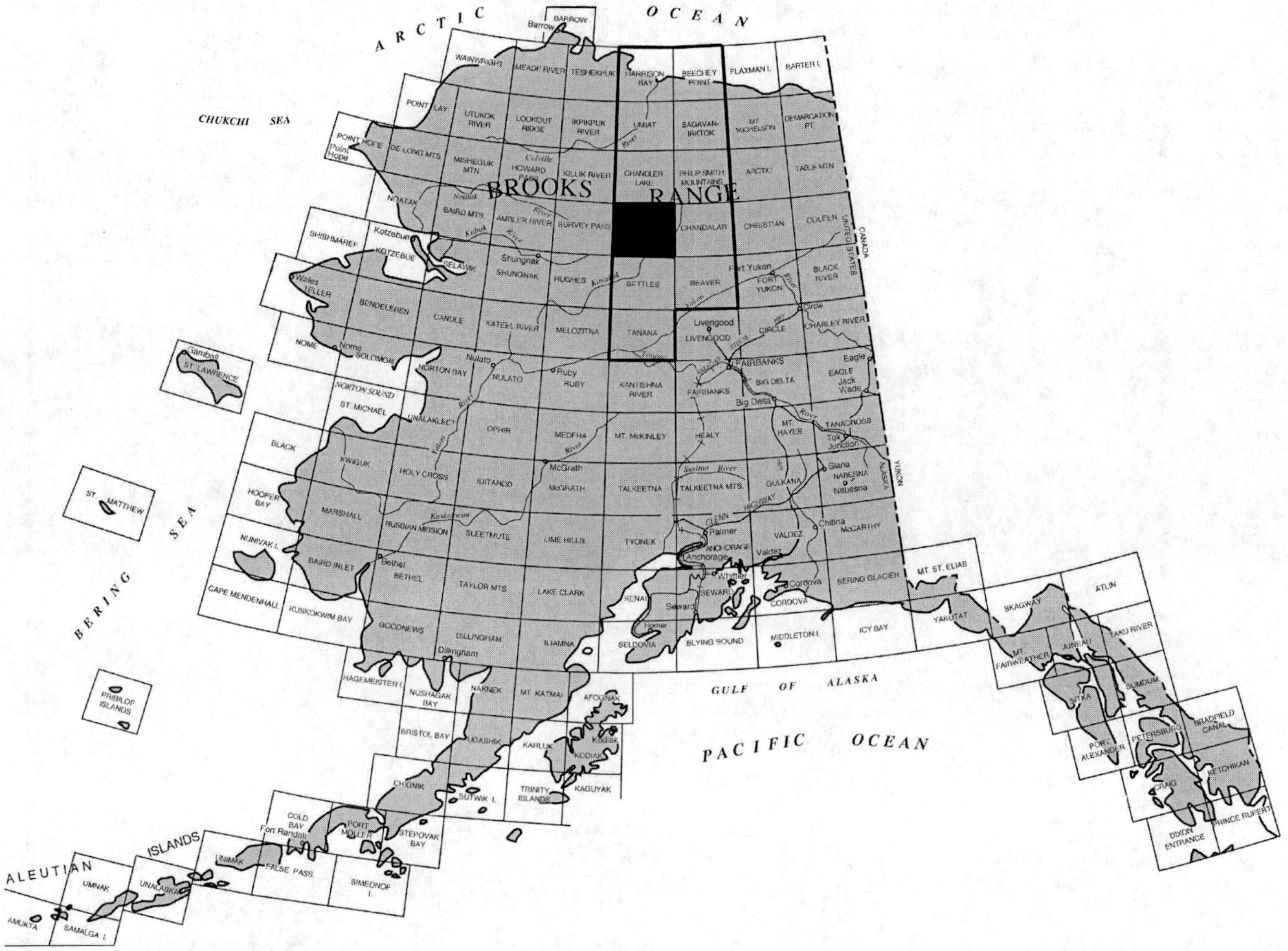

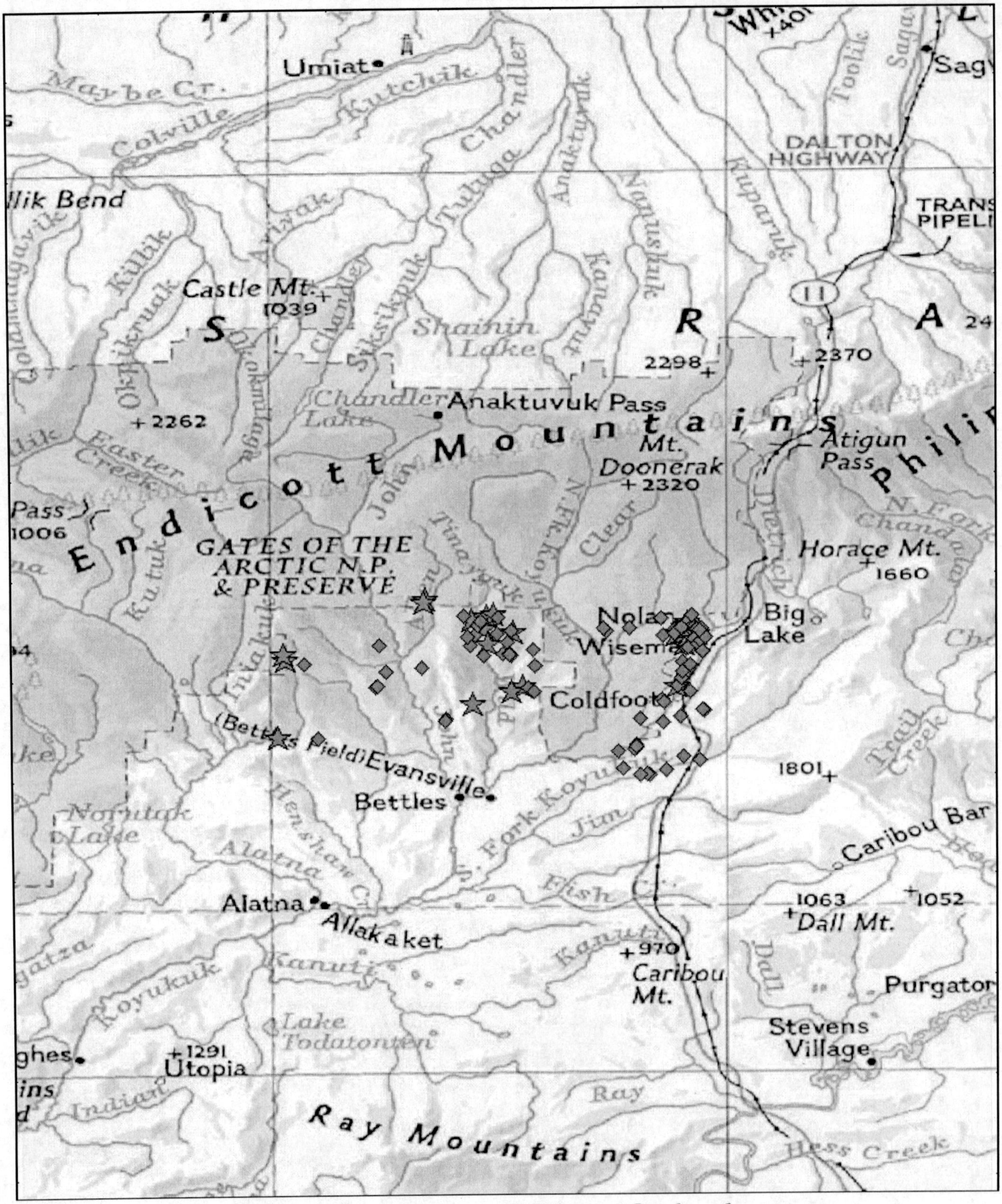

Gold Placer Locations in the Wiseman Quadrangle

♦ Diamonds shows placer gold locations. ★ Stars show gold-bearing hard rock locations. Hard rock occurrences are not identified by name nor is a description of the occurrence included in this book. They are provided so the reader can see where the hard rock sources for the placer deposits occur. Areas are located by latitude and longitude WGS 84 coordinates.

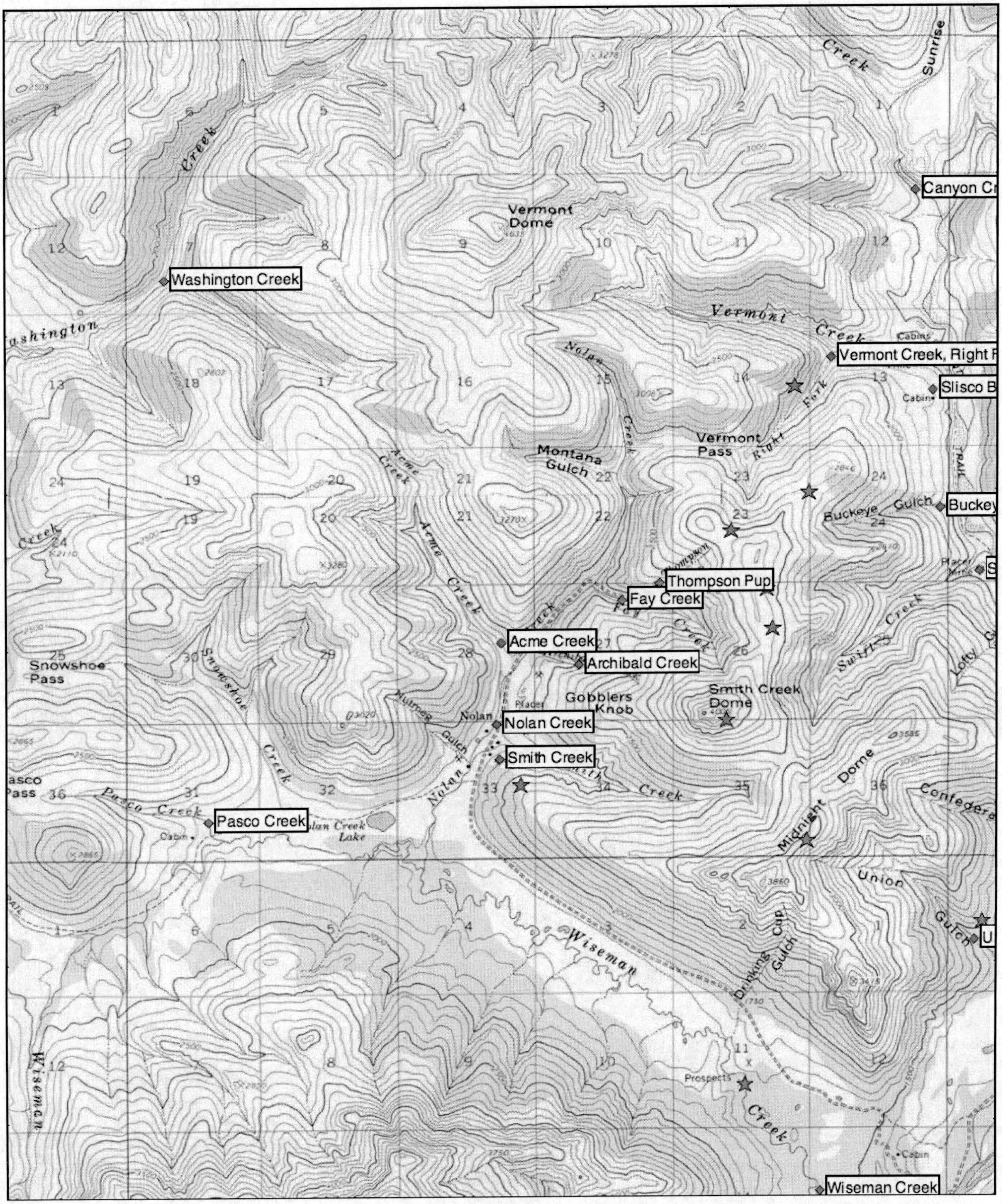
Canyon Cr
Vermont Dome
Washington Creek
Vermont Creek, Right F
Slisco B
Vermont Pass
Montana Gulch
Buckeye Gulch
Buckey
Thompson Pup
Fay Creek
Acme Creek
Archibald Creek
Snowshoe Pass
Gobblers Knob
Smith Creek Dome
Nolan Creek
Smith Creek
Pasco Pass
Pasco Creek
Midnight Dome
Union Gulch
Drinking Cup Gulch
Wiseman Creek
Prospects

Map 5-1

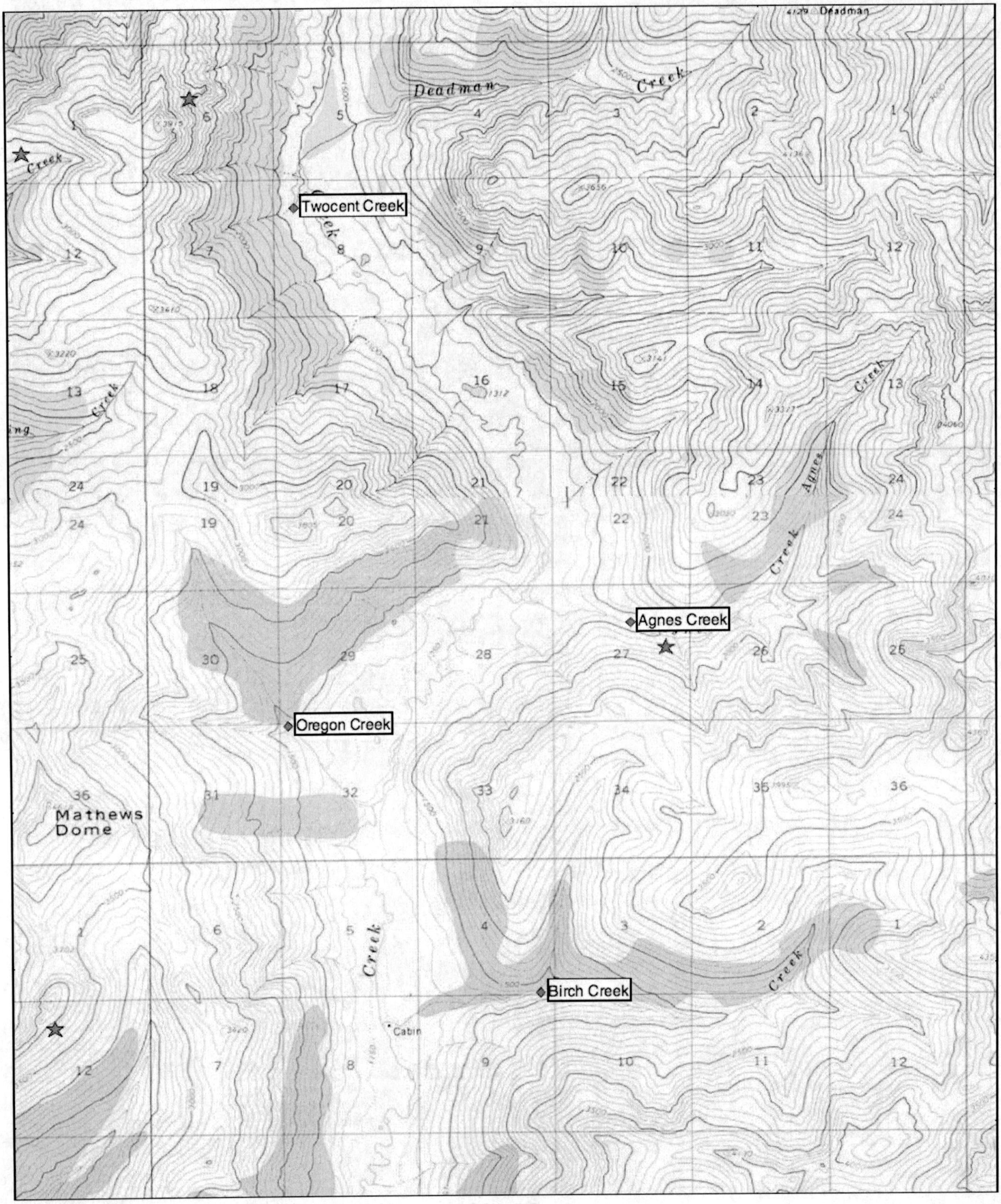
Deadman
Deadman
Creek
Twocent Creek
Agnes Creek
Oregon Creek
Birch Creek
Mathews
Dome
Cabin

Map 5-2

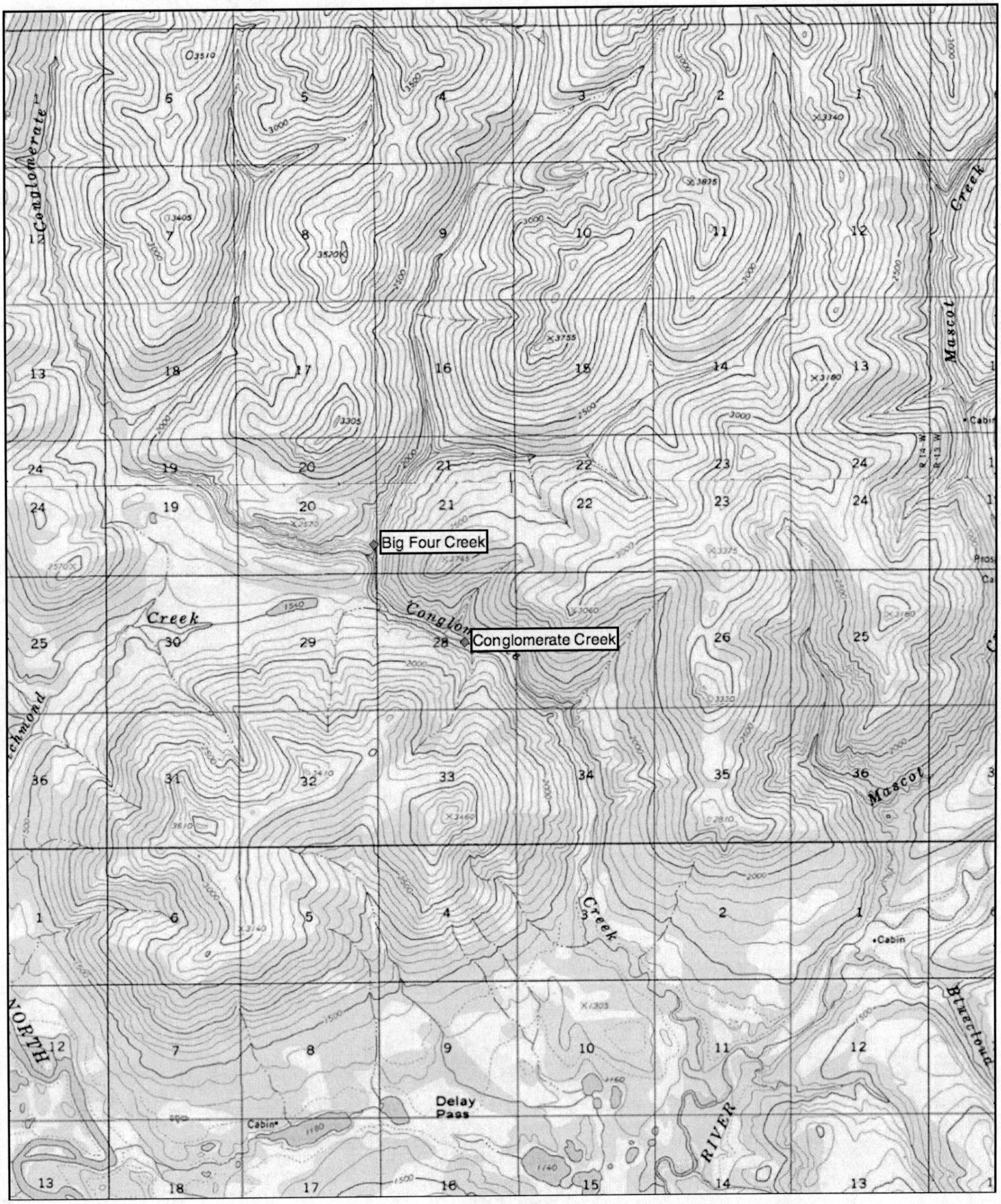
Big Four Creek
Conglomerate Creek
Conglomerate
Creek
Mascot
Creek
Delay
Pass
Cabin
RIVER
NORTH
Bluecloud

Map 5-3

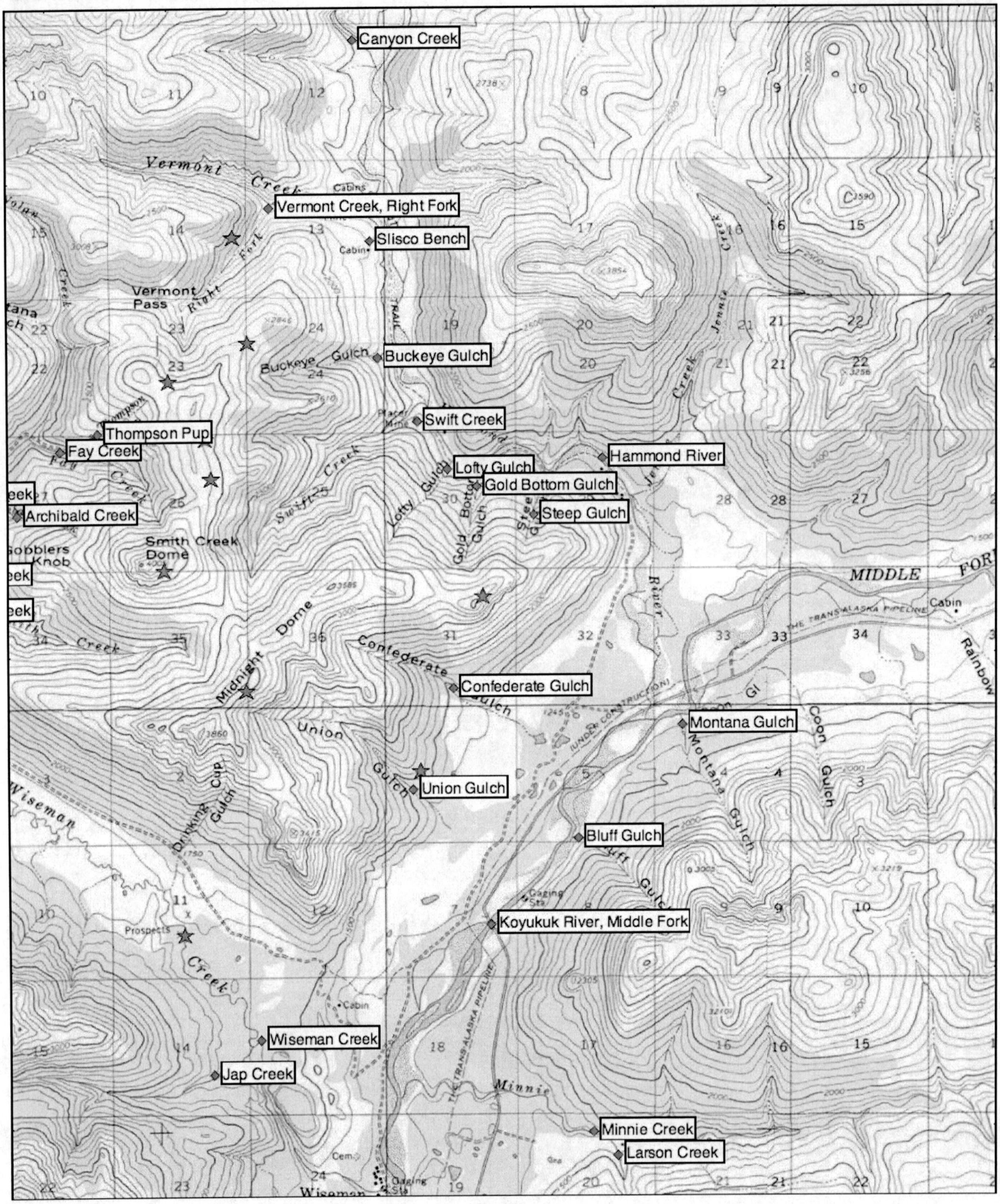

Canyon Creek
Vermont Creek, Right Fork
Slisco Bench
Buckeye Gulch
Swift Creek
Thompson Pup
Fay Creek
Hammond River
Lofty Gulch
Gold Bottom Gulch
Steep Gulch
Archibald Creek
Confederate Gulch
Montana Gulch
Union Gulch
Bluff Gulch
Koyukuk River, Middle Fork
Wiseman Creek
Jap Creek
Minnie Creek
Larson Creek
Vermont Creek
Vermont Pass
Smith Creek Dome
Gobblers Knob
Midnight Dome
MIDDLE FORK
THE TRANS-ALASKA PIPELINE
Wiseman

Map 5-4

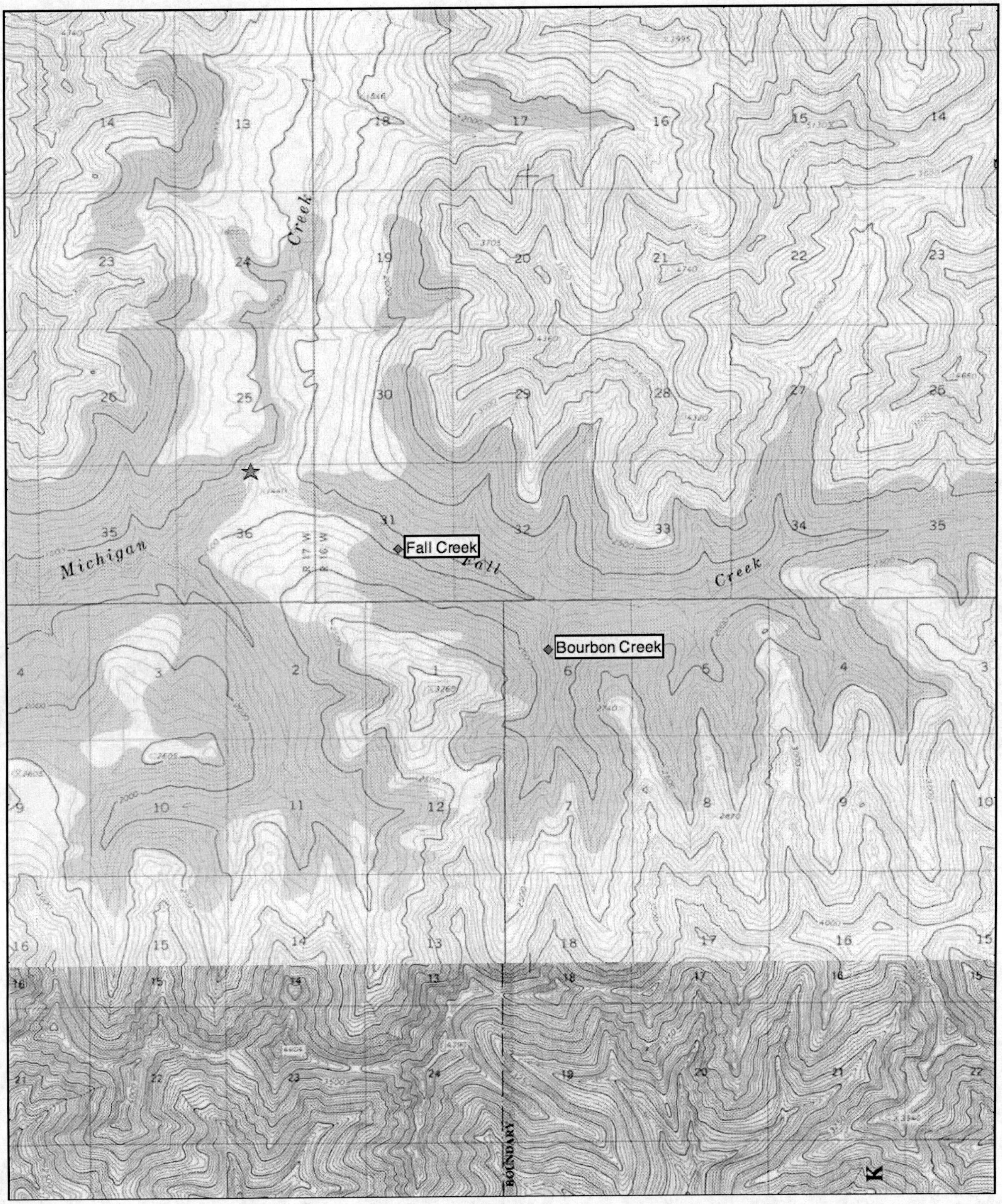
Creek
Michigan
Fall Creek
Fall
Creek
Bourbon Creek
BOUNDARY
R 17 W
R 16 W
K

Map 5-5

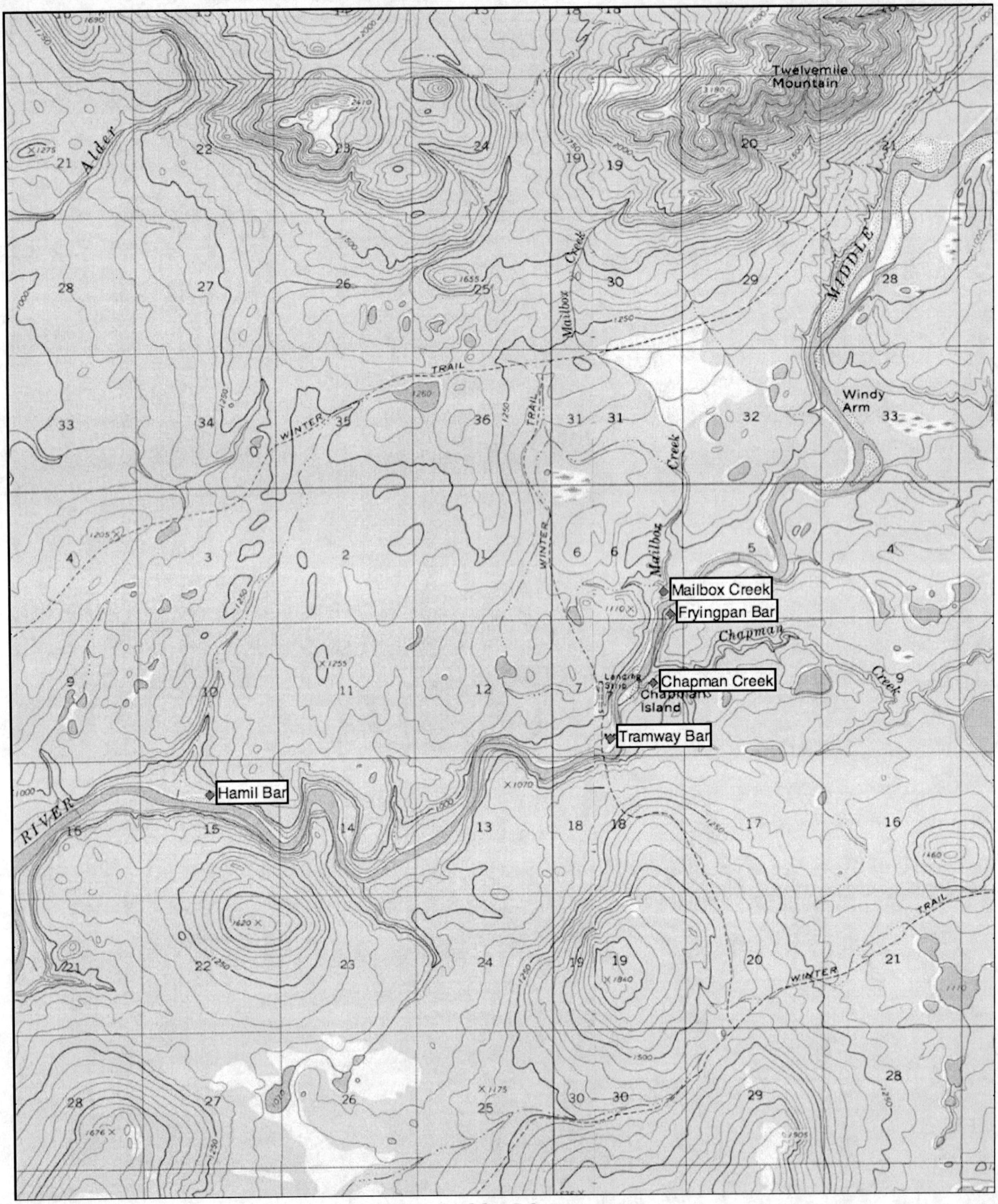
Twelvemile Mountain
Alder
Mailbox Creek
MIDDLE
Windy Arm
WINTER TRAIL
Mailbox Creek
Mailbox Creek
Fryingpan Bar
Chapman Creek
Chapman Creek
Landing Strip
Chapman Island
Tramway Bar
Hamil Bar
RIVER
WINTER TRAIL

Map 5-6

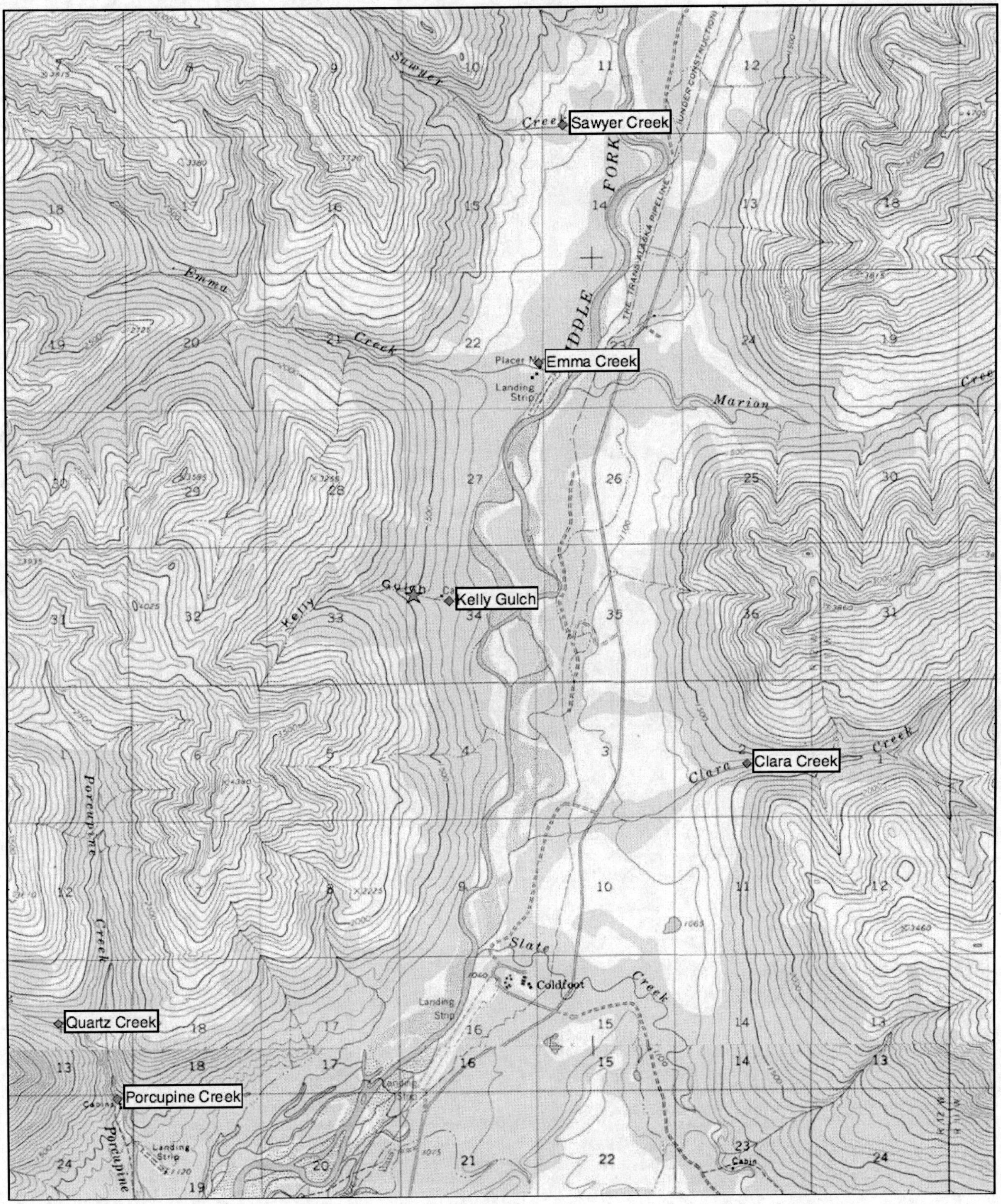
Sawyer Creek
Emma Creek
Kelly Gulch
Clara Creek
Quartz Creek
Porcupine Creek
Coldfoot
Landing Strip
Marion
Slate
Creek
MIDDLE FORK
THE TRANS-ALASKA PIPELINE
UNDER CONSTRUCTION

Map 5-7

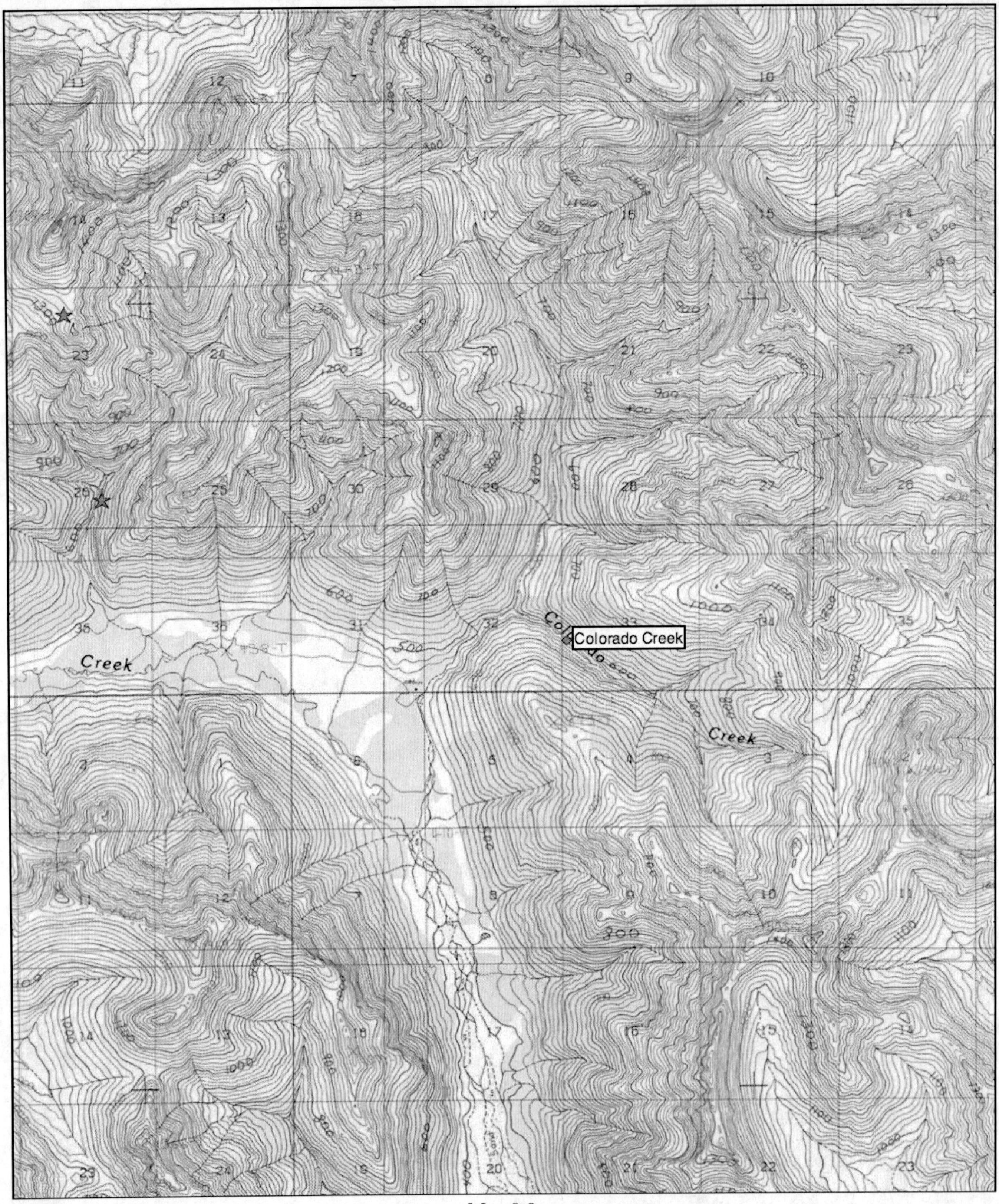

Map 5-8

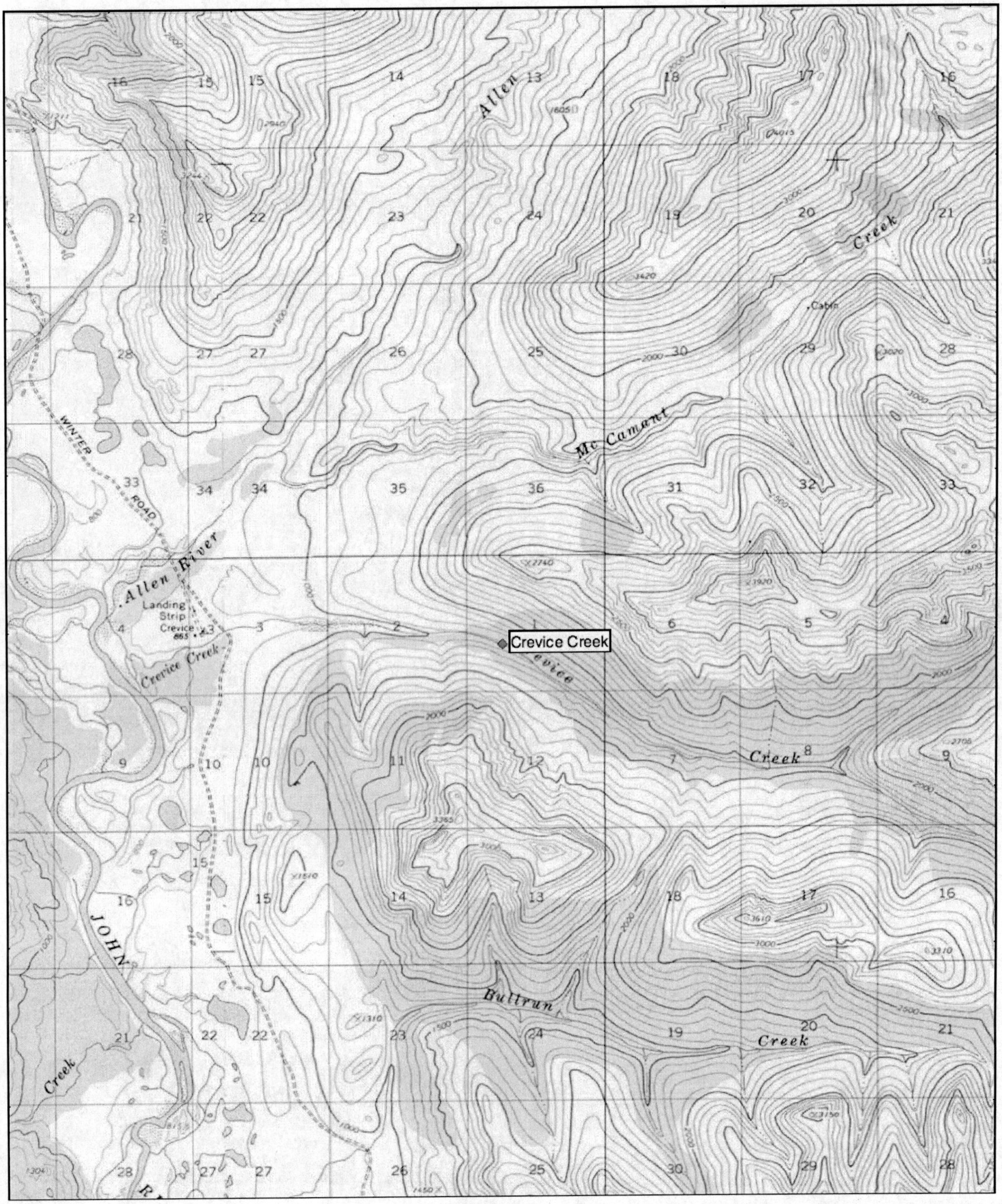
Allen
Creek
Cabin
Mc Camant
WINTER ROAD
Allen River
Landing Strip
Crevice
Crevice Creek
Crevice Creek
Creek
Bullrun
Creek
JOHN
Creek

Map 5-9

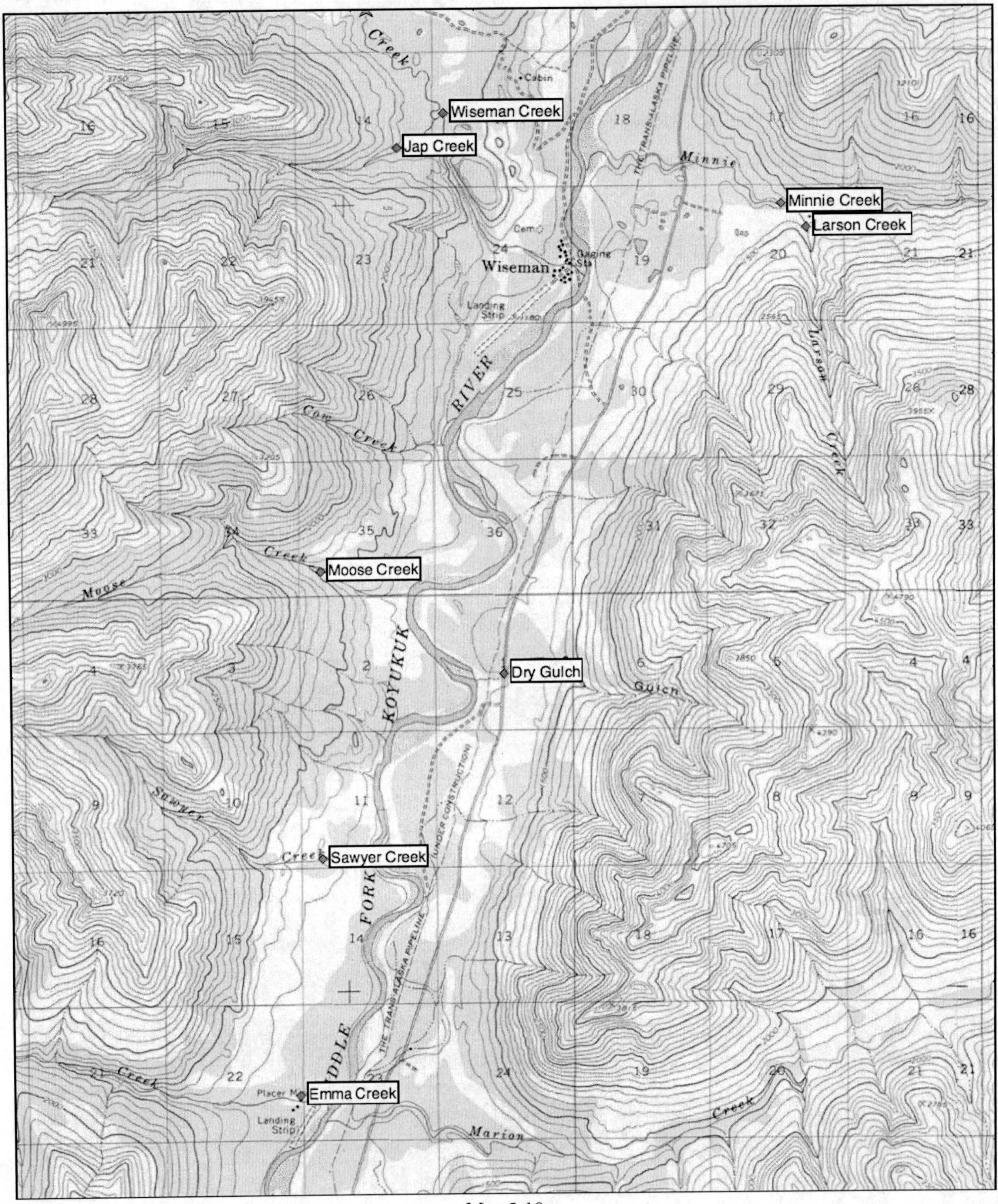
Wiseman Creek
Jap Creek
Minnie Creek
Larson Creek
Wiseman
Moose Creek
Dry Gulch
Sawyer Creek
Emma Creek
KOYUKUK
RIVER
MIDDLE FORK
THE TRANS-ALASKA PIPELINE
Landing Strip
Placer Mine
Marion Creek

Map 5-10

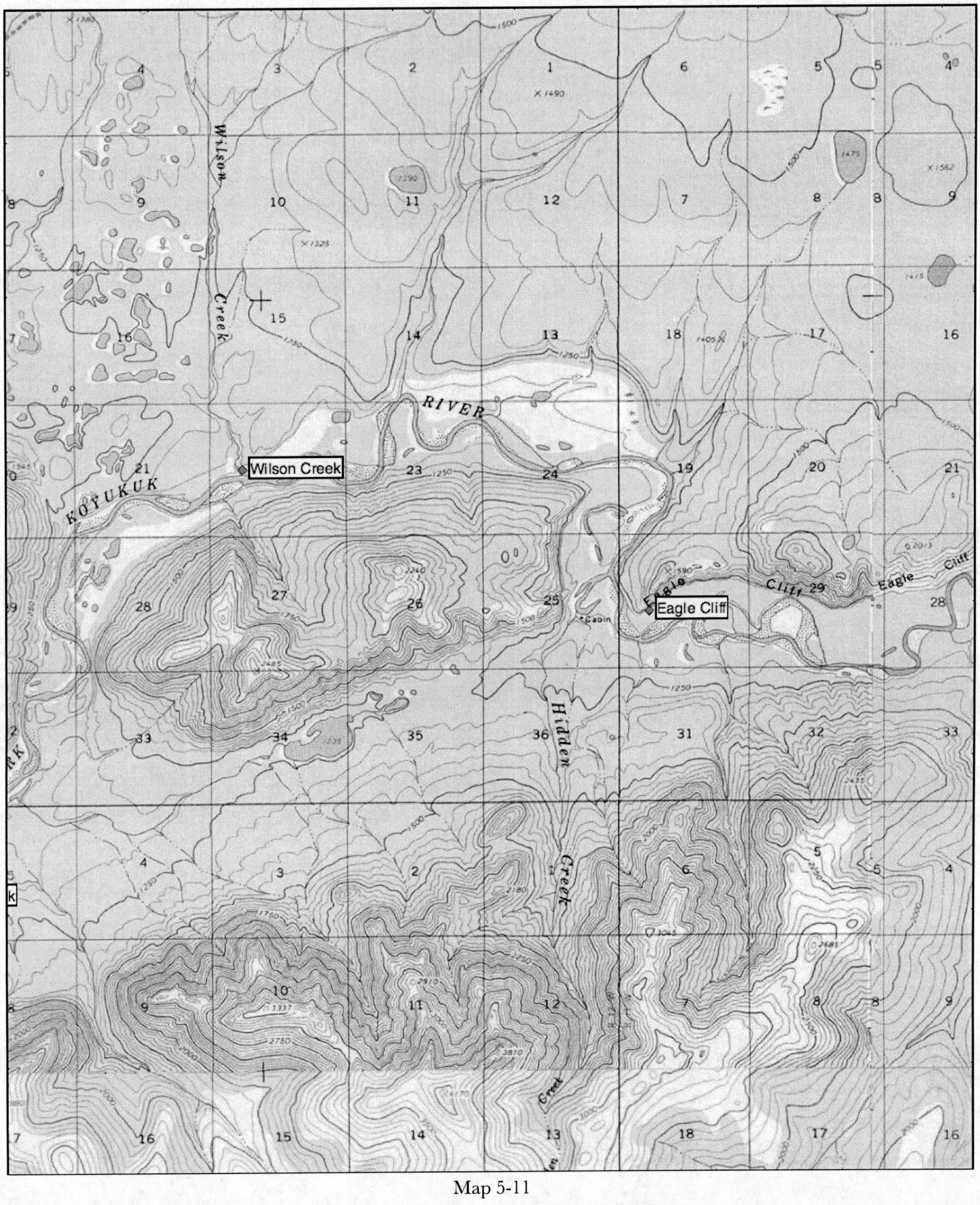
Wilson Creek
KOYUKUK
RIVER
Eagle Cliff
Eagle Cliff
Eagle Cliff
Hidden Creek
Cabin

Map 5-11

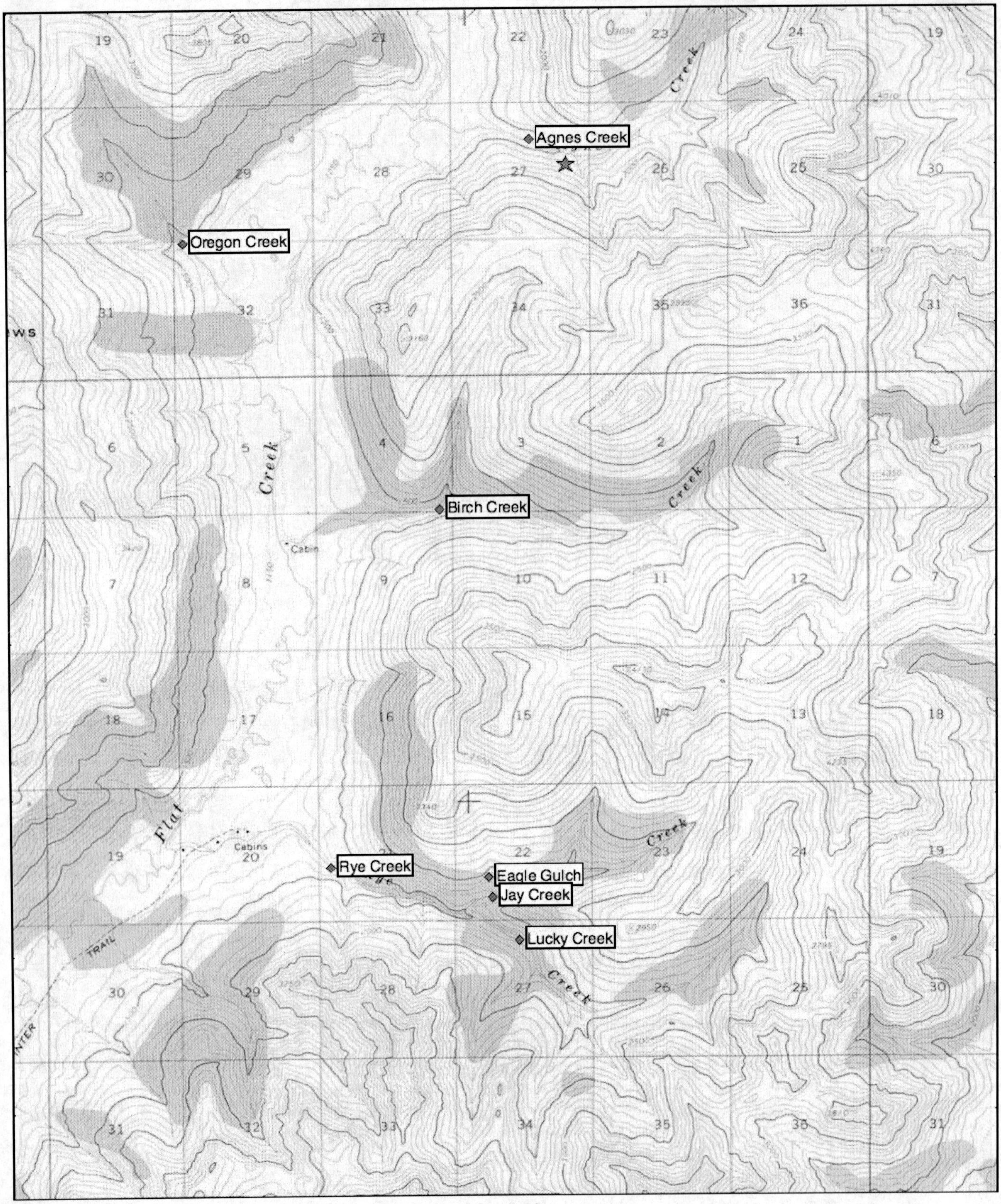
Agnes Creek
Oregon Creek
Birch Creek
Creek
Cabin
Flat
Cabins
Rye Creek
Eagle Gulch
Jay Creek
Lucky Creek
TRAIL

Map 5-12

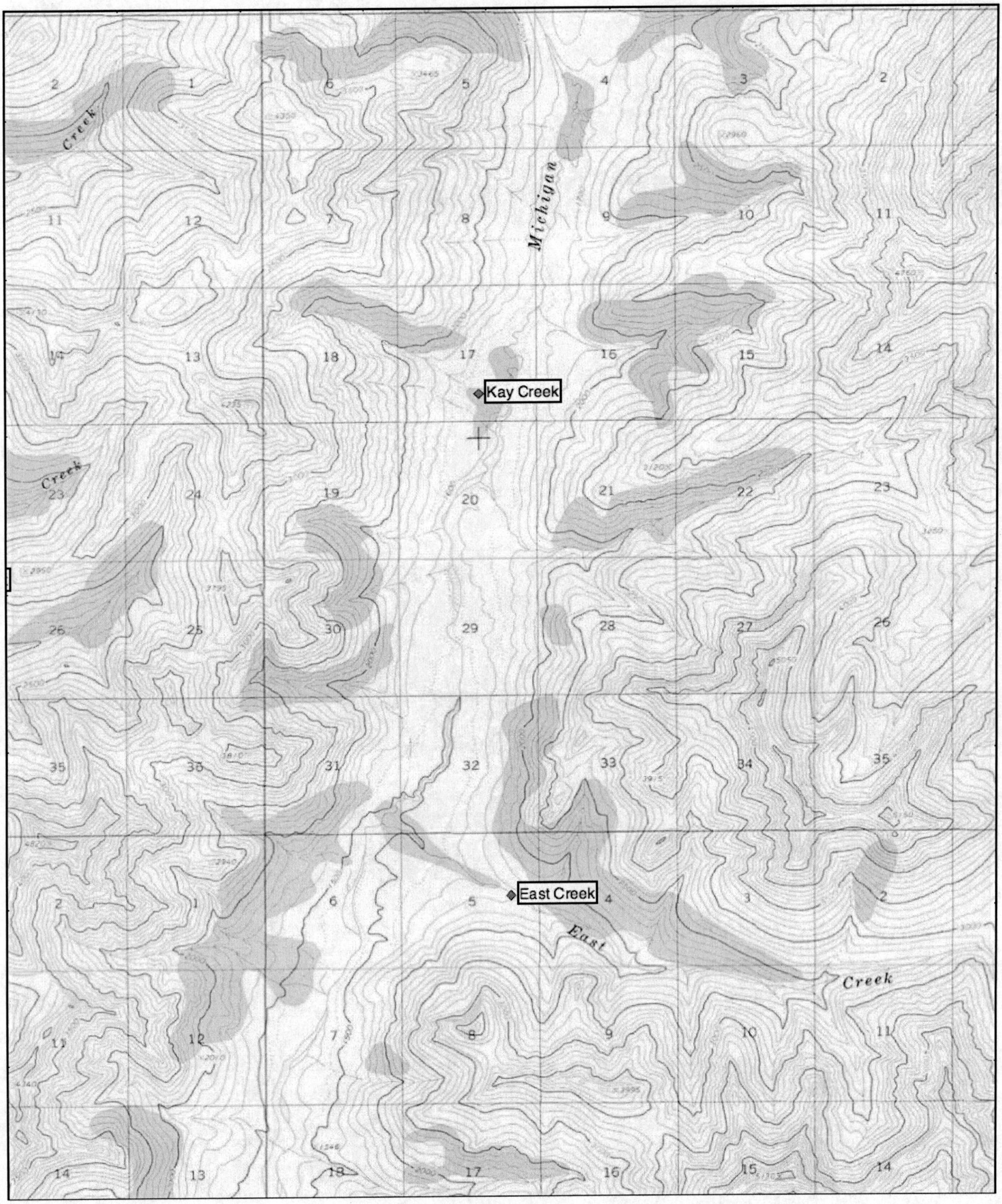

Map 5-13

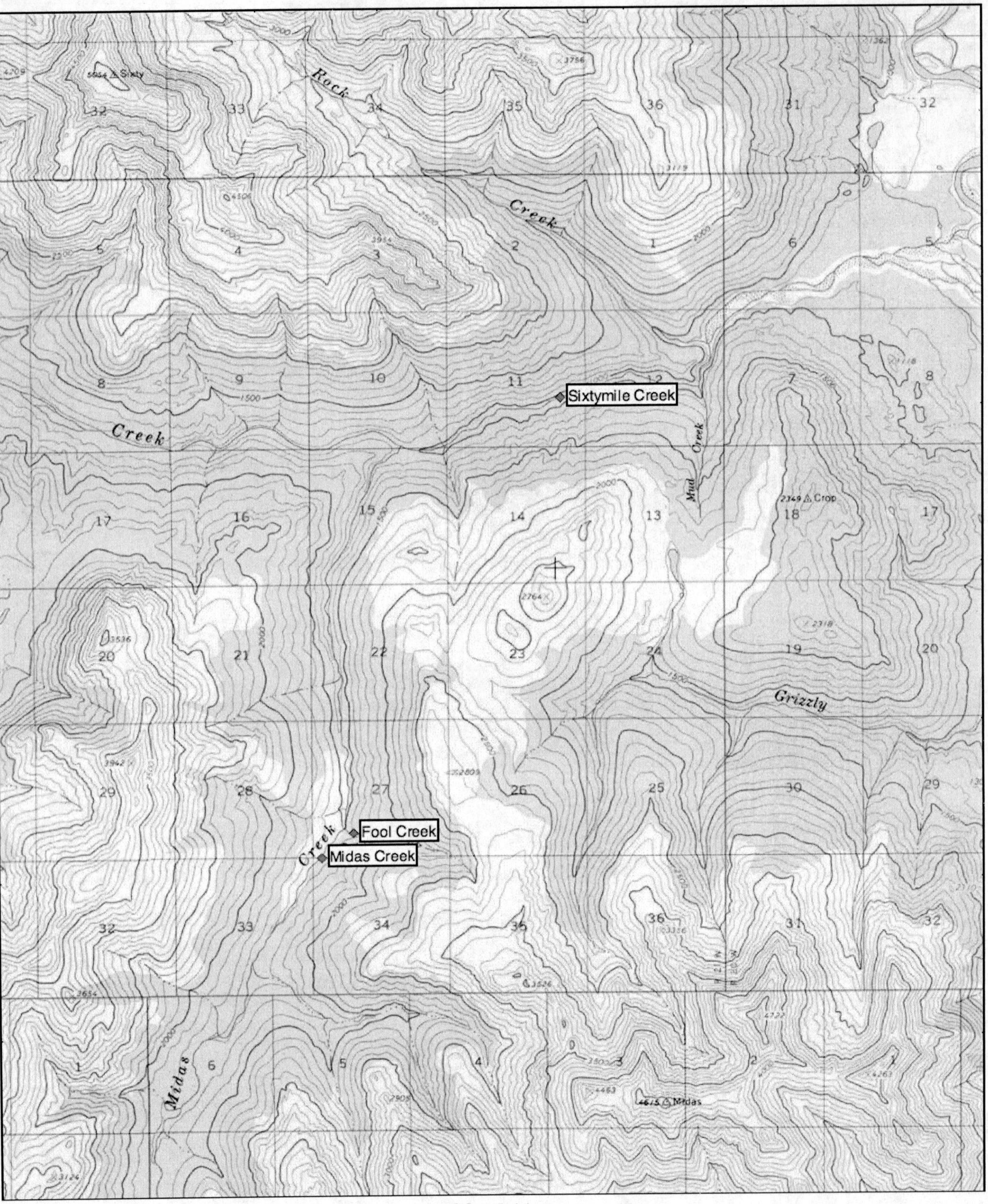
Sixtymile Creek
Fool Creek
Midas Creek
Sixty
Rock
Creek
Creek
Mud Creek
Crop
Grizzly
Creek
Midas
Midas

Map 5-14

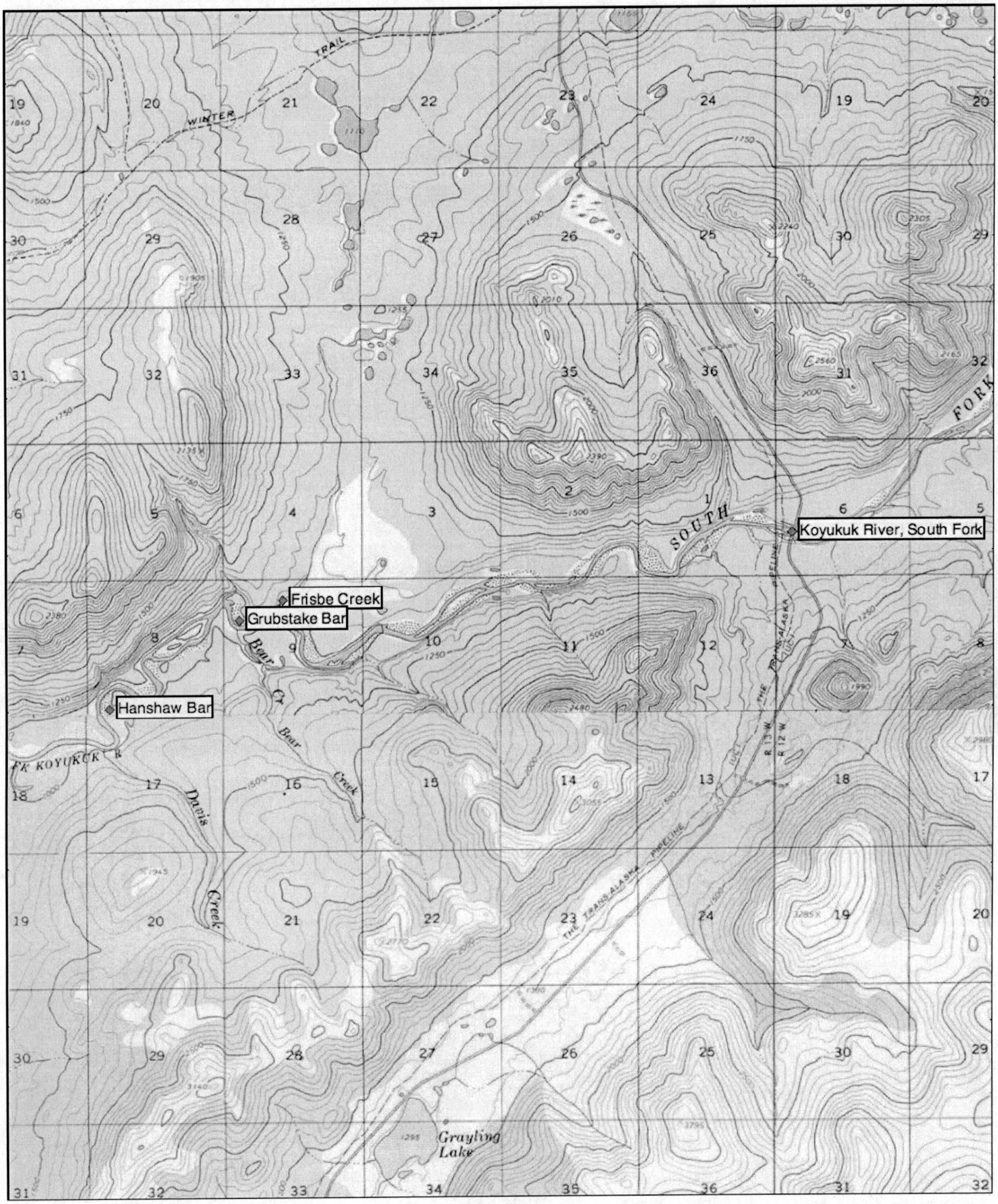

Koyukuk River, South Fork
Frisbe Creek
Grubstake Bar
Hanshaw Bar
WINTER TRAIL
SOUTH FORK
FK KOYUKUK R
Bear Cr
Bear Creek
Davis Creek
THE TRANS-ALASKA PIPELINE
Grayling Lake
R 13 W
R 12 W

Map 5-15

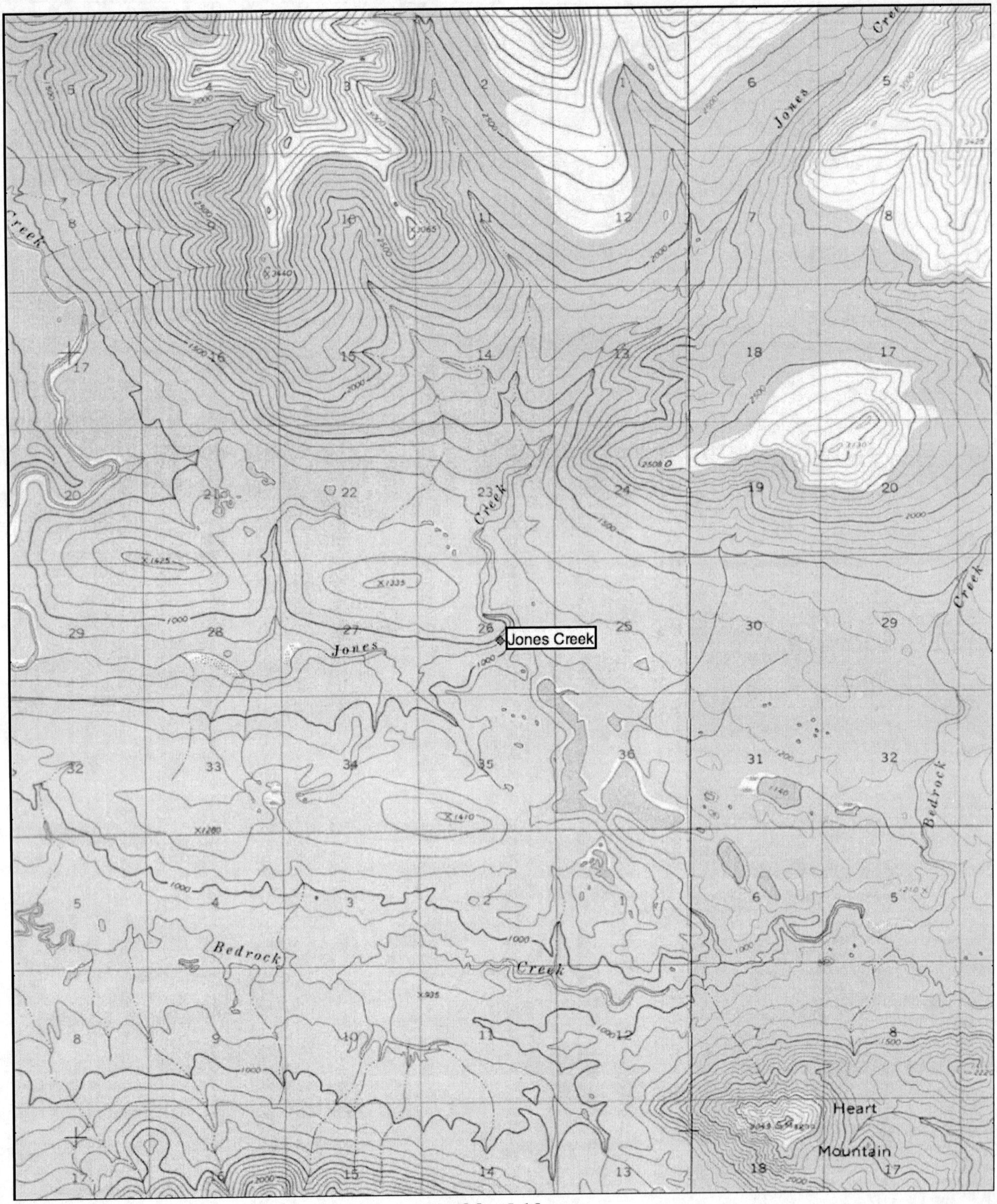
Jones Creek
Jones
Creek
Jones
Creek
Bedrock
Creek
Bedrock
Creek
Heart
Mountain

Map 5-16

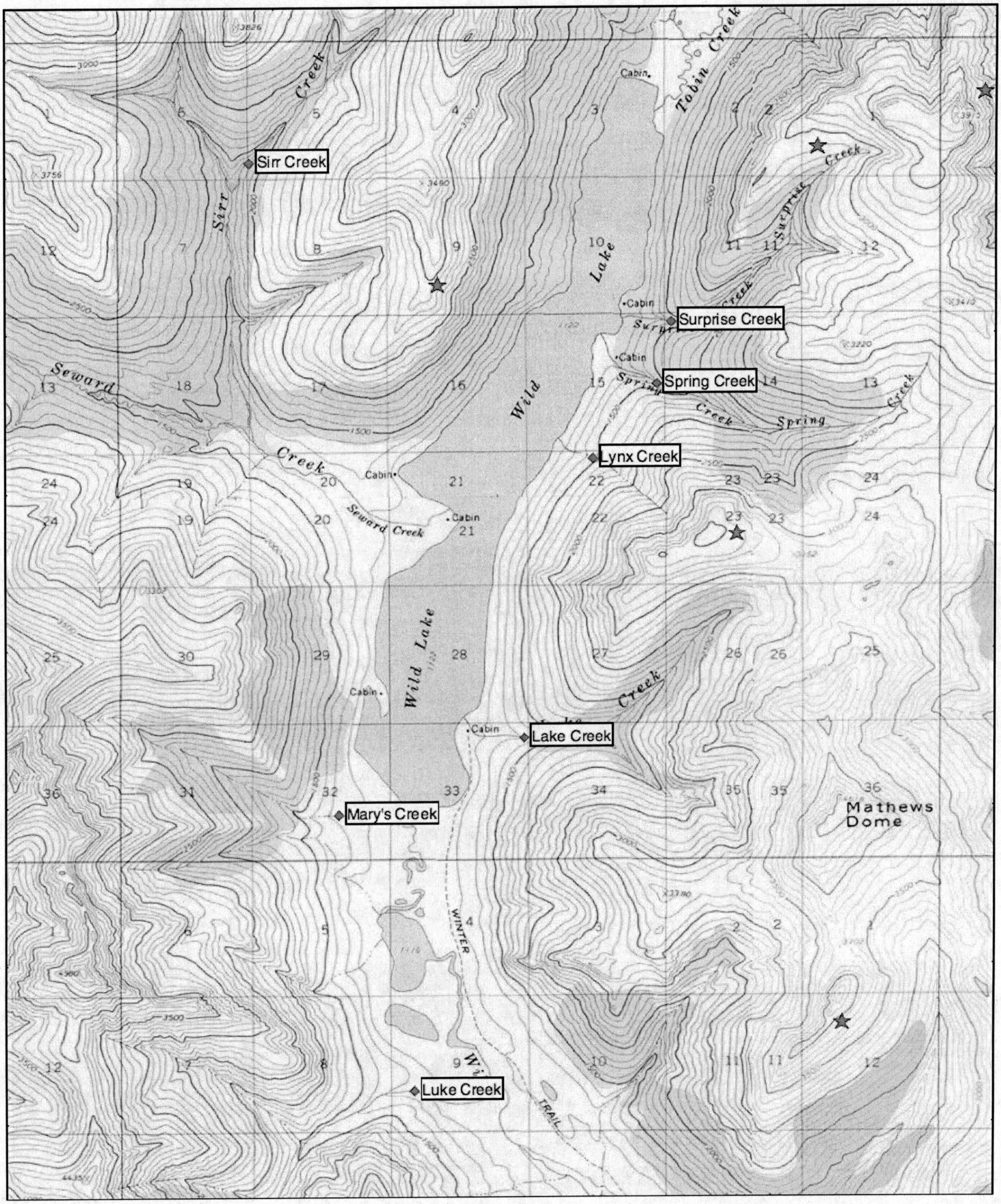
Sirr Creek
Surprise Creek
Spring Creek
Lynx Creek
Lake Creek
Mary's Creek
Luke Creek
Wild Lake
Tobin Creek
Seward Creek
Mathews Dome
WINTER TRAIL
Cabin

Map 5-17

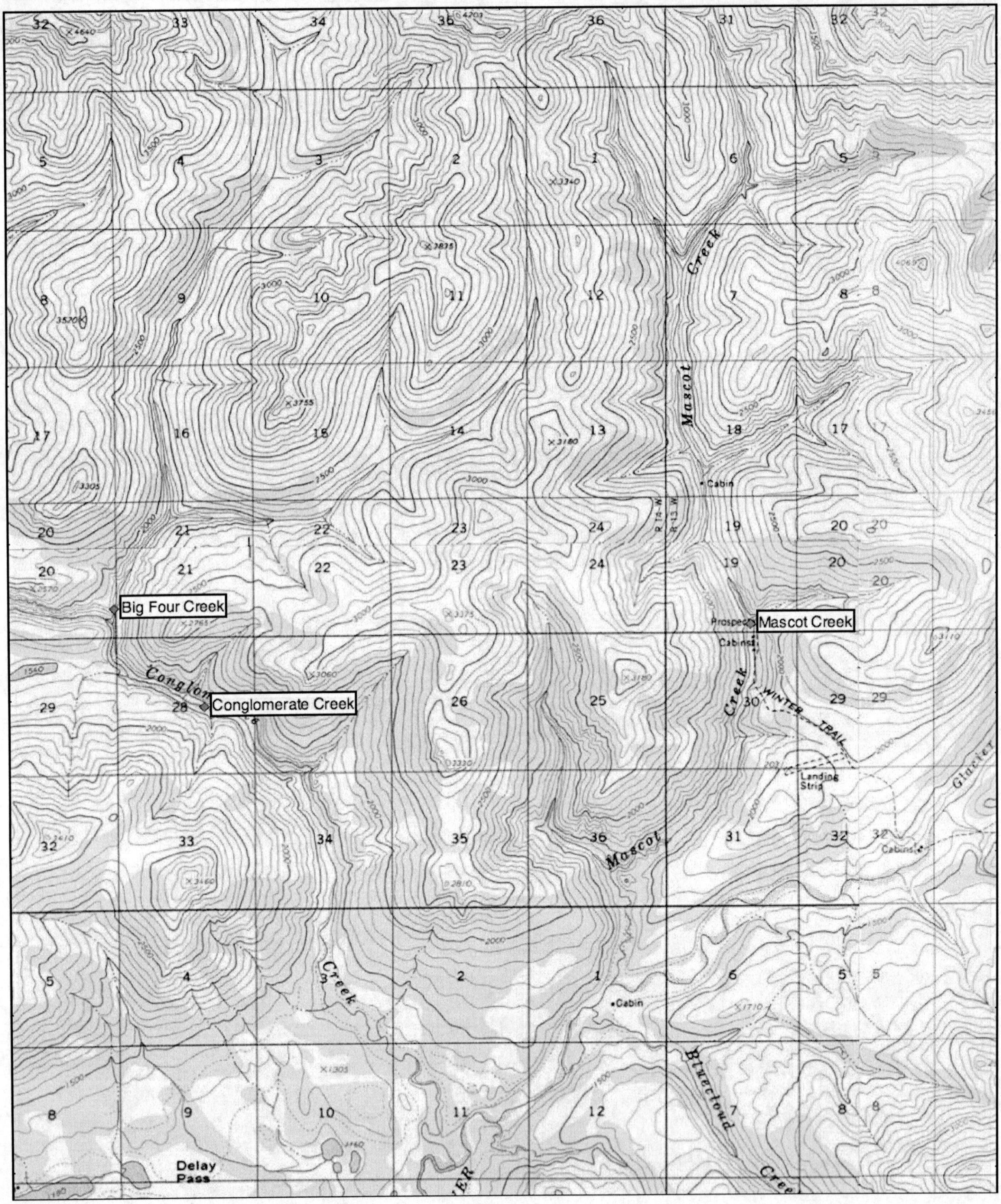
Big Four Creek
Conglomerate Creek
Mascot Creek
Mascot Creek
Prospect
Cabins
Cabin
Conglomerate
WINTER TRAIL
Landing Strip
Cabins
Mascot
Creek
Cabin
Bluecloud Creek
Delay Pass
Glacier
R 14 W
R 13 W

Map 5-18

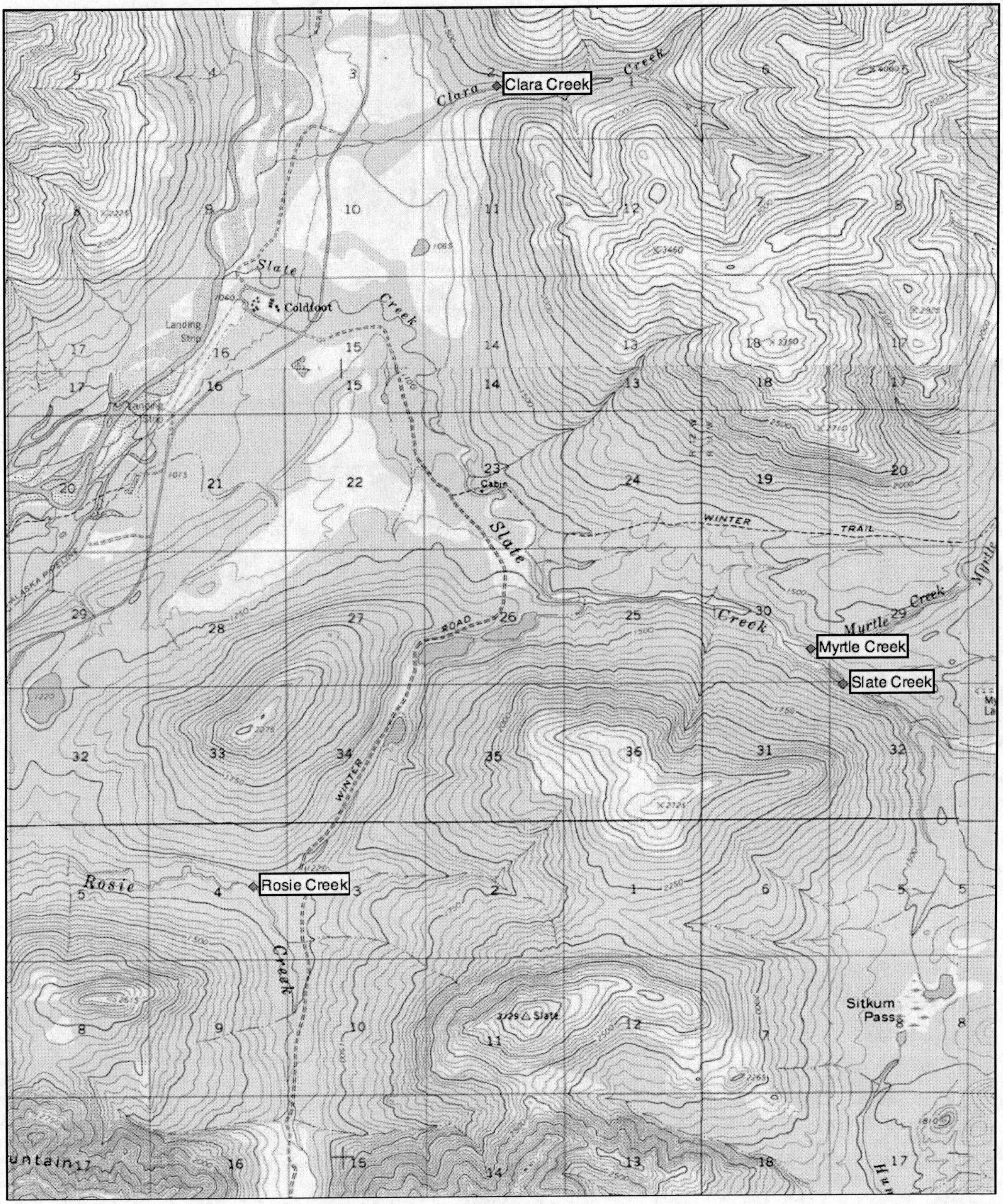
Clara Creek
Clara
Creek
Slate
Creek
Coldfoot
Landing Strip
Cabin
Slate
Creek
WINTER
TRAIL
Myrtle
Creek
Myrtle Creek
Slate Creek
ROAD
WINTER
Rosie
Rosie Creek
Creek
Slate
Sitkum Pass
ALASKA PIPELINE

Map 5-19

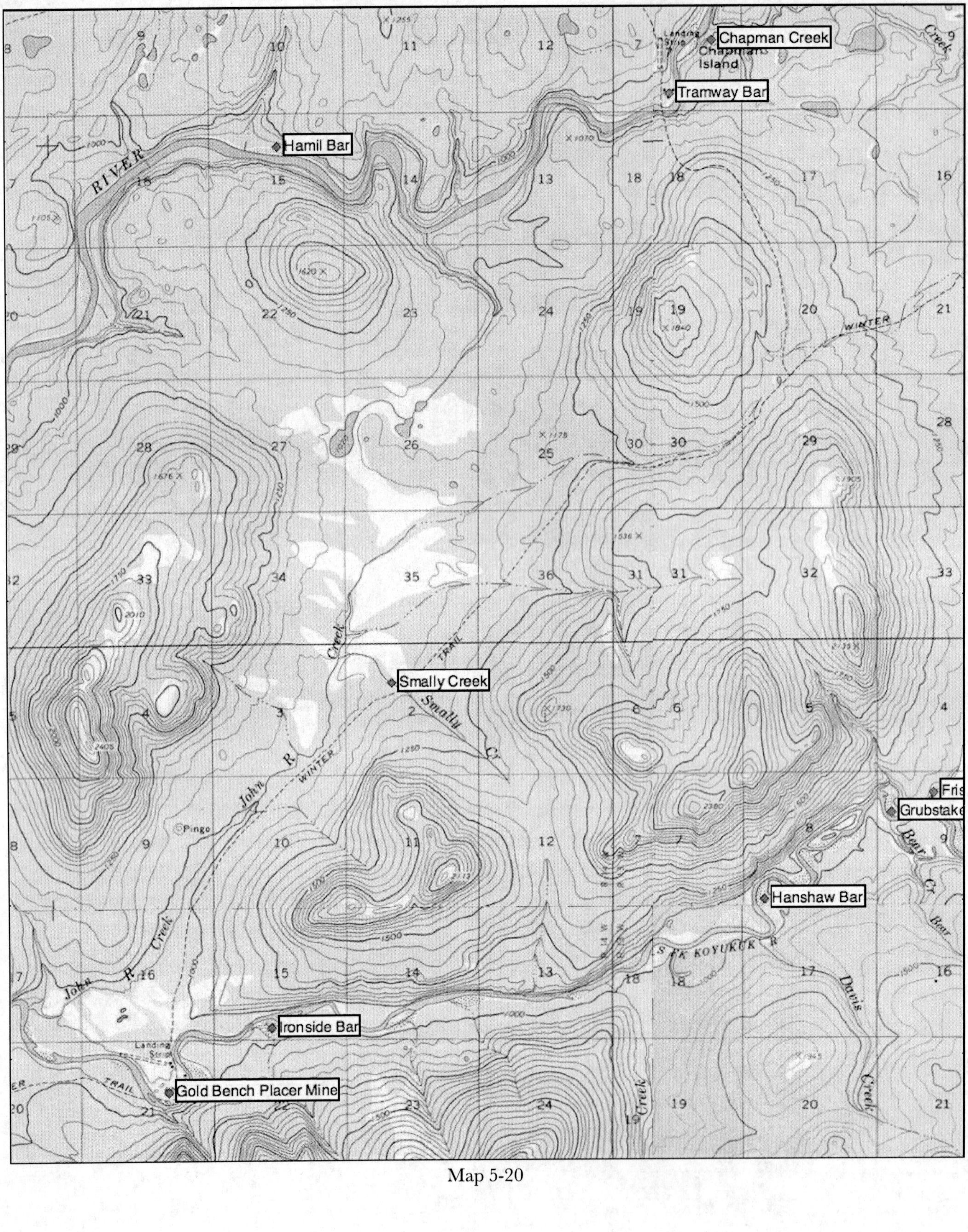
Chapman Creek
Chapman Island
Landing Strip
Tramway Bar
Hamil Bar
RIVER
WINTER
TRAIL
Smally Creek
Smally Cr
Creek
John R
Pingo
Hanshaw Bar
S FK KOYUKUK R
Davis Creek
Bear Cr
John R Creek
Ironside Bar
Landing Strip
Gold Bench Placer Mine

Map 5-20

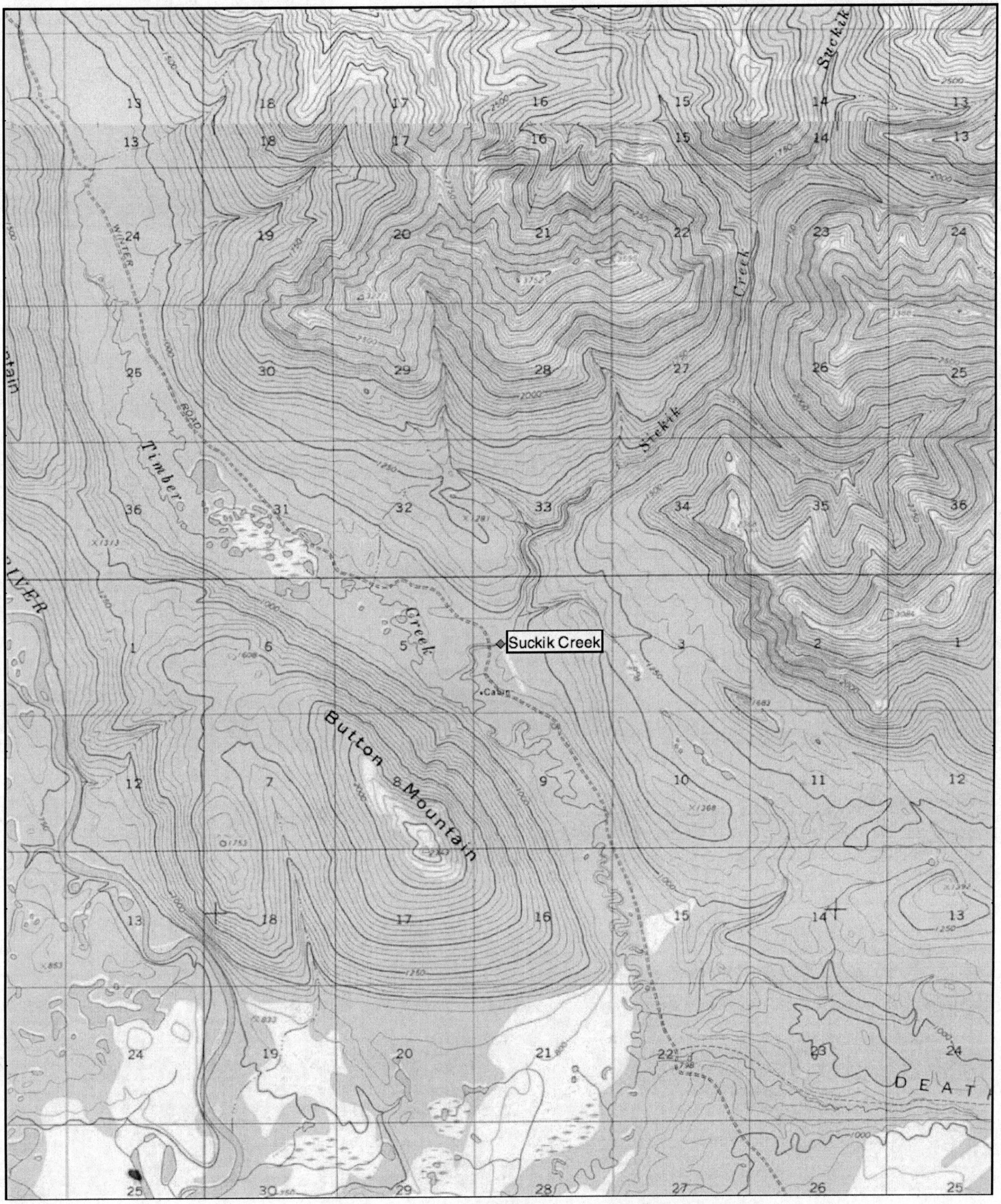
Suckik Creek
Button Mountain
Timber Creek
Suckik Creek
Cabin

Map 5-21

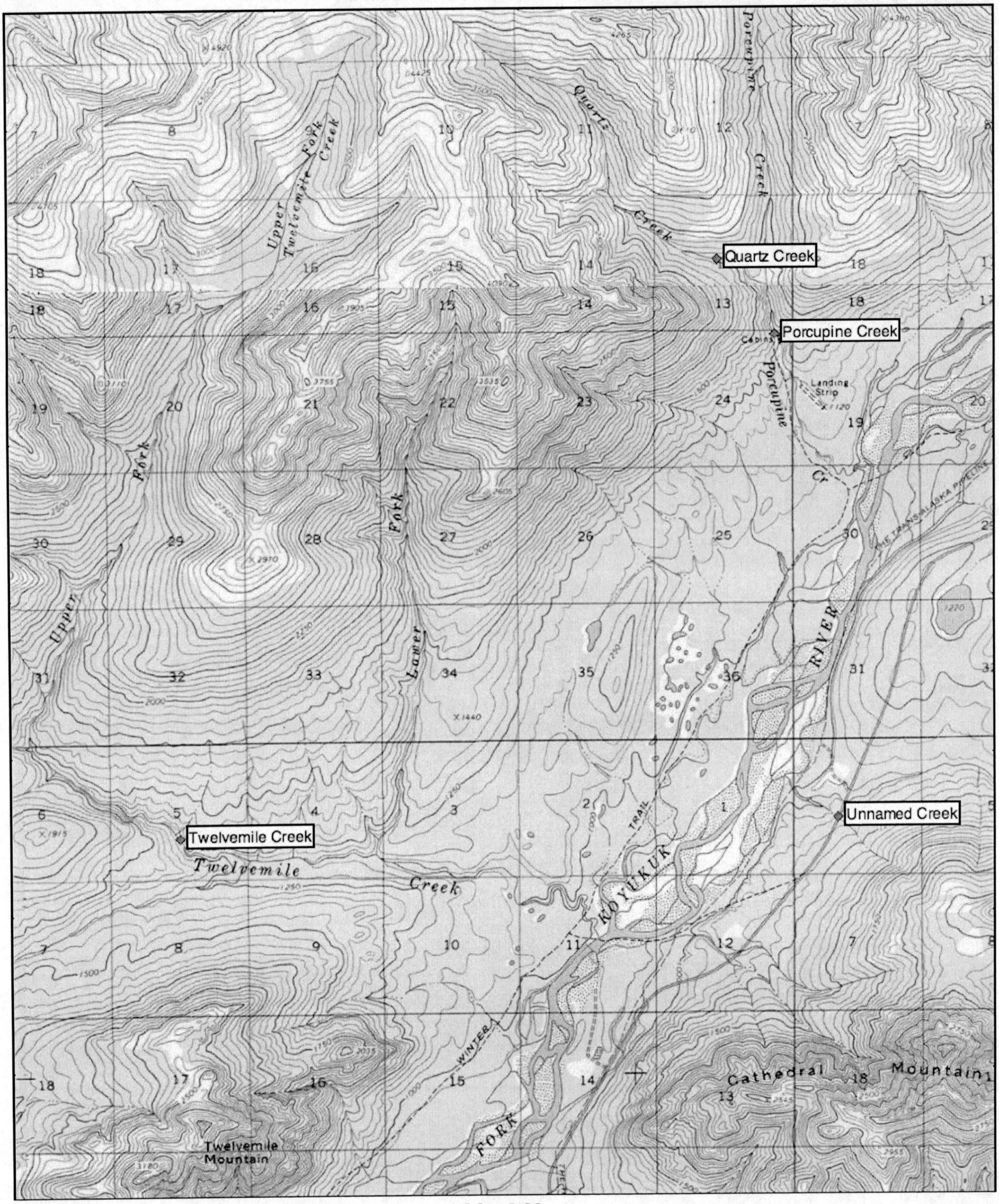
Quartz Creek
Porcupine Creek
Unnamed Creek
Twelvemile Creek
Upper Twelvemile Fork Creek
Quartz Creek
Porcupine Creek
Porcupine Cr
Cabins
Landing Strip
Upper Fork
Lower Fork
Twelvemile Creek
Koyukuk River
Fork
Winter Trail
Cathedral Mountain
Twelvemile Mountain

Map 5-22

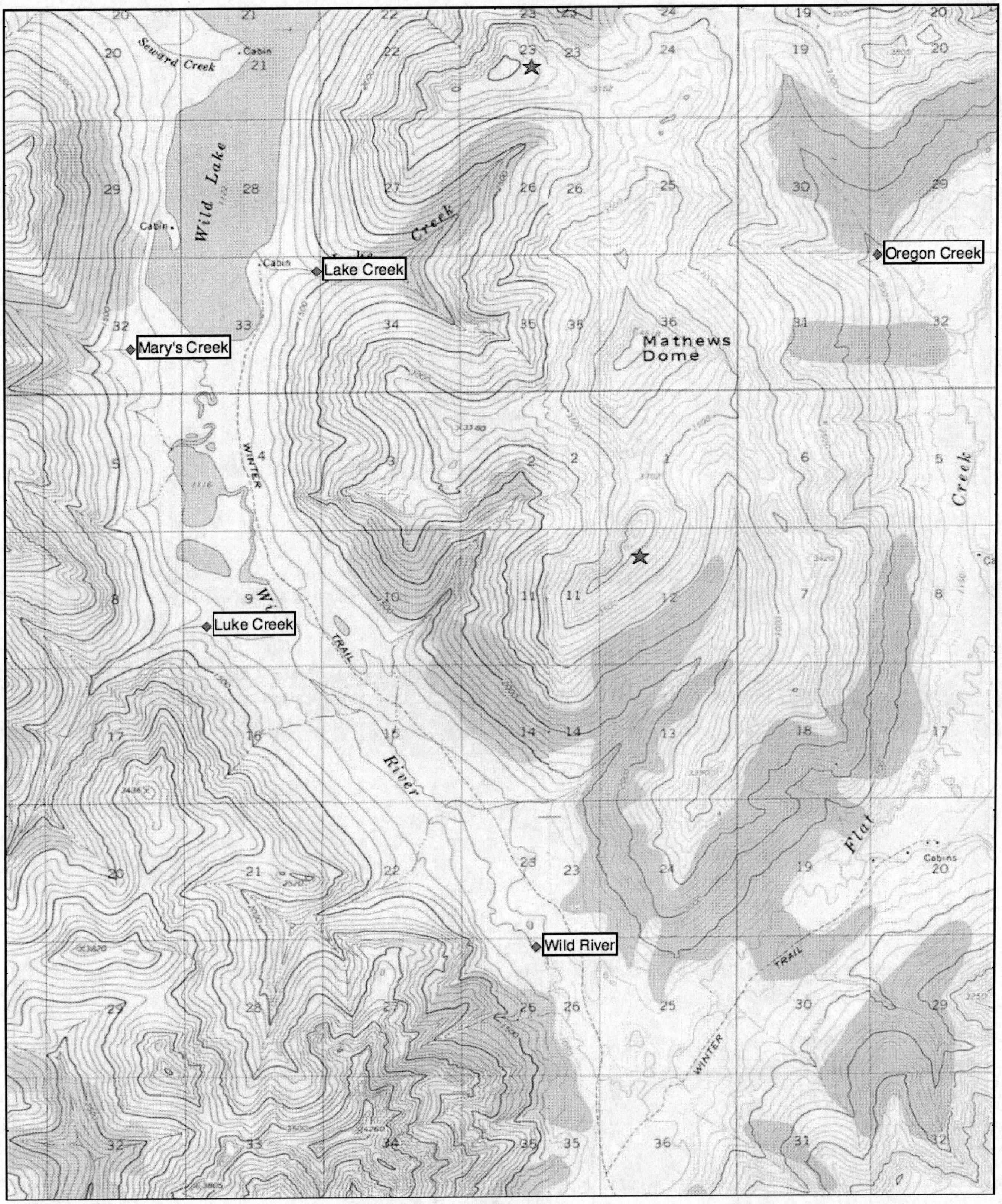
Seward Creek
Cabin
Wild Lake
Cabin
Cabin
Creek
Lake Creek
Oregon Creek
Mary's Creek
Mathews Dome
WINTER
TRAIL
Luke Creek
River
Creek
Flat
Cabins
Wild River
TRAIL
WINTER

Map 5-23

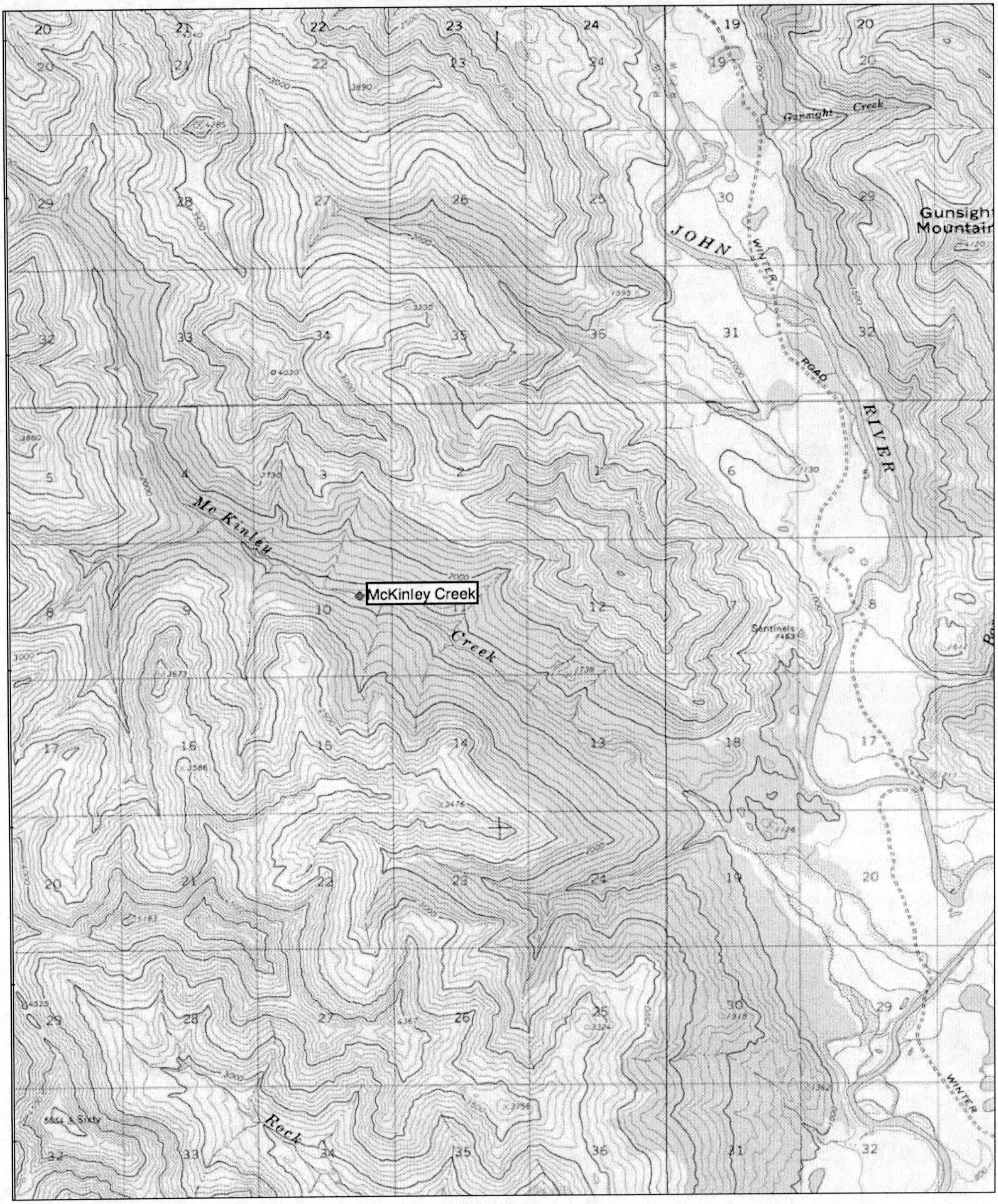
McKinley Creek
Mc Kinley
Creek
JOHN
RIVER
WINTER
ROAD
Gunsight Creek
Gunsight
Mountain
Sentinels
1453
Rock
Sixty

Map 5-24

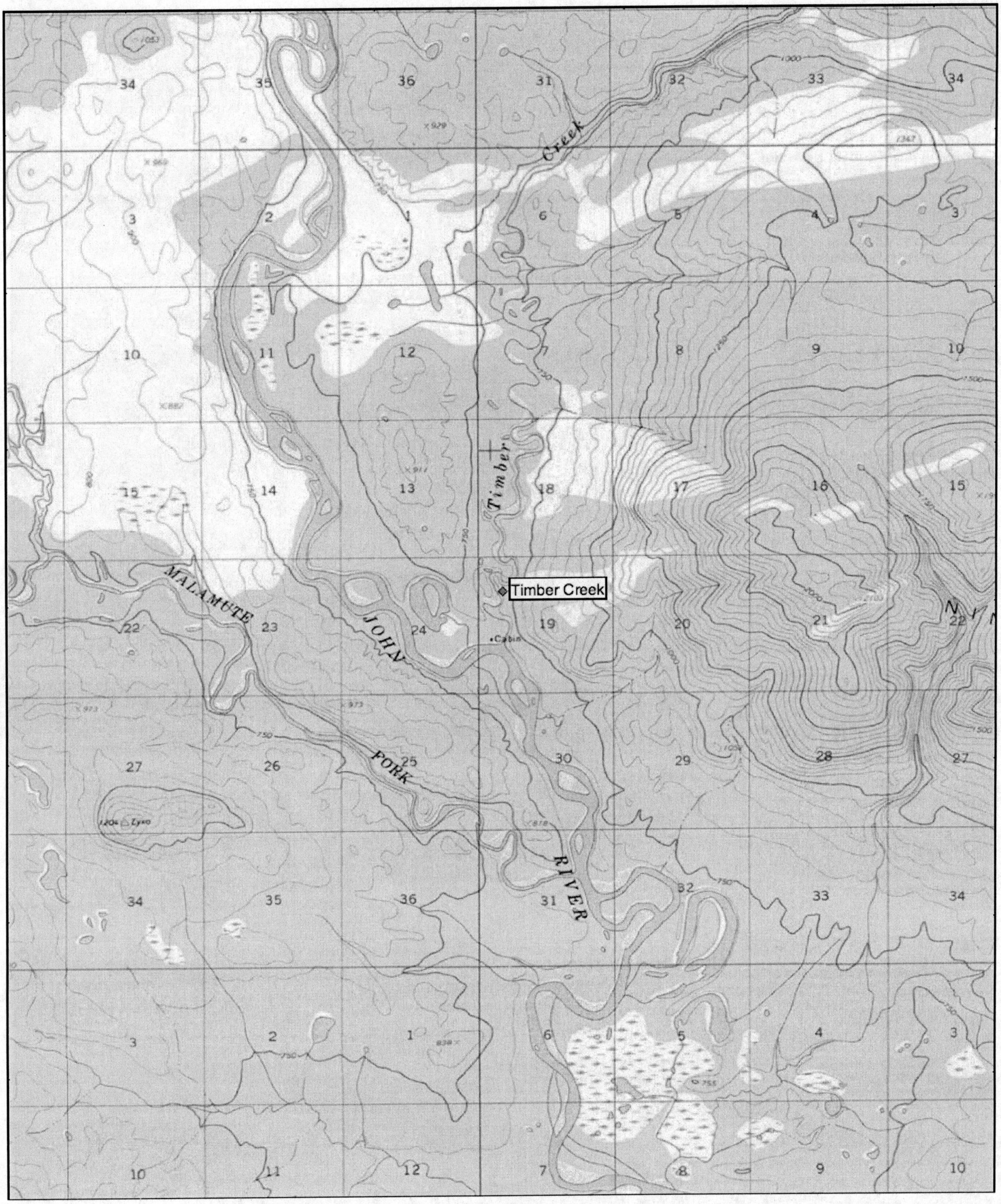
Timber Creek
Timber
Creek
MALAMUTE
JOHN
FORK
RIVER
Cabin
Zyxo

Map 5-25

Acme Creek

(Gold) **Quadrangle:** Wiseman (B1); **Coordinates:** 67.4839°N, 150.2292°W; **Map:** 5-1; **Worked:** 1900s-; **Access:** Unimproved road from Wiseman, Alaska; **Summary:** Tributary of Nolan Creek. Reed (1938) reported that a small amount of fine gold was mined on Acme Creek from a deep channel 70 to 110 feet below the surface, a few hundred feet upstream from the junction of the Acme Creek deep channel and the deep channel of Nolan Creek. The gold from Acme Creek was fine and bright and differs from that in Nolan Creek. Reed (1938) noted that gold had not been found in either the present or high channels of Acme Creek. (Cobb, 1976, OF 76-340. p. 83). (The Alaska Resource Data File, Wiseman Quadrangle).

Agnes Creek

(Gold) **Quadrangle:** Wiseman (B3); **Coordinates:** 67.4862°N, 151.3188°W ; **Map:** 5-2, 5-12; **Worked:** 1900s-; **Summary:** Tributary of Flat Creek. Extensive prospecting and some mining in early 1900s but mining records were lost. (Cobb, 1976, OF 76-340. p. 84). (The Alaska Resource Data File, Wiseman Quadrangle).

Archibald Creek

(Antimony, Gold) **Quadrangle:** Wiseman (B1); **Coordinates:** 67.48165°N, 150.2084°W; **Map:** 5-1, 5-4; **Worked:** 1909-1999; **Purity:** 905 Au; **Yield:** 0.0126-0.0603 oz/ft^2; **Production:** 1900 to 1999=5429.90 ozs; **Access:** Unimproved road from Wiseman, Alaska; **Summary:** Tributary of Nolan Creek. Short, steep east side tributary of Nolan Creek cut in soft schist. Coarse, subangular frozen gravel contains smooth, flattened fine gold and coarse, porous gold. Gold in bench deposits gold is fine and smooth. In old channel depth to bedrock is as much as 25 feet, decreasing upstream. Bench gravels mined by hand and hydraulic methods; deep gravels were drift mined. A stibnite vein is reported to have been exposed by drift mining; another reported on spur between Archibald and Smith Gulches. Mining before 1909, in 1927, 1933-39, and 1974 was reported. Production, 1900-09, was 290 fine ounces (Cobb, 1976, OF 76-340. p. 86). **Geology:** Placer gold has been mined both from the modern channel and a deep channel on Archibald Creek. The modern channel was mined near the mouth of the creek in the early 1900's and was being mined about 3/8 mile upstream in 1937. The gravel in the modern channel is coarse and subangular; it contains many boulders. Five to twelve feet of frozen gravel, covered by 3 feet of muck and a foot of moss, overlie soft, schist bedrock. The gold is distributed throughout the gravel and occurs both as fine particles worn flat and smooth, and as coarse particles that are rough and porous. The average fineness of the gold is 905 and the value of the ground is about $0.44 per square foot of bedrock (gold at $35 per ounce) (Reed, 1938). The deep channel which is 5 to to 14 feet wide is on the north side of the creek at a depth of about 25 feet near the mouth of the creek; it rises upstream to coalesce with the modern channel. Seven to nine feet of frozen, coarse, subangular gravel overlies bedrock described as alternating bands of black graphitic and yellow schist (Reed, 1938). The gravel is overlain by 16 to 18 feet of muck and a foot of moss. Unlike the modern channel, the deep channel is free of large boulders. The gold in the deep channel occurs both as coarse, worn and rounded particles, and as coarse, porous and unworn particles. Reed (1938) reports values of approximately $2.11 per square foot of bedrock (gold at $35 per ounce) and an average fineness of 904. A stibnite vein was reported during drift mining on Archibald Creek (Ebbley and Wright, 1948. **Workings:** Eden (2000) reports a total of 5429.90 ounces of gold were produced from Archibald Creek between 1904 and 1999. Most of the production from Archibald Creek has been from the buried channels. Tri-Con Mining mined on Archibald Creek as late as 1991 (Bundtzen and others, 1992).(The Alaska Resource Data File, Wiseman Quadrangle).

Big Four Creek

(Gold) **Quadrangle:** Wiseman (B2); **Coordinates:** 67.4929°N, 150.7066°W; **Map:** 5-3, 5-18; **Worked:** 1902-1937; **Summary:** Tributary of Conglomerate Creek. Mining in 1902 reported (Cobb, 1976, OF 76-340. p. 87) Schrader (1904) reported mining in 1902. Reed (1938) noted that mining in the early days was small scale, but reports that there was no activity in 1937. No additional information is available. (The Alaska Resource Data File, Wiseman Quadrangle).

Birch Creek

(Gold) **Quadrangle:** Wiseman (B3); **Coordinates:** 67.4471°N, 151.3439°W; **Map:** 5-2, 5-12; **Worked:** 1904-1998; **Production:** Early 1900s=965-1,450 ozs; **Summary:** Tributary of Flat Creek which flows through a schist belt. Gold discovered in 1904. Mined for about a mile near mouth of Rue Creek. Mining has been in creek bed and in deposits about 20 feet below a low left-limit bench. Total production 965 -1,450 fine ounces of gold in early 1900's and an unknown amount in the early 1930's. Last mining was in 1933 (Cobb, 1976, OF 76-340. p. 88). **Geology:** Gold was discovered on Birch Creek in 1904 by Louis Rue (Reed, 1938). He reportedly recovered about $1,800 in gold in 1905 (gold at $20.67 per ounce) (Reed, 1938). The modern channel of Birch Creek has been mined for about 1/2 mile above and below the mouth of Rue Creek, a name reported by Reed (1938), but not shown on the modern topographic map. Rue Creek is a south-flowing tributary to Birch Creek that enters Birch Creek about 1.25 miles above the mouth of Birch Creek. The gravel is coarse and waterworn in Birch Creek and contains many boulders. An area along the left limit of Birch Creek below Rue Creek was drift mined. The depth of the drift workings is probably about 20 feet, similar to the depth to bedrock in the creek bed (Reed, 1938). Several shafts were also sunk in the alluvial fan where Birch Creek enters the valley of Flat Creek but the results of this work are not known. Birch Creek was mined by shoveling-in and drifting. Reed, (1938) noted that mining had not occurred on this creek since 1933. Reed estimated the total production from the early 1900's until about 1933 to be $20,000 to $30,000 (at the old price of gold at $20.67 per ounce). A placer operation on Birch Creek in 1998 was mining stream gravel with dozers and then allowed paying customers to pan, sluice, dredge or use gold detectors to recover gold from the gravels (Kurtak and others, 1999). A sample of quartz-mica schist from the ridge on the north side of upper Birch Creek contains 35 ppb gold and 1,767 ppm arsenic (Kurtak and others, 1999). (The Alaska Resource Data File, Wiseman Quadrangle).

Bluff Gulch

(Gold) **Quadrangle:** Wiseman (B1); **Coordinates:** 67.4472°N, 150.0543°W; **Map:** 5-4; **Yield:** 0.0025 oz/yd^3; **Access:** From the Dalton Highway at 67.4626°N, 150.0238°W; **Summary:** Tributary of Middle Fork of the Koyukuk River. Average yield of gravel is estimated to be 0.0025 oz/yd^3 of fine gold.

Bourbon Creek

(Gold) **Quadrangle:** Wiseman (B3); **Coordinates:** 67.2827°N, 151.1643°W; **Map:** 5-5; **Worked:** 1900s-1937; **Summary:** Tributary of Fall Creek. Was extensively mined in "early days." In 1937 was considered to have been mined out (Cobb, 1976, OF 76-340. p. 90). Bourbon Creek is not named on the Wiseman B-3 topographic map. Reed (1938) describes Bourbon Creek as a left-limit tributary of Fall Creek. Later publications (Cobb (1972 [MF469]); Maas, 1987; Bliss and others, 1988) placed Bourbon Creek about 2.5 miles upstream from the mouth of Fall Creek although at least two other tributaries farther upstream could correspond to Reed's description. This mine is somewhat arbitrarily located near the mouth of the tributary in the NW1/4 section 6, T. 28 N., R. 16 W., of the Fairbanks Meridian. (The Alaska Resource Data File, Wiseman Quadrangle).

Buckeye Gulch

(Gold) **Quadrangle:** Wiseman (B1); **Coordinates:** 67.4981°N, 150.1094°W; **Map:** 5-4; **Worked:** 1900s-1937; **Nugget Size:** 50 ozs; **Access:** Unimproved road from Wiseman, Alaska; **Summary:** Small, steep gully on left limit of Hammond River. Some of channel near mouth was mined out in early 1900's. Gold from Buckeye Gulch was probably from bench gravels of Hammond River. Much of gold was very coarse with nuggets over 50 ounces reported. Concentrates contain pyrite, some exhibiting free gold. Placers work continuously from discovery to World War II. Inactive in 1974 (Cobb, 1976, OF 76-340. p. 127). (The Alaska Resource Data File, Wiseman Quadrangle).

Canyon Creek

(Gold) **Quadrangle:** Wiseman (C1); **Coordinates:** 67.5316°N, 150.1162°W; **Map:** 5-4; **Access:** Unimproved road from Wiseman, Alaska; **Summary:** Tributary of the Hammond River. Traces of gold in canyon in lower reaches of creek; upper valley also appears favorable (Cobb, 1976, OF 76-340. p. 92). Canyon Creek is long, southeasterly-flowing tributary to Hammond River. Its mouth is about 2.75 miles southwest of Jennie Creek Lake. This gold occurrence is on lower Canyon Creek, in the NE1/4 section 12, T. 31 N., R. 12 W., of the Fairbanks Meridian. The location is accurate within 1 mile. Mulligan (1974) reported that trace amounts of gold had been found along Lower Canyon Creek, and that the upper valley appears favorable for placer gold. (The Alaska Resource Data File, Wiseman Quadrangle).

Chapman Creek

(Gold) **Quadrangle:** Wiseman (A1); **Coordinates:** 67.0939°N, 150.4864°W; Dalton Hyw Xing, 67.0966°N, 150.3492°W; **Map:** 5-6, 5-20; **Worked:** 1908-; **Access:** Trail from landing strip at 67.0922°N, 150.5007°W or from Dalton Highway at 67.0797°N, 150.3508°W; **Summary:** Tributary of the Middle Fork of the Koyukuk River. Holes 9 - 14 feet deep 1/4 mile to 1 miles above mouth found some gold in schist gravel on clay. Two miles above mouth bench gravels 20 - 30 feet above creek also contained gold (0.002 - 0.005 ounces per pan) in a layer of blue schist gravel 1 - 3 feet thick. Not enough water for sluicing (Cobb, 1976, OF 76-340. p. 93). **Geology:** Gold was found in shafts sunk in 1908-09 along a stretch of Chapman Creek 1/4 to 1 mile above its mouth (Maddren, 1913). The workings included thirteen or fourteen shafts, 9 to 14 feet deep. A layer of blue, gravel composed mainly of schist that is below barren sand and reddish gravel produced gold values of 2-3 cents per pan (with gold at $20.67 per ounce). Maddren (1913) reported that gold values of 5 to 10 cents per pan also occurred in bench deposits about 2 miles above the mouth of the creek. The gold in the bench deposits also was in a 1- to 3 foot-thick layer of blue, gravel composed mainly of schist. (The Alaska Resource Data File, Wiseman Quadrangle).

Clara Creek

(Gold) **Quadrangle:** Wiseman (B1); **Coordinates:** 67.2793°N, 150.1269°W, 67.2724°N, 150.1710°W (Dalton Hwy

Xing); **Map:** 5-7, 5-19; **Worked:** 1900-; **Purity:** 940 Au; **Yield:** 0.0025 oz/yd^3; **Nugget Size:** 18 ozs; **Production:** 1900-01=145 ozs; **Access:** From Dalton Highway at 67.2724°N, 150.1710°W; **Summary:** Tributary of the Middle Fork of the Koyukuk River. Gold discovery in 1900 followed those on Slate and Myrtle Creeks. In 1900 a nugget weighing 18 ounces was found. Fineness of gold was about 940 Au. Mining was in 1900-1901 and 1934. Production, 1900-01, was 145 fine ounces (Cobb, 1976, OF 76-340. p. 94). Clara Creek has been designated as a **Recreation Mining Area** by the State of Alaska upstream 0.5 miles on the east side of the Dalton Highway at mile marker 176 (67.2724°N, 150.1710°W). Average yield of gravel is estimated to be 0.0025 oz/yd^3 of fine gold. Reed (1938) reported no evidence of mining on the creek since the early 1900's, although Smith (1936) reported mining in 1934. (The Alaska Resource Data File, Wiseman Quadrangle).

Colorado Creek

(Gold) **Quadrangle:** Wiseman (B6); **Coordinates:** 67.3800°N, 152.7201°W; **Map:** 5-8; **Worked:** 1937-; **Summary:** Tributary of Mettenpherg Creek. A prospector reported that fine gold with abundant magnetite and few garnets was found in gravel on schist bedrock. In 1937 there was small-scale mining. Includes references to Mecklenberg Creek (Cobb, 1976, OF 76-340. p. 95). **Geology:** Bliss and others (1988) speculate that gold in Colorado Creek may be derived from unspecified skarns in the headwaters of the Colorado Creek. **Workings:** Small-scale mining by booming and shoveling in from a small cut in 1937 (Reed, 1938). (The Alaska Resource Data File, Wiseman Quadrangle).

Confederate Gulch

(Gold) **Quadrangle:** Wiseman (B1); **Coordinates:** 67.4631°N, 150.0885°W; **Map:** 5-4; **Worked:** 1900s-; **Access:** Trail from Wiseman, Alaska; **Summary:** Coarse gold present, but not in paying quantities (Cobb, 1976, OF 76-340. p. 96). Confederate Gulch is a small drainage west of the Middle Fork, Koyukuk River; it is about 3.5 miles north-northeast of Wiseman. Placer prospecting extends from the coordinates given above where the creek enters the valley of the Middle Fork, Koyukuk River, and continues upstream for about 4,000 feet. The location is accurate. **Geology:** Confederate Gulch was prospected extensively in the early 1900's and some coarse gold was recovered during prospecting (Maddren, 1913). The amount of gold, however, was insufficient to justify intensive mining. (The Alaska Resource Data File, Wiseman Quadrangle).

Conglomerate Creek

(Gold) **Quadrangle:** Wiseman (B2, C2); **Coordinates:** 67.4825°N, 150.6821°W; **Map:** 5-3, 5-18; **Worked:** 1900s-1980s; **Summary:** Tributary of Glacier Creek. Good prospects and some mining reported from early days (Cobb, 1976, OF 76-340. p. 97). Conglomerate Creek is a southeast-flowing tributary to Glacier River. Its mouth is approximately 2 miles east of Delay Pass and 5.5 miles northwest of Bluecloud Mountain. There is little information to locate this placer occurrence along Conglomerate Creek and the site is arbitrarily plotted about 0.2 mile east of the center of section 28, T. 31 N., R. 14 W., of the Fairbanks Meridian, based on the location shown in Dillon (1982). **Geology:** Although he did not visit Conglomerate Creek, Reed (1938) reported that very good prospects were found on it in the early 1900's. Bliss and others (1988) indicated that there was some activity in the 1980's. The mining is s hampered by large conglomerate boulders in the creek. Mosier and Lewis (1986) analyzed six samples of yellow gold from Conglomerate Creek. Panned samples collected by Kurtak and others (1999) contained no visible gold although one sample contained 250 ppb gold. Only one pan contained any metallic minerals, chiefly traces of magnetite and some pyrite. (The Alaska Resource Data File, Wiseman Quadrangle).

Crevice Creek

(Gold, Copper, Lead) **Quadrangle:** Wiseman (B3, B4); **Coordinates:** 67.3652°N, 151.9274°W; **Map:** 5-9; **Worked:** 1904-1998; **Purity:** 958 Au; **Production:** Through 1904=121 ozs; **Access:** Trail from Crevice Creek landing strip at 67.3684°N, 152.0104°W, or trail from the Dalton Highway at 66.7915°N, 150.6878°W to Bettles, Alaska then from Bettles along the John River to Crevice Creek landing Strip; **Summary:** Tributary of the John River. Small gold production. Stream said to run on bedrock and gold to be in crevices and potholes in schist-marble unit. Upstream from placer mine galena and copper minerals occur in Skagit Limestone (Devonian) about 1/2 mile from a small body of mafic rock. Production in 1904 reported to be 121 fine ounces (Cobb, 1976, OF 76-340. p. 99). **Geology:** Although he did not visit Crevice Creek in 1937, Reed (1938) noted that Crevice Creek flowed principally on bedrock and that the gold that was mined from potholes and crevices in the bedrock. Some of the gold-bearing crevices were several feet deep. Dillon (1982) mapped the rocks in the area as a schist-marble unit, and tentatively identified stibnite in the placer concentrates. Occasional native silver nuggets have been recovered with the gold (Kurtak and others, 2002). Panned-concentrate samples from Crevice Creek contain visible gold grains and abundant magnetite; a sample from a sluice-box concentrate has over 8,000 ppb gold (Kurtak and others, 1999). Bliss and others (1988) speculate that the gold is derived from sulfide-bearing quartz veins in the Devonian Skajit limestone. Analyses of seven gold samples from Crevice Creek gave median values of 4.2% silver, 0.14% copper, 0.008% lead, and 0.47% mercury, and a median fineness of 958. Three of these analyses showed detectable levels (but less than 2 ppm) of Pd (Mosier and Lewis, 1986). **Workings:** Mining was reported as early as 1904 and the property was sluice mined in the 1980's (Bliss and others, 1988). The last year of significant mining on the creek was in 1994, followed by only minor suction dredging on the lower creek in 1998 (Kurtak and others, 2002). Approximately $1,800 in gold (at $20.67 per ounce) was said to have been produced in 1904 (Maddren, 1913). Dillon (1982) reports production of 30 ounces of gold per year, but it is unclear whether this production was historical or current as of 1982. (The Alaska Resource Data File, Wiseman Quadrangle).

Dry Gulch

(Gold) **Quadrangle:** Wiseman (B1); **Coordinates:** 67.3665°N, 150.1260°W, 67°21'59.7°N, 150°07'25.9°W (Dalton Hwy Xing); **Map:** 5-10; **Yield:** 0.0025 oz/yd^3; **Access:** From Rampart Dalton Highway at 67.3665°N, 150.1260°W; **Summary:** Tributary of the Middle Fork of the Koyukuk River. Dry Gulch has been designated as a **Recreation Mining Area** by the State of Alaska upstream on east side of the Dalton Highway at mile marker 184 (67.3665°N, 150.1260°W) all the way upstream. Average yield of gravel is estimated to be 0.0025 oz/yd^3 of fine gold.

Eagle Cliff

(Gold) **Quadrangle:** Wiseman (A1); **Coordinates:** 67.0472°N, 150.0694°W ; **Map:** 5-11; **Worked:** 1899-1937; **Production:** Through 1910=121 ozs; **Access:** By boat, upstream on the South Fork of the Koykuk River from Dalton Highway at 67.0186°N, 150.2885°W; **Summary:** Well-washed gravel on a rock-cut bench 10 - 12 feet above low-water stage of South Fork, Koyukuk River contains rounded pieces of gold about the size of a grain of wheat. Said to be similar to Ironside Bench and Gold Bench in Bettles quadrangle. Through 1910 production was worth about 121 fine ounces. No mining in 1937 (Cobb, 1976, OF 76-340. p. 100). The Eagle Cliff placer mine is on the north bank of the South Fork, Koyukuk River, about 12 miles eastsoutheast of Tramway Bar. The mine is near the west end of Eagle Cliff, a geographic feature shown on the Wiseman A-1 topographic map (1975 ed.) in sections 29 & 30, T. 26N., R. 11W., of the Fairbanks Meridian. The extent of the placer is uncertain but it may extend a mile or so to the east along the valley floor at the base of the Eagle Cliff bluff. The location is accurate to within 1 mile. **Geology:** Gold in coarse, wheat-sized rounded grains, was mined from well-washed gravel on top of a bench cut into rock about 10-12 feet above the low-water level of the South Fork Koyukuk River (Maddren, 1913). Reed (1938) noted that Eagle Bluff was said to be a high channel that resembled the Ironside Bench and Gold Bench placers (both in the Bettles quadrangle) about 15-20 miles downstream. **Workings:** A waterwheel was used to raise water to work the placer in 1899 (Schrader, 1900). Four or five men worked the placer in 1910. Reed (1938) noted that no one was working the site in 1937, although there were plans to work there in 1938. Maddren (1913) noted that $1,000 in gold was produced from 1900 to 1909, and an additional $1,500 in gold was produced in 1910. (The Alaska Resource Data File, Wiseman Quadrangle).

Eagle Gulch

(Gold, Copper, Antimony) **Quadrangle:** Wiseman (B3); **Coordinates:** 67.4081°N, 151.3305°W; **Map:** 5-12; **Worked:** 1904-1937; **Summary:** Small tributary of Jay Creek. A little gold was found near mouth (Cobb, 1976, OF 76-340. p. 101).

East Creek

(Gold) **Quadrangle:** Wiseman (B3); **Coordinates:** 67.3679°N, 151.1609°W; **Map:** 5-13; **Worked:** 1900s-; **Summary:** Tributary of Michigan Creek. Bedrock is schist. Very good prospects found in early days. Location on creek not given (Cobb, 1976, OF 76-340. p. 102). East Creek originates near Ipnek Mountain and flows west to Michigan Creek. The junction of East and Michigan Creeks is about 5.2 miles northwest of Ipnek Mountain. The location of the placer occurrence along East Creek is not known and this occurrence is arbitrarily placed about a mile upstream from the mouth of the creek. (The Alaska Resource Data File, Wiseman Quadrangle).

Emma Creek

(Gold) **Quadrangle:** Wiseman (B1); **Coordinates:** 67.3219°N, 150.1831°W; **Map:** 5-7, 5-10; **Worked:** 1900-2000; **Purity:** 905 Au; **Nugget Size:** Pennyweight; **Production:** Through 1984=7,700-9,700 ozs; **Access:** Trail from landing strip on the west side of the Middle Fork of the Koyukuk River at 67.3186°N, 150.18211°W; **Summary:** Tributary of the Middle Fork of the Koyukuk River. Gold discovered in 1900. Near mouth a canyon 100 feet deep and 1/2 mile long has been cut through crystalline limestone and schist in the bottom of an older and wider valley. Some gold in old gravels on top of canyon walls, but most is in stream gravel in and near canyon. Present channel has been mined for about 2-3/4 miles. A high channel about 30 feet above the present stream was also mined near the mouth of Frisbie Creek. Bedrock is schist with bands of limestone. Many granite and limestone boulders in gravel. In old channel depth to bedrock is from 5 - 91 feet. Gold is coarse. Rich gold deposit formed by steep-gradient stream in narrow, boulder-laden gulch. Mining has been small scale; most recently reported in 1974, but most was probably before 1930. Total production was probably between 7,700 - 9,700 fine ounces (Cobb, 1976, OF 76-340. p. 103). **Geology:** Gold was discovered on Emma Creek in 1900. The gold occurs in the modern channel and on a high bench. The productive portion of the modern channel is centered in a steep canyon cut through crystalline limestone. The canyon begins about a mile above the mouth of Emma Creek and extends upstream 1/2 mile or more. Most of the placer gold in this part of Emma Creek was concentrated at the upper end of the canyon, both within the canyon and in a gravel fan at its lower end. The richest ground was at the lower end of the canyon (Maddren, 1913). Gold also occurs in a high bench channel on the south side of Emma Creek about a half mile from the mouth. This channel lies about 30 feet above the modern creek bed and is approximately 150 feet wide. Toward the Middle Fork, it diverges from the modern creek bed and swings to the southeast. The bench gravel is small and waterworn and contains numerous granite and limestone boulders (Reed, 1938). The depth to bedrock varies from 5 to 91 feet. From the top, the section consists of a foot of moss, 0-50 feet of sand and clay, and 4-40 feet of gravel. All of the gravel, except the lower 4-5 feet is frozen. The gold is coarse and angular and commonly attached to quartz (Bliss and others, 1988). The gold on bedrock was in small nuggets valued at $0.25 to $1 (gold at $20 per ounce.) The bedrock in the area around the high channel is micaceous and graphitic schist with bands of limestone (Reed, 1938). Bliss and others (1988) suggested that the source of the gold may be the numerous quartz veins that occur in the drainage area. Three samples of gold from Emma Creek have a median fineness of 905 and median trace elements contents of 0.047% copper, 50 ppm lead, 80 ppm antimony, and 0.31% mercury (Mosier and Lewis, 1986). **Workings:** Early mining of the modern channel was by sluicing, and of the high bench by drifting and open cuts. Mining in the modern channel was complicated by numerous large boulders, 3 to 10 feet in diameter, above and below the canyon. Small-scale mining was reported in 1974 (Mulligan, 1974). According to Kurtak and others (2002) a suction dredge has been used on Emma creek to clean out potholes in the bedrock. Mining on Emma Creek has been regularly reported during the 1980's, 1990's (Eakins and others, 1983 and 1985; Bundtzen and others, 1984, 1986, 1990, and 1992; Green and others, 1989; Swainbank and others, 1993), and as recently as 2000 (Szumigala and others, 2001). Most of the production from this creek occurred prior to 1930. The value of the gold produced from 1900 to 1909 was estimated to be $160,000 (gold at $20/ounce), most of which was produced from 1900 to 1902 (Maddren, 1913). According to Reed (1938) as much as $12,000 (gold at $35/ounce) was produced from the high channel through 1928. (The Alaska Resource Data File, Wiseman Quadrangle).

Fall Creek

(Gold) **Quadrangle:** Wiseman (B3); **Coordinates:** 67.2935° N, 151.2049° W; **Map:** 5-5 **Summary:** Fall Creek is a west-flowing tributary to Michigan Creek; its mouth is about 6 miles southwest of Ipnek Mountain. The exact locations of placer prospects on Fall Creek are unknown and the site is arbitrarily placed approximately 1 mile upstream from its mouth. **Geology:** Reed (1938) did not visit visit Fall Creek but was told that it flowed entirely over schist and that very good prospects had been found on it in the early 1900's. No other information is available on these prospects. (The Alaska Resource Data File, Wiseman Quadrangle).

Fay Creek

(Gold) **Quadrangle:** Wiseman (B1); **Coordinates:** 67.4883°N, 150.1965°W; **Map:** 5-1, 5-4; **Worked:** 1901-1999; **Purity:** 842 Au; **Production:** Through 1901-1999=1,865.41 ozs; **Access:** Light duty road from Wiseman, Alaska; **Summary:** Tributary of Nolan Creek that drains mineralized area. By 1937 the present channel and a deep (20-foot) channel near Nolan Creek were considered mined out. A bench 275 feet higher than and 1,000 feet upstream from mouth of creek was prospected in 1929. Another part of high channel, farther upstream, is reported to have been mined. No mining in 1937. Gold recovered from frozen ground is rough and angular with quartz attached. Early (1901-1909) production was 1,451 fine ounces. Some recent (probably post-World War II) mining reported. See also Smith Creek Dome (Cobb, 1976, OF 76-340. p. 106). **Geology:** Placer gold in Fay Creek occurs in the modern channel, in a deep channel, and in a high channel. The discovery in 1901 of gold in the shallow gravels near the mouth of the modern channel of Fay Creek sparked the initial interest in the Nolan Creek valley. These shallow gravels were first mined beginning in 1903 but were exhausted within a few years. Gold in portions of a high channel along the north side of Fay Creek was mined prior to 1930. These high channel placers begin about 1000 feet upstream from the mouth of the creek and are about 275 higher in elevation (Reed, 1938). A gold-bearing deep channel near the mouth of the creek, which in fact was a deep channel of Nolan Creek, was mined out in the early 1900's. This deep channel was about 20 feet below the surface. Maddren (1913) noted that the placer gold in Fay Creek appeared not to have moved very far from its bedrock source as it generally occurs in rough and angular grains attached to quartz. There are reports of lode gold occurrences nearby on Smith Creek Dome, near the head of Fay Creek, and as well on on Thompson Pup, a tributary of Fay Creek (Brosge and Reiser, 1972). A sample of gold from Fay Creek is 842 fine, and contains 16% silver, 94 ppm copper, 16 ppm lead, and 0.625% mercury (Mosier and Lewis, 1986). **Workings:** The modern and deep channels of Fay Creek were mined in the early 1900's, but there is no record of the mining methods. Sections of the high channel were mined in the early 1900's and the high channel was being prospected in 1929 (Reed, 1938). Brosge and Reiser (1960) noted probable post-World War II mining activity and Dillon (1982) noted reports of recent mining activity. Fay Creek was the first creek in the Nolan Creek valley to have been mined. Reed (1938) indicated that mining began in 1901, whereas Maddren (1913) reported that it began in 1903. According to Maddren (1913) Fay Creek produced about $30,000 of gold (about 1,500 ounces) between 1900 and 1909. Eden (2000) reports that a total of 1865.41 ounces of gold were produced from Fay Creek between 1904 and 1999. The modern stream channel and the deep channel were considered to have been mined out by 1937. (The Alaska Resource Data File, Wiseman Quadrangle).

Fool Creek

(Gold) **Quadrangle:** Wiseman (B5); **Coordinates:** 67.3050°N, 152.2247°W; **Map:** 5-14; **Worked:** 1922-; **Summary:** Tributary of Midas Creek. Encouraging prospects have been found. Mining in 1922 was reported. See also Midas Creek. Includes reference to Sixtymile River (Cobb, 1976, OF 76-340. p. 108).

Frisbe Creek

(Gold) **Quadrangle:** Wiseman (A1); **Coordinates:** 67.0117°N, 150.4253°W; **Map:** 5-15; **Worked:** 1900s-; **Access:** By boat downstream about 4 miles from the Dalton Highway at 67.0186°N, 150.2885°W; **Summary:** Right-limit tributary of South Fork Koyukuk River near Grubstake Bar. Prospecting and mining in early days reported. Cobb, 1976, OF 76-340. p. 109). Frisbe Creek was described by Reed (1938) as a right-limit tributary to the South Fork, Koyukuk River; it is about 2.5 miles upstream from Davis Creek and just upstream of Grubstake Bar. Based on Reed's description, Frisbe Creek is inferred to be the south-flowing tributary to the South Fork whose mouth is approximately 0.3 mile upstream from the mouth of Bear Creek. The location of placer activity along the creek is unknown; its location is arbitrarily placed near the mouth of the creek.The location of placer activity along the creek is unknown; its location is arbitrarily placed near the mouth of the creek. **Geology:** Reed (1938) did not visit Frisbe Creek. He reported that the Frisbe brothers had mined and prospected the creek in the early 1900's, and that no one had mined it since that time. Reed did not know the extent of the mining. (The Alaska Resource Data File, Wiseman Quadrangle).

Fryingpan Bar

(Gold) **Quadrangle:** Wiseman (A1); **Coordinates:** 67.1013°N, 150.4820°W ; **Map:** 5-6, 5-20; **Worked:** 1908-; **Access:** Trail from landing strip on the west bank of rhe Middle Fork of the Koyukuk River at 67.0922°N, 150.5010°W, or by trail from the Dalton Highway at at 67.0798°N, 150.3506°W; **Summary:** Segment of old high channel of Middle Fork Koyukuk River. Was mined in 1908 with water pumped from river using local coal for fuel (Cobb, 1976, OF 76-340. p. 110). Fryingpan Bar is on the east side of the Middle Fork, Koyukuk River opposite the mouth of Mailbox Creek; it is about 1 mile north-northeast of Tramway Bar. **Geology:** According to Reed (1938), the gold in Fryingpan Bar is derived from the erosion of the high channel of the Middle Fork of the Koyukuk. This channel is on the south side of the river on top of the wall of the Middle Fork Canyon; the channel was forced southward by the action of Mailbox Creek. The placer was mined in 1908 but the results of the mining are unknown. (The Alaska Resource Data File, Wiseman Quadrangle).

Gold Bottom Gulch

(Gold, Lead) **Quadrangle:** Wiseman (B1); **Coordinates:** 67.4844°N, 150.0819°W; **Map:** 5-4; **Worked:** 1904-2000; **Nugget Size:** 50 ozs; **Access:** Light duty road from Wiseman, Alaska; **Summary:** Tributary of the Hammond River River. Gold from Gold Bottom Gulch was probably from bench gravels of Hammond River. Much of gold was very coarse with nuggets over 50 ounces reported. Concentrates contain pyrite, some exhibiting free gold. Placers work continuously from discovery to World War II (Cobb, 1976, OF 76-340. p. 114-115). Gold Bottom Gulch is a short, steep, south tributary to the lower Hammond River. It drains the northeast flank of Midnight Dome between Lofty Gulch and Steep Gulch. This mine is about 5 miles north-northeast of Wiseman. Most placer mining on Gold Bottom Gulch was near the mouth but it extended approximately 0.5 mile upstream. **Geology:** Maddren (1913) reported that 600 to 700 feet of the modern channel near the mouth of Gold Bottom Gulch was mined; a $400 nugget was recovered in 1904 (gold at $20.67 per ounce). These shallow gravels were mined out in the early 1900's. Reed (1938) noted mining of a high channel and a deep channel in 1937 near the mouth of the creek. Bedrock had not yet been reached in the deep channel and the only production was of gold scattered through the gravel. There were two types of gold in the deep channel (Reed, 1938). Most was a coarse-grained, rounded gold similar to that found in the Hammond River deep channel; this was accompanied by lesser amounts of very fine and flaky gold. The gravel in the deep channel is not frozen. The high channel is on the west side of Gold Bottom Gulch, about 50 feet above the bottom of the cut in which the deep channel was being mined in 1937 (Reed, 1938). The high channel was prospected in 1937, but not mined. From 1978 to the present (2000), mining has been by dozer, excavator and washing plant; production figures are not available. Dillon (1982) speculates that the gold in the lower parts of tributaries of the Hammond River, such as Gold Bottom Gulch, is mostly derived from bench gravels of the Hammond River. (The Alaska Resource Data File, Wiseman Quadrangle).

Grubstake Bar

(Gold) **Quadrangle:** Wiseman (A1); **Coordinates:** 67.0096°N, 150.4369°W ; **Map:** 5-15; **Worked:** 1900-; **Production:** 1900-1909=193 ozs; **Access:** By boat downstream about 4 miles from the Dalton Highway at 67.0186°N, 150.2885°W; **Summary:** Low bench deposit or bar of the South Fork of the Koyukuk River. A few thousand dollars worth of gold was sluiced in 1900-1901. Similar to Hanshaw Bar. Production, 1900-09, was 193 fine ounces (Cobb, 1976, OF 76-340. p. 112). Grubstake Bar is described by Reed (1938) as being on the right limit of the South Fork, Koyukuk River about 2 miles above Hanshaw Bar which is about 5.8 miles by river above Gold Bench (in the Bettles quadrangle). Maddren (1913) described the occurrence on the NW bank of South Fork about 9 miles above Gold Bench. The site is here plotted on a sand bar on the north bank of the South Fork, opposite the mouth of Bear Creek, in the northwest corner of section 9, T. 25 N., R. 13 W., of the Fairbanks Meridian. **Geology:** Maddren (1913) described Grubstake Bar as a low-bench deposit. Reed (1938) described it as a superficial deposit in the present river channel, similar to Hanshaw Bar. (The Alaska Resource Data File, Wiseman Quadrangle).

Hamil Bar

(Gold) **Quadrangle:** Wiseman (A2); **Coordinates:** 67.0825°N, 150.6067°W ; **Map:** 5-6, 5-20; **Worked:** 1900s-; **Access:** By boat downstream from the Dalton Highway at 67.0186°N, 150.2885°W; **Summary:** Located on the Middle Fork of the Koyukuk River below Tramway Bar. Alluvial fan or delta of small stream that flows into Middle Fork Koyukuk River, was hand mined in early days. No data on production. Gold probably concentrated from old high channel of main stream (Cobb, 1976, OF 76-340. p. 113).

Hammond River

(Gold, Lead) **Quadrangle:** Wiseman (B1, C1); **Coordinates:** 67.4873°N, 150.0476°W; **Map:** 5-4; **Worked:** 1900-Present; **Purity:** 845-961 Au; **Yield:** 0.0417 oz/ft^2; **Nugget Size:** 137 ozs; **Production:** 1900-1999=17,255.90 ozs; **Access:** Trail from Rampart landing strip on the Yukon River at 65.5085°N, 150.1425°W; **Summary:** Largest tributary of Middle Fork of Koyukuk River. Coarse gold discovered 2 miles above mouth in 1900. Gold found only in lower 13 miles where bedrock is schist. There is limestone farther upstream. Most of gold came from deep (as much as 115 feet; pay gravel 8-10 feet thick) channel that extends into valley of Middle Fork, where it is truncated, probably by a glacier that came down Middle Fork. Pay gravel also in an old channel 40 - 50 feet higher than the present stream. Deep, frozen gold-bearing gravels (in places thawed on bedrock) were drift mined near mouth. Gold from the mouths of tributaries such as Buckeye and Goldbottom Gulches and Steep Creek was probably from bench gravels of Hammond River though some also came from mineralized area near heads of tributaries. Much of gold was very coarse with nuggets over 50 ounces reported. Concentrates contain pyrite, some exhibiting free gold. Winter mining in 1902-03. In 1902 a nugget containing 53.21 fine ounces was found. In 1913 a 137 ounce nugget was recovered. Very rich placer ground reported to have been found at depths of from 60 to 120 feet. Some deep mines cannot be worked because of water. Pumps brought in to try to control water in some of deep mines. A quartz vein near mouth of Swift Creek contains galena. Includes references to Buckeye Gulch, Goldbottom Gulch, Steep Creek. See also Swift Gulch and Vermont Creek. Production, 1900-1905, was 1,800 fine ounces. Output decreased in 1914 because abnormally rich ground had been worked in 1913. Largest producer in district in 1915. Placers work continuously from discovery to World War II (Cobb, 1976, OF 76-340. p. 114-115). The Hammond River produced Alaska's second largest gold nugget weighing 137 ounces with many others over 40 ounces. These large nuggets were usually found under large boulders. The Hammond River is a large, northern tributary of the Middle Fork, Koyukuk River. It begins near Alhamblar Mountain, nearly 35 miles north of its junction with the Middle Fork. The mouth of the Hammond River is located approximately 4 miles northeast of Wiseman and 3 miles east of Midnight Dome. **Geology:** Gold was discovered on the Hammond River about 1900 (Maddren, 1913). The gold is mainly along the lower 13 miles of the river, although most mining has occurred within 3 miles of its mouth. The gold is in the modern channel, a deep channel, and a high channel. Most of the gold has come from the deep channel. There has been intermittent mining of the gravels in the modern channel, mostly prior to 1910. This mining was principally in areas where the river has cut to bedrock from just below Buckeye Gulch to above the mouth of the Hammond River. According to Maddren (1913), this mining in the modern channel produced about $8,000 to $10,000 in gold (gold at $20.67 per ounce) by 1913. Gravel in the deep channel have been mined from near the mouth of Buckeye Gulch to about a mile or so above the mouth of the Hammond River. The lower end of the deep channel extends into the valley of the Middle Fork, Koyukuk River, but is truncated in valley, presumably by glacial scouring. The deep channel is approximately 30 feet wide and lies more than 115 feet below the modern river level at its lower end, rising to about 35 feet below the modern channel near its upper end. The gravels of the deep channel were said to become richer downstream. The gold in the deep channel occurs directly on bedrock; it is coarse grained, rounded, and waterworn. Galena and pyrite are locally associated with the gold. The deep channel, except for a few areas overlooked by the early miners, was thought to be mined out by 1937 (Reed, 1938). Workings in one of these overlooked spots in 1936 and 1937 found a section 8-10 thick of paying gravel. The channel was 35 feet wide and the depth to bedrock was 115 feet. The gravel consisted of small boulders and large cobbles, separated by sand and fine gravel. The gravel was frozen to bedrock (although this was not always the case at other locations along the deep channel) and the bedrock was described as hard schist (Reed, 1938). The gold included many nuggets, was waterworn, and had a fineness of 961. According to Reed (1938), the ground was valued at about $1.46 per square foot of bedrock (gold at $35 per ounce). Gold also occurs in sections of a high channel preserved along the lower Hammond River. These sections appear to be 40-50 feet above the modern river bed and were first encountered in locations where the high channel is cut by tributary streams such as at Buckeye Gulch. Little additional information is available about these bench deposits. Much coarse gold has been produced from the Hammond River, including several nuggets valued at $800 to $1100 (gold at $20.67 per ounce.) and a 137 ounce nugget (Maddren, 1913; Dillon, 1982). Mosier and Lewis (1986) reported that gold in 14 samples from Hammond River varies from from 845 to 929 fine with a median fineness of 900. See also: Slisco Bench. **Workings:** Workings included extensive drift mining on the deep channel, beginning around 1913 and continuing until around 1950. Drift mining of the deep channel was often impeded by water problems in thawed gravels. There was limited mining of the modern stream bed, generally by hand methods, in the early 1900's. Some mining took place on preserved sections of a high channel early 1900's. Mining has continued along the Hammond River intermittently through at least the 1990's. This work included open cut mining near Wiseman in 1993 (Bundtzen and others, 1994) and open-pit mining as late as 2000 (Szumigala and others, 2001). **Production:** The total placer gold production figures for the Hammond River are uncertain, but most reports agree that it has been substantial. Bliss and others (1988) indicate that total production may have been in the range of 1.2 to 1.8 tonnes of gold (although these figures likely included production from tributaries, including Gold Bottom Gulch, Buckeye Gulch, Swift Creek, and Steep Creek). Eden (2000) reported that only 17,255.90 ounces of gold has been produced from the Hammond River between 1904 and 1999. (The Alaska Resource Data File, Wiseman Quadrangle). The Hammond River has been designated as a **Recreation Mining Area** by the State of Alaska upstream 1 mile upstream from the confluence with the Middle Fork of the Koyukuk River. Access is by the Dalton Highway at mile marker 190.5 (67.4609°N, 150.0348°W). Average yield of gravel is estimated to be 0.0025 oz/yd^3 of gold. It is closed to all suction dredging from July 15 to May 20 for salmon spawning.

Hanshaw Bar

(Gold) **Quadrangle:** Wiseman (A1); **Coordinates:** 67.0002°N, 150.4720°W ; **Map:** 5-15, 5-20; **Worked:** 1937-; **Purity:** 907 Au; **Yield:** 0.12 oz/yd^3; **Access:** By boat downstream from the Dalton Highway at 67.0186°N, 150.2885°W; **Summary:** The location of Hanshaw Bar is described by Reed (1938) as being on the left limit of the South Fork, Koyukuk River about 4.8 miles above Ironside Bench (in the Bettles quadrangle) and about 0.25 mile above Davis Creek (in the Bettles quadrangle). As plotted here, the mine is on a sand bar on the east bank of the South Fork, in the southwest corner of section 8, T. 25 N., R. 13 W. of the Fairbanks Meridian. Location is accurate within one-half mile. High-water bar on false bedrock in left-limit channel of South Fork of Koyukuk River was being mined in 1937. Averaged about 0.12 ounces per yd^3 in fine, flat, flaky gold. Gold found in 2 - 3 feet of gravel and sand (Cobb, 1976, OF 76-340. p. 116). **Geology:** Gold, as fine, flat flakes, is distributed throughout 2 to 3 feet of medium- to fairly fine-grained gravel and sand on a high-water bar in the channel of the South Fork, Koyukuk River. The paying gravel and sand lies on a false bedrock of clayey sand. Reed (1938) noted that the average value of the ground was $0.23 per square foot of bedrock that the gold had a fineness of 907. Mining in 1937 was by shoveling into sluice boxes and washing with water from the South Fork (Reed, 1938). (The Alaska Resource Data File, Wiseman Quadrangle).

Jap Creek

(Gold) **Quadrangle:** Wiseman (B1); **Coordinates:** 67.4223°N, 150.1548°W; **Map:** 5-4, 5-10; **Access:** Light duty road from Wiseman, Alaska; **Summary:** Tributary of Wiseman Creek. Gold has been found but not enough to encourage mining (Cobb, 1976, OF 76-340. p. 118). Two stream-sediment samples contain 700 ppm and 3,000 ppm antimony (Brosge and Reiser, 1972). (The Alaska Resource Data File, Wiseman Quadrangle).

Jay Creek

(Gold, Copper, Antimony) **Quadrangle:** Wiseman (B3); **Coordinates:** 67.4060°N, 151.3297°W; **Map:** 5-12; **Worked:** 1904-1998; **Purity:** 970 Au; **Yield:** 0.0144 oz/ft^2; **Production:** Through 1935=9,671 ozs; **Access:** Trail from Wild Lake; **Summary:** Tributary of Rye Creek. Gold discovered in 1904, but mining did not begin until 1912. Bedrock is alternating bands of graphitic, micaceous, and greenstone schist. Richest creek in Wild River area. Production to 1935 was about 9,671 fine ounces. Lower part of stream flows in narrow canyon, 25 - 50 feet wide, cut in schist, with 3 - 12 feet of coarse, angular gravel over bedrock. Pay streak said to be 30 feet wide and 4 feet thick. Gold is coarse and unworn (Cobb, 1976, OF 76-340. p. 119). Although the exact location of the placer workings on Jay Creek is not known, Reed (1938) noted that the claims extended from No. 1 Below Discovery (which reached about 500 feet into the Rye Creek valley) to No. 5 Above Discover, and that the creek was mined from the mouth to the lower reaches of No. 5 Above Discovery claim. **Geology:** Jay Creek is reputed to be the richest creek in the Wild River district. Jay Creek is in a canyon that averages about 50 feet wide. Brooks (1916) reported that the paystreak was 30 feet wide and 4 feet thick. Reed (1938) describes the gravel as thawed, very coarse grained with subangular sediments, 3 to 12 feet thick , and contains numerous erratic, conglomerate boulders. The gold occurs throughout the gravel and in the upper foot of the bedrock. The gold is coarse grained, unworn, very dark colored, and about 970 fine. Dillon (1982) reports a stibnite-quartz vein upstream from the placer, and Bliss and others (1988) speculated that the source of the gold is stibnite (?)-quartz veins within the drainage. The rocks in the vicinity consist of alternating bands of graphitic schist, micaceous schist, and greenstone schist (Reed, 1938). A little gold was found in Jay Creek near the mouth of Eagle Gulch, a locally named, small, right-limit tributary to Jay Creek (Reed, 1938). A chip sample of a quartz vein with pyrite and chalcopyrite in the upper drainage of Jay Creek contains 21 ppb gold and 82 ppm arsenic (Kurtak and others, 1999; sample no. 95). See also Rye Creek. **Workings:** From 1912 through at least 1935, the creek was mined by booming and shoveling into sluice boxes. Mining took place over 5 or 6 claim lengths from the mouth of the creek to the lower part of No. 5 Above Discovery. A mechanized mining operation was active on Jay Creek in 1998. **Production notes:** The estimated total production from Jay Creek through 1935 is about $200,000 (approximately 10,000 oz.). The ground that was mined had an average value of about $0.505 per square foot of bedrock (gold at $35 per ounce) (Reed, 1938). There is no information available about recent mining. (The Alaska Resource Data File, Wiseman Quadrangle).

Jones Creek

(Gold) **Quadrangle:** Wiseman (A6); **Coordinates:** 67.1350° N, 152.5530° W **Map:** 5-16 **Worked:** 1976-; **Summary:** This placer occurrence is located on Jones Creek, an east tributary to lower Mettenpherg Creek. The occurrence is about 3 miles upstream of the mouth of Jones Creek in the SW 1/4 of section 26, T. 7 N., R. 23 W. of the Fairbanks Meridian. **Geology:** Pan samples taken from two separate sites on lower Jones Creek (Kurtak and others, 2002) contained visible placer gold. Schist in the creek contained 1-2% disseminated pyrite. Regional geologic mapping indicates the lower Jones Creek area is underlain by Devonian(?), quartz-mica schist, mica schist, phyllite, and quartzite (Brosge and Reiser, 1971). Placer claims were located on Jones Creek in 1976 (Kurtak and others, 2002). (The Alaska Resource Data File, Wiseman Quadrangle).

Kay Creek

(Gold) **Quadrangle:** Wiseman (B3); **Coordinates:** 67.4209°N, 151.1692°W; **Map:** 5-13; **Worked:** 1910-; **Summary:** Tributary of Michigan Creek that heads against Jay Creek. Some mining and extensive prospecting in 1910 (Cobb, 1976, OF 76-340. p. 123).

Kelly Gulch

(Gold) **Quadrangle:** Wiseman (B1); **Coordinates:** 67.2967°N, 150.2078°W; **Map:** 5-7; **Worked:** 1901-1937; **Production:** 1901=24 ozs; **Access:** Trail from Emma Creek landing strip on the west bank of the Middle Fork of the Koyukuk River at 67.3185°N, 150.1826°W or possibly from the Dalton Highway at 67.2930°N, 150.1620°W then across the river by boat; **Summary:** Short tributary on west side of Middle Fork of Koyukuk River. About 24 ounces of gold was mined in 1901. Work in 1937 was probably not successful. Includes references to Kelly's Gulch, Kelly's Mistake Creek (Cobb, 1976, OF 76-340. p. 124). (The Alaska Resource Data File, Wiseman Quadrangle).

Koyukuk River, Middle Fork

(Gold) **Quadrangle:** Wiseman (A1, A2, A3, B1); **Coordinates:** 67.4380°N, 150.0781°W (Dalton Hwy Xing); **Map:** 5-4; **Yield:** 0.0025 oz/yd^3; **Access:** From the Dalton Highway at 67.4380°N, 150.0781°W; **Summary:** Tributary of the Koyukuk River. Average yield of gravel is estimated to be 0.0025 oz/yd^3 of gold.

Koyukuk River, South Fork

(Gold) **Quadrangle:** Wiseman (A1, B1); **Coordinates:** 67.0186°N, 150.2885°W (Dalton Hwy Xing); **Map:** 5-15; **Worked:** 1930s-; **Yield:** 0.1 oz/yd^3; **Access:** From Dalton Highway Bridge at 67.0186°N, 150.2885°W; **Summary:** Tributary of the Middle Fork of the Koyukuk River. Prospect drilling (1930) and mining (1935, 1936) reported. Operations probably were on Eagle Cliff, Grubstake Bar, or some other bar or small tributary (Cobb, 1976, OF 76-340. p. 125). The South Fork of the Koyukuk River between 1.5 miles upstream and 2 miles downstream from the Dalton Highway, at mile marker 156.3 (67.0186°N, 150.2885°W) has been designated as a **Recreation Mining Area** by the State of Alaska, except for pipeline right-of-way. Average yield of gravel is estimated to be more than 0.1 oz/yd^3 of gold. It is closed to all suction dredging from July 15 to May 20 for salmon spawning.

Lake Creek

(Gold, . Platinum, Antimony, Bismuth, Copper, Tungsten) **Quadrangle:** Wiseman (B4); **Coordinates:** 67.4740°N, 151.5651°W; **Map:** 5-17, 5-23; **Worked:** 1903-Present; **Yield:** 0.0531-0.0734 oz/ft^2; **Nugget Size:** 21 ozs; **Production:** Total=1,258 ozs; **Access:** Trail from Wild Lake; **Summary:** Lake Creek drains the area between Mathews Dome and Wild Lake; it empties into the east side of Wild Lake about 0.5 mile north of the south end of the lake. Mining on Lake Creek has mainly been in an area near the head of the alluvial fan on the lower part of the creek and on an area above a fork of the creek about 1.5 miles east of Wild Lake. The two areas are about 0.8 mile apart and the coordinates are given for the midpoint between them.Tributary of Wild Lake. Bedrock is various kinds of schist cut by at least one greenstone dike; many quartz stringers near head of creek. Bedrock at head of delta into Wild Lake is 60 - 95 feet deep. Gravel in stream bed above delta is only 3 feet thick. Concentrates contain gold, large pieces of stibnite, scheelite, native copper and bismuth, hematite, pyrite, and a little magnetite. A sample of vein quartz contained a little chalcopyrite. Gold, both coarse and very fine, was mined intermittently from 1904 until at least as recently as 1938. In 1932 mining was handicapped by damaging floods in middle of season. In 1903-04, 97 fine ounces were taken from a claim on a headwater gulch. Gold was coarse with 4.35 to 7.25 ounce nuggets. Total recorded production was 1,258 fine ounces, which is probably many thousand dollars too low because the stream is still being worked today. (Cobb, 1976, OF 76-340 p. 126). On September 6, 1995 a 21 ounce nugget was found by Tim Chadwick, in a tailings pile, while working at the Henry Wojtusik placer mine located on Lake Creek about 45 miles west of Wiseman. The placer recovery system installed at the mine is designed to capture gold up to about an inch in size. Anything larger is discarded along with the rest of the stream gravel. The sources for the gold on

Lake Creek has not been found yet. **Geology:** Gold was discovered on Lake Creek in 1903 (Maddren, 1913) and mining has continued intermittently through at least 1996 (Kurtak and others, 1999). Mining has been concentrated near the head of the alluvial fan where the creek enters Wild Lake and in a narrow portion of the creek valley about 1.5 miles east of the lake. Near the head of the alluvial fan, the depth to bedrock is 60 to 95 feet. In the upper valley of the creek, the depth to bedrock is only 3 feet. Reed (1938) reported production from early mining from 1904 to 1937 to be at least $26,000 (about 1,300 ounces of gold) but he speculated that the extent of the workings indicated the production was many thousands of dollars more. No information is available on production since 1937. The rocks in the area are predominantly muscovite-chlorite schist that contains numerous quartz lenses and veinlets (Kurtak and others,1999). Some of the quartz is parallel to the cleavage or schistosity, and some cross cuts it. Reed (1938) describes the bedrock about 200 feet above the last of the old workings (about 2 miles from the lake) as alternating micaceous and graphitic schist cut by a 300-feet-wide, greenstone schist dike. This greenstone dike was not observed by Kurtak and others (1999). Reed (1938) also noted soft reddish schist containing numerous quartz stingers at the head of Lake Creek. A sample of vein quartz from the Lake Creek drainage near the head of the alluvial fan contained minor chalcopyrite and tourmaline (Chipp, 1972, sample 133). The gravels on Lake Creek are coarse and waterworn with numerous large schist boulders. The upper part of the gravel contains erratics of conglomerate and black chert. The gold occurs as coarse, waterworn nuggets, and as fine flour gold. Maddren (1913) noted recovery of nuggets valued at $90 to $150 (about 4.7 to 7.9 ounces). Two estimates of the value of the ground in two separate locations along the creek are $1.86 and $2.50 per square foot of bedrock (Reed, 1938). Joesting (1943) lists several minerals in the placer concentrates from Lake Creek , including two, inch-long pieces of stibnite, scheelite, native bismuth, native copper, hematite, pyrite and magnetite. Sluice concentrates collected by the Kurtak and others (1999) contains up to 5,930 ppb platinum, 976 ppm tungsten, 976 ppm tungsten, 0.44% bismuth, 1,750 ppm arsenic, and 3,043 ppb gold. The sluice concentrates did not contain visible gold. Samples of quartz veins and the schist bedrock contain no valuable metals. **Workings:** Intermittent extensive mining by sluicing and drifting from 1904 through to at least 1996 (Kurtak and others, 1999). In 1986 a placer reserve of 10,000 ounces of gold was announced by the owners (Bundtzen and others, 1987). (The Alaska Resource Data File, Wiseman Quadrangle).

ERIK HILL / Anchorage Daily News
Wally Gordon, right, found this 21-troy-ounce gold nugget in a tailings pile while working at Henry Wojtusik's mining operation this summer.

Larson Creek

(Gold) **Quadrangle:** Wiseman (B1); **Coordinates:** 67.4136°N, 150.0433°W; **Map:** 5-4, 5-10; **Worked:** 1900-; **Yield:** 0.0025 oz/yd^3; **Production:** 1900-09=48 ozs; **Access:** Side road from Dalton Highway at 67.4163 °N, 150.0768°W; **Summary:** Tributary of Minnie Creek. Has been much prospecting. Some coarse gold has been recovered. Gravel has so much live water that prospecting and mining are difficult. Heavy shot gold on bedrock. Much prospecting and a little drift mining in early days. Unsuccessful prospecting in 1937. Production, 1900-09, was 48 fine ounces (Cobb, 1976, OF 76-340. p. 138). Larson Creek has been designated as a **Recreation Mining Area** by the State of Alaska for its entire length. Access is by the Dalton Highway at 67.4163"N, 150.0768°W. Average yield of gravel is estimated to be 0.0025 oz/yd^3 of fine gold. (The Alaska Resource Data File, Wiseman Quadrangle).

Lofty Gulch

(Gold) **Quadrangle:** Wiseman (B1); **Coordinates:** 67.4862°N, 150.0902°W; **Map:** 5-4; **Worked:** 1900s-; **Nugget Size:** 50 ozs; **Access:** Light duty road from Wiseman, Alaska; **Summary:** Small, steep gully on left limit of Hammond River. Some of channel near mouth was mined out in early 1900's. Gold from Lofty Gulch was probably from bench gravels of Hammond River. Much of gold was very coarse with nuggets over 50 ounces reported. Concentrates contain pyrite, some exhibiting free gold. Placers work continuously from discovery to World War II. Inactive in 1974 (Cobb, 1976, OF 76-340. p. 127). Placer mining has occurred from near the mouth of Lofty Gulch for a distance of about 1,500 feet upstream. The mine is plotted at an elevation of about 1,350 feet near the mouth of the gulch. **Geology:** The modern channel of Lofty Gulch was mined near its mouth in the early 1900's (Reed, 1938). There has been no mining on the upper parts of the creek, where it is narrow, steep and contains only limited amounts of gravel in which gold could occur. According to Reed (1938), the creek was mined out. (The Alaska Resource Data File, Wiseman Quadrangle).

Lucky Creek

(Gold) **Quadrangle:** Wiseman (B3); **Coordinates:** 67.4015°N, 151.3224°W; **Map:** 5-12; **Worked:** 1900s-; **Access:** Trail from Wild Lake; **Summary:** Tributary of Rye Creek. Small amounts of gold have been found. Location on creek not given; old name for upper part of Rye Creek (Cobb, 1976, OF 76-340. p. 128).

Luke Creek

(Gold) **Quadrangle:** Wiseman (B4); **Coordinates:** 67.4366°N, 151.5952°W; **Map:** 5-17, **Worked:** 1900s-; **Access:** Trail from Wild Lake; **Summary:** Small creek that runs into the Wild River from west. Reed (1938) reported placer gold prospects on Luke Creek. No other information is available. (Cobb, 1976, OF 76-340. p. 129). (The Alaska Resource Data File, Wiseman Quadrangle).

Lynx Creek

(Gold) **Quadrangle:** Wiseman (B4); **Coordinates:** 67.5035°N, 151.5465°W; **Map:** 5-17; **Worked:** 1900s-; **Access:** Trail from Wild Lake; **Summary:** Small gulch that flows west into Wild Lake. Seems to occupy a slide basin. Good prospects said to have been found near mouth. Reed (1938) notes reports of good placer gold prospects near the mouth of Lynx Creek although no mining had been done. In Reed's judgment, the appearance of the creek suggested that it was unlikely that any significant concentration of gold had taken place. No other information is available. (Cobb, 1976, OF 76-340. p. 130).

Mailbox Creek

(Gold) **Quadrangle:** Wiseman (A1); **Coordinates:** 67.1036°N, 150.4839°W; **Map:** 5-6, 5-20; **Worked:** 1930s-1999; **Purity:** 898 Au; **Yield:** 0.037 ozs/ft^2; **Access:** Trail from landing strip on the west bank of the Middle Fork of the Koyukuk River at 67.0924°N, 150.5007°W or trail from the Dalton Highway at 67.0800°N, 150.3509°W; **Summary:** Tributary of the middle Fork of the Koyukuk River. Stream has cut canyon in faulted conglomerate. About 2,000 feet above mouth 2 men mined by groundsluicing and shoveling in. Mining in middle 1930's recovered fine, flaky gold from stream gravel and coarser gold from near bedrock, which is clay and fine conglomerate with some coal. Stream gravel derived from conglomerate. Ground mined in 1937 ran about 0.037 ounces per ft^2 bedrock. No data on thickness of gravel (Cobb, 1976, OF 76-340. p. 131). **Geology:** Placer workings about a half mile upstream from the mouth of Mailbox Creek have recovered gold from fairly fine gravel that contains numerous erratic boulders. The fine gravel is derived mainly from conglomerate bedrock in the area. Fine, flaky gold occurs throughout the gravel; coarser gold is in a layer of reddish gravel near the bedrock surface. The value of the ground being worked in the mid-1930's was reported to be about $0.75 per square foot of bedrock. The fineness of the gold is 898. The State of Alaska reported that a Mining License had been issued for an operation on Mailbox Creek in 1999 (Swainbank and others, 2000). Its issuance, however, does not mean that any mining took place. **Workings**: The placer was mined in the 1930's by ground sluicing and shoveling in from an open cut. (The Alaska Resource Data File, Wiseman Quadrangle).

Mary's Creek

(Gold) **Quadrangle:** Wiseman (B4); **Coordinates:** 67.4659°N, 151.6157°W; **Map:** 5-17; **Access:** Trail from Wild Lake; **Summary:** Tributary of The wild River. Good prospect said to have been found; no mining reported (Cobb, 1976, OF 76-340. p. 132).

Mascot Creek

(Gold) **Quadrangle:** Wiseman (B2, C2); **Coordinates:** 67.4908°N, 150.5321°W; **Map:** 5-18; **Worked:** 1902-1990s; **Purity:** 942-977 Au; **Yield:** 0.19 oz/ft^2; **Nugget Size:** 5 ozs; **Production:** Through 1910=7,282 ozs; **Access:** Trail from Rampart landing strip on the Yukon River at 65.5085°N, 150.1425°W; **Summary:** Tributary of Glacier River. Gold discovered in 1902 and mined intermittently until as recently as 1950's. Thin (3 ft. or less) creek gravels carry up to 0.19 ounces in gold per ft^2 bedrock. Bedrock is mica schist with quartz stringers and a few porphyry dikes. Gold is coarse and is in base of gravel, on bedrock, and in bedrock cracks and crevices. Some gold has been mined from old high channel, some of which is 30 feet above creek. Some nuggets weighing 5 fine ounces. Production through 1910 was 7,282 fine ounces. Includes reference to Koyukuk River, North Fork (Cobb, 1976, OF 76-340. p. 133). **Geology:** Most of the mining on Mascot Creek has been along a 3-mile stretch of the creek, downstream from the forks located approximately 7 miles above the mouth. Mining and prospecting above and below this section of the creek has generally proved to be unproductive. The gold occurs both in the modern stream channel and in high channels. Most of the gold has been taken from the modern stream channel. The gravel is waterworn and fine with few boulders; none of the gravel is frozen. The gravel in the modern channel generally is 0.5 to 3 feet thick. The relative ease with which these thin gravels could be mined made Mascot Creek one of the more profitable placer creeks in the district. The gold in the modern channel is just above and below the bedrock surface; it typically is coarse and well worn. Nuggets with values up to $100 (5 oz) have been recovered (Maddren, 1913). In one location near the upper end of the mined area, the gold was 90% coarse and 10% fine. Fineness ranged from about 942 to 977. Reed (1938) estimated that the value of the ground being mined in the modern channel in 1938 was $1.25 to nearly $7 per square foot of bedrock. Gold also occurs in remnants of a high channel on Mascot Creek. These remnants typically are along the right limit (west side) of the creek and a few tens of feet above the modern channel. Reed (1938) described one of these remnants as about 100 feet long and 10 feet wide; another was about 30 feet above the modern stream. According to Reed (1938), the bedrock in Mascot Creek is graphitic to quartz schist. The quartz schist is reddish from iron staining; its relation to the placer gold is unknown. According to Schrader (1904), mica schist on bedrock contains quartz stringers and a few porphyry dikes. Bliss and others (1988) speculated that the gold might be derived from these quartz stringers. Kurtak and others (1999) described the bedrock of Mascot Creek as quartz-mica schist, graphitic schist, phyllite, siliceous mudstone and schistose quartzite. **Workings:** Most mining was by ground sluicing and shoveling in. Although bedrock beneath the placer gravels in this part of the creek was thought to be about 20 feet deep, thawed gravel apparently prevented the prospect shafts from reaching bedrock. The last extensive mining reported on the creek was in the 1980's (Kurtak and others, 1999). The creek is considered to be essentially mined out except for perhaps a few small pockets. Some very small scale mining (hand methods?) directed at these pockets appears to have continued into the mid-1990's (Kurtak and others, 1999). (The Alaska Resource Data File, Wiseman Quadrangle).

McKinley Creek

(Gold) **Quadrangle:** Wiseman (B5); **Coordinates:** 67.44092° N, 152.20777° W Map: 5-24 **Worked:** 1900-; **Access:** Trail from Crevice Creek landing strip at 67.3684°N, 152.0104°W, or trail from the Dalton Highway at 66.7915°N, 150.6878°W to Bettles, Alaska then from Bettles along the John River to Crevice Creek landing Strip; **Summary:** McKinley Creek is a long, southeast-flowing tributary to the John River. The mouth of McKinley Creek is about 2 miles north of the junction of the Allen and John Rivers. The location and extent of placer prospects on McKinley Creek are unknown; the site is arbitrarily located at an elevation of about 1,500 feet on McKinley Creek, i.e. about in the center of this long creek. **Geology:** Schrader (1900) reported a paystreak of considerable width on McKinley Creek. Cobb (1976 [OF76- 340]) in summarizing Schrader's report on the McKinley Creek occurrence, indicated that Schrader's location was uncertain and that it might instead be a description of Midas Creek or Crevice Creek. Cobb, however, provides no evidence to support his suggestion. Kurtak and others (1999) collected 3 panned concentrate samples from McKinley Creek in the vicinity of this site. One of these samples contained 2 particles of coarse gold, with abundant magnetite and pyrite. The concentrate contains 625 ppb gold. The other 2 samples contained only moderate to minor amounts of magnetite and sulfides. Kurtak and others (1999) did not find any evidence of mining on McKinley Creek. **Workings:** Schrader (1900) noted that 50 men were mining in the area in 1900. This, however, may have referred to creeks other than McKinley Creek, as Kurtak and others (1999) found no evidence of mining. (The Alaska Resource Data File, Wiseman Quadrangle).

Midas Creek

(Gold) **Quadrangle:** Wiseman (B5); **Coordinates:** 67.3024°N, 152.2335°W; **Map:** 5-14; **Worked:** 1905-1952; **Summary:** Tributary of the Sixtymile Creek. Good prospects found in 1905. In schist belt (Cobb, 1976, OF 76-340. p. 137). **Geology:** Maddren (1913) reported that good placer gold prospects had been found on Midas Creek in 1905, although these had not been developed by 1913. A small one-man bulldozer-hydraulic mining operation was underway in 1952. No other information is available. (The Alaska Resource Data File, Wiseman Quadrangle).

Minnie Creek

(Gold) **Quadrangle:** Wiseman (B1); **Coordinates:** 67.4162°N, 150.0501°W, 67.4204°N, 150.07609°W (Dalton Hwy Xing); **Map:** 5-4, 5-10; **Worked:** 1900s-; **Production:** Through 1900-1909=48 ozs; **Access:** Trail from Rampart landing strip on the Yukon River at 65.5085°N, 150.1425°W; **Summary:** Tributary of the Middle Fork of the Koyukuk River. Has been much prospecting. Some coarse gold has been recovered. Gravel has so much live water that prospecting and mining are difficult. Heavy shot gold on bedrock. Much prospecting and a little drift mining in early days. Unsuccessful prospecting in 1937. Production, 1900-09, was 48 fine ounces (Cobb, 1976, OF 76-340. p. 138). **Geology:** Minnie Creek reportedly was heavily prospected and locally drift mined in the early 1900's, but it is unclear as to exactly where on Minnie Creek this activity occurred (Reed, 1938). Water problems hindered excavation to bedrock and the early mining efforts were not very successful (Reed, 1938). A shaft to bedrock in 1904 reportedly produced $500 of 'shot' gold from a short drift before it was flooded. Drift mining was concentrated in a deep channel, although some gold was taken out at higher levels. No prospects were reported on the modern channel or on a high channel, although several high benches indicate the presence of a high channel. Limited evidence suggests that most mining activity on Minnie Creek has been on the middle and upper parts of the creek in the Chandalar quadrangle, although it is likely that at least some prospecting has occurred along the lower portion of the creek in the Wiseman quadrangle. A panned concentrate sample from upper Minnie Creek (in the Wiseman quadrangle) contained 6,899 ppb gold (Kurtak and others, 1999). Minnie Creek has been designated as a **Recreation Mining Area** by the State of Alaska upstream 1.5 miles on Minnie Creek and the entire length of Larson Creek, one of its tributaries. Access is by the Dalton Highway at mile marker 187.2 (67.4163°N, 150.0766°W). Average yield of gravel is estimated to be 0.0025 oz/yd^3 of fine gold. (The Alaska Resource Data File, Wiseman Quadrangle).

Montana Gulch

(Gold) **Quadrangle:** Wiseman (B1); **Coordinates:** 67.4591°N, 150.0259°W; **Map:** 5-4; **Access:** 67.4624°N, 150.0245°W (Dalton Hwy Xing); **Summary:** Tributary of the Middle Fork of the Koyukuk River. Average yield of gravel is estimated to be 0.0025 oz/yd^3 of fine gold.

Moose Creek

(Gold) **Quadrangle:** Wiseman (B1); **Coordinates:** 67.3774°N, 150.1766°W; **Map:** 5-10; **Access:** Possible from Dalton Highway at 67.3709°N, 150.12556°W, crossing the Middle Fork of the Koyukuk River by boat; **Summary:** Tributary of the Middle Fork of the Koyukuk River. Reported to contain prospects, but is full of large boulders and has not been worked. Near Emma Creek (Cobb, 1976, OF 76-340. p. 139). **Geology:** Maddren (1913) described Moose Creek as a small gulch containing gold prospects, but numerous large boulders prevented any work from being done. No other information is available on this occurrence. (The Alaska Resource Data File, Wiseman Quadrangle).

Myrtle Creek

(Gold) **Quadrangle:** Wiseman (A1); **Coordinates:** 67.2196°N, 150.0424°W; **Map:** 5-19; **Worked:** 1899-1998; **Purity:** 866-914 Au; **Yield:** 0.0169-0.0581 oz/ft^2; **Nugget Size:** 38 ozs; **Production:** 1900-1909=8,805 ozs; **Access:** Trail from Myrtle Creek landing strip at 67.2152°N, 149.9869°W or by trail from the Dalton Highway at 67.0807°N, 150.3500°W; **Summary:** Tributary of Slate Creek. Bedrock is schist cut by at least one greenstone dike. Lower part of stream is in gravel flats of Slate Creek. Creek gravels were mined for 6 or more miles upstream from mouth. Bench gravels were mined also. Elevation of benches above stream increases downstream. Much of early mining was by drifting. Ditch built and hydraulicking began in 1909. Gravels are 2 - 4 feet deep with gold mainly on or in crevices in bedrock. One nugget weighing more than 23 ounces was recovered in the early days and another weighing over 38 ounces was recovered in 1940. One of the major producing creeks of the district. Mining began in 1899 and continued intermittently until as recently as 1953. No mining in 1974. Includes references to Gold Myrtle Creek. See also Myrtle Creek, Chandalar quadrangle. Production, 1900-09, was 8,805 fine ounces (Cobb, 1976, OF 76-340. p. 140). **Geology:** Gold was discovered in 1899 on Myrtle Creek about 2 1/2 miles above its mouth (Maddren, 1913). The most productive ground was 2 to 3 miles above the mouth (in the Chandalar quadrangle). The gold occurs in gravels in the modern channel and in bench gravels along the lower 2 to 3 miles of the creek. According to Maddren (1913), the width of the gravels in the modern channel is 100 to 300 feet . Reed (1938), however, puts the width at 30-150 feet. The gold was distributed throughout the modern stream gravels. The gravels averaged 2 to 4 feet thick but may be as much as 7 feet thick. Maddren (1913) described the gold as 'shot' or 'wheat' (sized) gold, although he speculated that much of the fine gold was not being recovered at the time. The gold was coarse, clean, and somewhat flattened (like melon seeds) and included occasional nuggets up to about 1 ounce. Smith (1942) reports recovery of a 23-ounce nugget. The gold generally is concentrated on schist bedrock and in crevices in the bedrock as much as 3 feet below the gravel/bedrock contact. Reed (1938) reports active mining in 1937 both on the modern channel and on benches. In at least two locations, Reed described the gold as both fine and flaky and coarse and well-worn. At one of these locations, a bench about 2 3/4 miles from the mouth, Reed noted that the fine and flaky gold was present throughout the gravel; coarser, worn gold (with occasional small nuggets) was on or in the top 3 feet of, bedrock. The gravel consist mainly of slabby schist with varying numbers of greenstone boulders. The bedrock is graphitic to siliceous and slate, that is cut by at least one greenstone dike (Reed, 1938). **Workings:** Mining in the early days was by open cut and later by drifting, principally on the benches. Hydraulicking was begun in 1909. Nearly all evidence of the early mining had disappeared by 1937 (Reed, 1938). Mining using a dragline and bulldozers began in 1940 and continued at least into the early 1950's (Heiner and Wolff, (1968); Cobb, 1973 [B1374]). The gravels are thawed near the modern channel but frozen near the benches. Gold from three locations along the creek was 886 fine at one location and 914 fine at the other two (Reed, 1938). There was exploration activity in 1984 (Eakins and others, 1985) and development activity as late as 1992 (Swainbank and others, 1993). Myrtle Creek (in both the Wiseman and Chandalar quadrangles) has been one of the largest placer gold producers in the Koyukuk district. **Production:** The following values reflect gold prices at the time of the early reports. The total production is not known but production through 1910 was estimated by Maddren (1913) to be $182,000 (about 9,000 ounces). Reed (1938) reported that at least $20,000 was mined in the early days (1900-1910?) from a short section above Kelly's Pup in one summer. Values reported by Reed (1938) ranged from $0.35 to $1.20 per square foot of bedrock. The most productive part of the stream was reported by Maddren (1913) to be at the mountain front, 2 to 3 miles above the mouth of Myrtle Creek. While the creek has been worked from the mouth up to claim No. 20 Above, the claims along the creek above No. 15 were not profitable. Mining continued sporadically from 1899 until at least 1998 (Kurtak and others, 2002). The most recent production was in 1995 (Bundtzen and others, 1996). el.(The Alaska Resource Data File, Wiseman Quadrangle).

Nolan Creek

(Gold) **Quadrangle:** Wiseman (B1, C1); **Coordinates:** 67.4754°N, 150.2306°W; **Map:** 5-1; **Worked:** 1903-Present;

Purity: 885-962 Au; **Yield:** 0.056-0.725 oz/ft^2; **Nugget Size:** ozs; **Production:** 1904-09=36,992 ozs, 1999=734 izs; **Access:** Light duty road from Wiseman, Alaska; **Summary:** Tributary of Wiseman Creek. Bedrock is schist and phyllite into which a narrow v-shaped valley was cut. During glaciation ice backed up into Wiseman Creek valley and lower part of Nolan Creek. Valley of Nolan Creek was filled and buried by sediments from a stream graded to a higher temporary base level. Richest and most productive creek in district. One fork heads against Vermont Creek. Tributaries from east have also been good producers. Gold found from Fay Gulch to Wiseman Creek flats. Most of the creek is in a narrow bedrock valley 4 miles long. The lower 2 miles is choked with unconsolidated deposits that slope to valley of Wiseman Creek. At Discovery (mouth of Fay Gulch) bedrock is 20 - 25 feet below the surface; 8 below (mouth of creek) it is 180 feet below the surface. Placer deposits are bench gravels 50 - 200 feet above the valley floor (auriferous between Smith and Fay Gulches); shallow gulch gravels (mined in tributary gulches from the east and on Discovery claim); and deep, frozen gravels in the old bedrock valley. Old valley floor uneven and has occasional large boulders. Creek gravels mined only at mouth of Fay Gulch. Valley is filled with muck below Fay Creek. Gold on false bedrock. Values ranged from 0.056 to 0.725 oz/ft^2 of bedrock. Gold coarse, but without large nuggets. Benches that merge with those of tributaries have been mined. Most production was from buried deep channel, 20 feet deep at Fay Gulch; and more than 200 feet deep, 3/4 mile below Smith Creek. Mining in 1901 and most succeeding years until World War II was reported. Mining in 1937 was on benches. Deep channel and benches are frozen. Has been more recent mining. Production, 1904-09, was 36,992 fine ounces. No data on total production, but may have been several times 1904 - 09 figures. See also Archibald Gulch, Fay Gulch, Smith Creek. (Cobb, 1976, OF 76-340. p. 141-142). **Geology:** Gold placers in Nolan Creek valley are in shallow gravels of the modern stream, in bench gravels, and in frozen gravels in a deep channel. The gravels of the deep channel have been the most productive by far. The present channel in Nolan Creek was mined just below the mouth of Fay Creek (near the Discovery claim) from 1903 to 1905. This channel together with shallow gravels on Smith and Fay Creeks produced a considerable amount of gold. The shallow gravels on Nolan Creek were about 3.5 feet thick and lay on false bedrock in the valley fill. Reed (1938) suggested that the gold in this placer was derived from erosion of gold-bearing gravels in Fay Creek. Recent information from the Internet page website of Silverado Gold Mines, the current property owner, (http://www.silverado.com) indicates that shallow placer deposits also occur on upper Nolan Creek for about 0.5 mile above Fay Creek. The bench deposits on Nolan Creek generally are between the mouth of Fay Creek to just below the mouth of Smith Creek. The configuration of several of these bench deposits suggests that at least some of them may be derived from the eastern tributaries to Nolan Creek. Most are located along the east side of Nolan Creek and 20 to more than 200 feet above the modern stream. The gravel in these deposits is frozen and coarse grained but contains few boulders. The gold is also coarse; it includes small nuggets and a few large nuggets. The gold fineness varies from 885 to 962. According to Reed (1938) these deposits varied in value from $0.25 to $4.20 per square foot of bedrock (gold at $35 per ounce). The most productive portion of the Nolan Creek placers has been the deep channel. This extends for about 2 miles along the Nolan Creek valley from near the mouth of Fay Creek to about 0.75 mile below the mouth of Smith Creek. At its upper end, the deep channel is about 20 feet below the surface; it gradually deepens to over 200 feet at its lower end. The average gradient of the deep channel is about 4.5 percent but there are a number of steeper gradients as the channel makes its way down valley. The gravel and overlying fill in Nolan Creek is frozen and waterworn and generally fairly fine, grained; however, large boulders commonly are scattered throughout the gravel. The gold in the deep channel lies directly on bedrock. It is rounded and waterworn and generally coarse grained; large nuggets are rare. Values, according to Reed (1938), range from $1.15 to as much as $15 per square foot of bedrock (gold at $35 per ounce). In addition to the principal deep channel, there are sections of other deep channels that have been preserved at levels above the principal channel. Reed (1938) speculated that these remnants may have been bench deposits formed either when the deep-channel stream was active, or when an aggrading Nolan Creek deposited gold in channels cut in the valley walls as it meandered across the valley floor. The rocks in the valley of Nolan Creek are predominantly black metasiltstone, phyllite, pelitic schist and black slate of the Middle to Upper Devonian, Beaucoup Formation (Eden, 2000). Nolan Creek has been extensively mining and prospected beginning in 1903 and continuing until the present (2000). Most early mining was by underground drifting. Modern mining has been by open-pit operation in the summer and underground drift mining in frozen gravel in the winter. Reed (1938) noted that by 1937 the deep channel was considered mined out. However, press releases from Silverado Gold Mines (the current property owner) indicate that mining is continuing in the Nolan Creek valley, and that further exploration on the deep channel is planned. (The Alaska Resource Data File, Wiseman Quadrangle). **Reserves:** A number of reserve figures for the Nolan Property (which includes Nolan Placer, Nolan Lode, Thompson's Pup, Dionne (Mary's Bench) and Smith Creek) are given on the website of the Silverado Gold Mines Ltd., the current property owner (http://www.silverado.com/) . A press release dated February 16, 2000 posted on the website indicated a gold reserve of 145,000 oz. of proven, probable, and drill-indicated reserves for the Nolan Gold Property. This figure is presumed to represent a total resource for the entire property. On the website, Silverado has also described a probable and possible gold resource for the Nolan Placer of 114,760 ounces. Although it is unclear from the information on the website what portion of the overall Nolan property this reserve figure represents, the information implies that it is for the Nolan deep channel.

Oregon Creek

(Gold) **Quadrangle:** Wiseman (B3); **Coordinates:** 67.4672°N, 151.3912°W; **Map:** 5-2, 5-12, **Summary:** Tributary of Flat Creek. Good prospects, but no mining. Bedrock schist (Cobb, 1976, OF 76-340. p. 144). Oregon Creek is not named on the Wiseman B-3 topographic map. It is a small right-limit tributary to Flat Creek. The mouth of Oregon Creek is located about 1.5 miles north of the junction of Birch and Flat Creeks. The location of placer prospects on Oregon Creek is not known and the site is plotted just upstream from the mouth of the creek in about the center of section 32, T. 31 N., R. 17 W., of the Fairbanks Meridian. **Geology:** Reed (1938) reported that good placer prospects had been found on Oregon Creek but that there had been no mining. Reed described the bedrock on Oregon Creek as schist; Dillon and others (1986) map the area as Upper Devonian(?), Hunt Fork Shale and Middle Devonian(?) siliceous clastic rocks. (The Alaska Resource Data File, Wiseman Quadrangle).

Pasco Creek

(Gold) **Quadrangle:** Wiseman (B1); **Coordinates:** 67.4652°N, 150.3092°W; **Map:** 5-1; **Worked:** 1900-; **Access:** Trail from Wiseman light duty road at 67.4692°N, 150.2365°W; **Summary:** Tributary of Wiseman Creek. Schrader (1900) reported placer gold on Pasco Creek but no other information is available. (Cobb, 1976, OF 76-340. p. 145).

Porcupine Creek

(Gold) **Quadrangle:** Wiseman (A1, B1); **Coordinates:** 67.2444°N, 150.2979°W; **Map:** 5-7, 5-22; **Worked:** 1901-2000; **Purity:** 854 Au; **Yield:** 0.0324 oz/ft^2; **Production:** 1900-1901=72 ozs; **Access:** Trail from Porcupine Creek landing strip on the Yukon River at 67.2380°N, 150.2883°W or from Dalton Highway at 67.2389°N, 150.2167°W; **Summary:** Tributary of the Middle Fork of the Koyukuk River. Above mouth of Quartz Creek small-scale mining in 1936-37 was encouraging, but not profitable. Near mouth of Quartz Creek the present stream gravel was mined from 1916 to 1937. Depth to bedrock is from 6 feet under the creek bed to 30 feet toward a bench on the left limit. Deposit is permanently frozen. Bedrock schist; most of gold in lower part of gravel rather than on bedrock. Gold is fine with a few nuggets. Deep channel (not frozen) was drift mined in winter. Mining has been intermittent from 1900 to 1974. Gravel said to contain many large boulder yielding 0.387 ounces per man per day. Coarse gold being recovered from gravel 20 feet thick in 1974. Has been drift mining in deep frozen gravel. Production during the period 1900-1901 was 72 fine ounces. No data on total production (Cobb, 1976, OF 76-340. p. 146). **Geology:** Placer gold mining on Porcupine Creek was first reported in 1901 when Maddren (1913) noted that four men were producing $8 per man per day. The placer gold on Porcupine Creek occurs both in the modern channel and in a deep channel. Reed (1938) reported mining on the modern channel for a distance of 600 feet below and 3 miles above the mouth of Quartz Creek. Below Quartz Creek, Porcupine Creek was mined from 1916 until the time of Reed's visit in 1937. Below Quartz Creek, the depth to bedrock was 6 feet in the modern creek bed and deepened to about 30 feet as the workings moved toward the left limit of the creek. The gravel was frozen and consisted of small worn pieces of schist mixed with coarse sand and many large boulders. The gold occurs in the lower 3 to 6 feet of the gravel, but only a little gold was on bedrock. The gold is generally fairly fine with an occasional nugget, and had a fineness of 854. The value of the ground (in 1938) was about $0.59 per square foot of bedrock. Reed (1938) described workings in a 20 to 25-foot-wide, deep channel about 30 feet below the surface. Although the overlying material was frozen, the deep channel itself was not. Reed reports that the value of the ground was $0.67 per square foot of bedrock. Early mining of the modern channel was by ground sluicing and shoveling in; the deep channel was drift mined. There are reports of mining into the 1980's using a dozer (Bundtzen and others, 1984). Mining was reported on the creek in 2000 (Kurtak and others, 2002). (The Alaska Resource Data File, Wiseman Quadrangle).

Quartz Creek

(Gold) **Quadrangle:** Wiseman (B1); **Coordinates:** 67.2524°N, 150.3138°W; **Map:** 5-7, 5-22; **Worked:** 1901-2000; **Production:** Through 1937=121 ozs; **Access:** Trail from Porcupine Creek landing strip on the Yukon River at 67.2380°N, 150.2883°W or from Dalton Highway at 67.2389°N, 150.2167°W; **Summary:** Heads in Emma Dome and flows into Porcupine Creek. Depth to bedrock about 3 feet. Hand mining in stream channel produced 121 fine ounces by 1937 (Cobb, 1976, OF 76-340. p. 147).

Rosie Creek

(Gold) **Quadrangle:** Wiseman (A1); **Coordinates:** 67.1949°N, 150.1932°W, 67.1949°N, 150.2794°W (Dalton Hwy Xing); **Map:** 5-19; **Worked:** 1900s-; **Access:** Trail from Dalton Highway at 67.2533°N, 150.1792°W; **Summary:** Tributary of the Middle Fork of the Koyukuk River. A few surface colors 2 miles above mouth. Ground said to be about 130 feet deep. Many colors on bedrock in 113-foot shaft about 3 miles above mouth. Includes reference to Rosie Creek (Cobb, 1976, OF 76-340. p. 149). **Geology:** A number of prospect shafts were sunk in the early 1900's on Rosie Creek and its tributary Lake Creek (Reed, 1938). The depth to bedrock varies from just over 100 feet to as much as 275 feet in one shaft near the head of Lake Creek. Maddren (1913) reported fine gold in surface gravels about 2 miles above the mouth of Rose(sic) Creek, and that a 113-foot shaft near the junction of Rose Creek and Lake Creek worked a thin layer of gravel on bedrock that produced numerous fine flakes of gold. Reed (1938) described several additional shafts from near the head of Lake Creek to about 0.2 mile below the junction of Rosie and Lake Creeks. Only the shafts on Lake Creek found any gold. This was fine flour gold, which occurred throughout the gravels from a depth of 20 feet to the top of the bedrock. Reed (1938) reported that the bedrock on Rosie Creek is greenstone and the bedrock on Lake Creek is schist. According to Dillon and others (1986), Rosie Creek is underlain by metagraywacke and phyllite, and upper Lake Creek is underlain by schist of the Hunt Fork Formation. (The Alaska Resource Data File, Wiseman Quadrangle).

Rye Creek

(Copper, Gold, Lead, Monazite, Tungsten, REE) **Quadrangle:** Wiseman (B3); **Coordinates:** 67.4092°N, 151.3744°W; **Map:** 5-12; **Worked:** 1915-Present; **Production:** Through 1937=2.261 ozs; **Access:** Trail from Wild Lake; **Summary:** Tributary of Flat Creek. Creek flows on limestone and schist and along contact (probably a fault) between them. High channel, present channel, and a deep channel. No gold in paying quantities above Jay Creek. Has been both drift and hand mining. Placers found in 1915. Production through 1937 was worth about 2,261 fine ounces. **Minerals:** Ilmenite, andalusite, kyanite, pyrite, zircon, chalcopyrite, monazite, galena, scheelite, and gold (Cobb, 1976, OF 76-340. p. 150). **Geology:** The gravel in Rye Creek is angular, very coarse grained, and contains numerous small boulders. Gold in paying amounts in Rye Creek has not been found above the mouth of Jay Creek. Reed (1938) reports that the bedrock on lower Rye Creek to about 0.5 mile below the mouth of Jay Creek is blue crystalline limestone; above this point, the bedrock is schist. He suspects a fault contact between these two units. Dillon and others (1986) map most of the area around Rye Creek as Skajit Limestone; the uppermost part of the creek flows across chloritic and carbonate rocks. **Workings:** Placer gold was found at Rye Creek in 1915 (Brooks, 1916) and subsequently gold was mined from the modern channel, a high channel, and a deep channel (Reed, 1938). The total production from Rye Creek as of 1938 was about $55,000 (approx. 2,700 oz.). Nearly half of that was recovered from a 300-foot section of the modern channel below the mouth of Jay Creek (Reed, 1938). Mining on the modern channel extended from near the mouth of Jay Creek downstream to where Rye Creek enters Flat Creek valley. A high channel occurs along the right limit of lower Rye Creek, about 0.2 mile north of the modern creek, but it apparently contained little gold. A deep channel also occurs along the right limit of Rye Creek. A shaft was sunk 85 feet to bedrock beside lower Rye Creek and the deep channel was drift mined for a short distance. Additional drift mining was conducted on the deep channel about 0.2 mile below Jay Creek, where the depth to bedrock is about 20 feet. (The Alaska Resource Data File, Wiseman Quadrangle).

Sawyer Creek

(Gold) **Quadrangle:** Wiseman (B1); **Coordinates:** 67.3470°N, 150.1766°W; **Map:** 5-7, 5-10; **Worked:** 1936-; **Access:** Trail from Emma Creek landing strip at 67.3188°N, 150.1821°W; **Summary:** Tributary of the Middle fork of the Koyukuk River. Gravel coarse with many large boulders. Some gold reported, but no mining (Cobb, 1976, OF 76-340. p. 151). **Workings:** Sawyer Creek was prospected for gold in 1936 (Reed, 1938). There is no record of any mining before then. Mosier and Lewis (1986) collected and analyzed five samples of gold from Sawyer Creek. However, there are no reports of actual mining on the creek, probably due in part to numerous large boulders in the creek gravels. (The Alaska Resource Data File, Wiseman Quadrangle).

Sirr Creek

(Gold) **Quadrangle:** Wiseman (C4); **Coordinates:** 67.5349°N, 151.6400°W; **Map:** 5-17; **Worked:** 1900s-; **Access:** Trail from Wild Lake; **Summary:** Tributary of Seward Creek which flows into Wild Lake. Has been some mining (as of 1937). Mountain at head of creek is schist with limestone cap (Cobb, 1976, OF 76-340. p. 152). **Geology:** Reed (1938) reported that there was extensive prospecting and a little placer mining on Sirr Creek in the early 1900's. The rocks in the area are Devonian phyllite and schist (Chipp, 1972). No other information is available. (The Alaska Resource Data File, Wiseman Quadrangle).

Sixtymile Creek

(Gold) **Quadrangle:** Wiseman (B5); **Coordinates:** 67.35106°N, 152.16774°W; Map: 5-14 **Worked:** 1900s-;**Production:** 83 ozs; **Access:** Trail from Crevice Creek landing strip at 67.3685°N, 152.0109°W; **Geologic description:** Maddren (1913) reported good prospects on 'Fool Creek' and its tributaries, and Wimmler (1922) reported coarse gold along benches of Sixtymile Creek. (which is probably another name for Fool Creek). According to Kurtak and others (2002), 83 ounces of gold have been recovered from Sixtymile Creek. The rocks in the area consist mostly of Devonian, Skajit Limestone, with subordinate Middle or Upper Devonian(?) chloritic and carbonate rocks and Proterozoic(?) granite gneiss (Dillon and others, 1986). (The Alaska Resource Data File, Wiseman Quadrangle).

Slate Creek

(Gold) **Quadrangle:** Wiseman (A1); **Coordinates:** 67.2159°N, 150.0338°W, 67.2546°N, 150.1770°W (Dalton Hwy Xing); **Map:** 5-19; **Worked:** 1899-1993; **Yield:** 0.0025 oz/yd^3; **Nugget Size:** ozs; **Production:** 1900-1909=145 ozs; **Access:** Trail from Mrytle Creek landing strip at 67.2153°N, 149.9864°W or trail from the Dalton Highway at 67.2533°N, 150.1789°W; **Summary:** Flows westward from a wide, gravel-floored pass for about 20 miles to Middle Fork of the Koyukuk. Has cut down into phyllite and mica schist on lower two thirds of its course. Bedrock cut by many quartz veins and altered dikes. Mining near mouth of Myrtle Creek in 1903 and for several miles upstream. Much of gold was brought in by Myrtle Creek, but there is some in bench gravel above mouth of Myrtle. Bedrock mica schist and slate. Gravels thin, with most of gold in cracks and crevices in bedrock. Gold probably locally derived from quartz veinlets; large veins are barren. Mining began in 1899 and was reported as recently as 1939. Production, 1900-09, was 145 fine ounces. Gold is reported to be flat pieces the size of shot or wheat (Cobb, 1976, OF 76-340. p. 153). **Geology:** The gold in Slate Creek occurs in the modern channel and in high channels both above and below the junction with Myrtle Creek (Reed, 1938). In the modern channel from the mouth of Slate Creek upstream to the junction with Myrtle Creek, the gravels are 10 to 40 feet thick. Reed (1938) indicates that it was possible for the early miners to recover $3 to $4 per day of flour gold (approx. 0.15 - 0.20 ounce of gold) by rocking. The high channels along Slate Creek were prospected in the early 1900's, but they contained insufficient gold to justify mining. The productive portion of Slate Creek above the junction with Myrtle Creek is probably in the adjoining Chandalar quadrangle. Bedrock along Slate Creek is mica schist, slate, or phyllite containing small bodies of quartz and cut by several altered diorite dikes (Maddren, 1913). Slate Creek was mined by hand and by mechanized methods including use of a dragline (Bliss and others, 1988). There was some scattered exploration drilling. Intermittent mining continued up through at least 1993 (Bundtzen and others, 1994). Maddren (1913) reported production of $3,000 (about 150 ounces of gold) from 1900 to 1909. No information is available for later production. (The Alaska Resource Data File, Wiseman Quadrangle). Slate Creek has been designated as a **Recreation Mining Area** by the State of Alaska upstream from 0.25 to 0.75 miles above the bridge on the Dalton Highway at mile marker 175.5 (67.2546°N, 150.1770°W). Average yield of gravel is estimated to be 0.0025 oz/yd^3 of fine gold.

Slisco Bench

(Gold) **Quadrangle:** Wiseman (A2); **Coordinates:** 67.51050°N, 150.11140°W Map 5-4 **Worked:** 1920s-1995; **Access:** Light duty road from Wiseman, Alaska; **Summary:** The Slisco Bench placer mine is located along the west side of Hammond River about 0.2 mile southwest of the mouth of Vermont Creek. The mine is 0.3 mile east-southeast of the center of section 13, T. 31 N., R. 11 W., of the Fairbanks Meridian. It marks the north end of a reported buried channel that may extend a mile to the south. **Geology:** The Slisco Bench deposit is one of a number of bench placers in the valley of the Hammond River. Silverado Mines, the current (2000) owner of most of the property in the vicinity, reports that the Slisco Bench was discovered in the late 1920's and that a small amount of mining was done in the years between the discovery and 1935. Additional mining also occurred in the 1960's. Drilling by Silverado in 1995 intersected gold-bearing gravel in a buried channel beneath 60 to 200 feet of frozen gravel (Silverado Mines website: www.silverado.com). Silverado speculates that the Slisco Bench channel may extend from its shallow north end just south of Vermont Creek, as much as 5000 feet to the south; however, much of this interval is unexplored. (The Alaska Resource Data File, Wiseman Quadrangle).

Smally Creek

(Gold) **Quadrangle:** Wiseman (A2); **Coordinates:** 67.0240°N, 150.5751°W; **Map:** 5-20; **Worked:** 1937-1980s; **Purity:** 853-933 Au; **Access:** Trail from Gold Bench landing strip on the South Fork of the Koyukuk River at 66.9831°N, 150.6435°W; **Summary:** Tributary of John R Creek. Mining in early days and in 1937. Mining in small area near forks; groundsluicing and shoveling in. Bedrock is black shale and quartz conglomerate. Depth to bedrock about 17 feet. Gold mainly fine and flat; mostly on bedrock. Conglomerate contains some gold; may have been source of gold in placers of area (Cobb, 1976, OF 76-340. p. 154). **Geology:** Placer gold on Smally Creek occurs in the modern stream bed. The alluvium is about 17 feet thick. The upper 4 to 7 feet consists of muck, and the remainder is coarse and waterworn gravel with numerous boulders. The gold occurs principally on bedrock, but also in the lower 6 feet of the gravel. The gold forms very fine, flat flakes; there are infrequent coarse pieces. Reed (1938) noted that the average fineness of the gold was 907. The fineness of five samples of gold from Smally Creek collected around 1985 varies from 853 to 933 (Mosier and Lewis, 1986). The bedrock on Smally Creek is black shale and quartz conglomerate; Reed (1938) suspects that the conglomerate was the source of the fine gold in the placer. The placer was mined in the early 1900's. Reed (1938) reports mining by ground sluicing and shoveling in in 1937, and notes that there were plans for hydraulic and/or mechanized mining. Numerous claims were active in the vicinity in the early 1980's. (The Alaska Resource Data File, Wiseman Quadrangle).

Smith Creek

(Gold, Antimony) **Quadrangle:** Wiseman (B1); **Coordinates:** 67.4716°N, 150.2299°W; **Map:** 5-1; **Worked:** 1903-; **Purity:** 925-986 Au; **Yield:** 0.015 oz/ft^2; **Production:** 1903-09=10,063 ozs; **Access:** Light duty road from Wiseman, Alaska; **Summary:** A major gold-producing tributary of Nolan Creek. Heads against Union Gulch. Bedrock is micaceous schist. Creek gravels are 6 - 20 feet thick and frozen. Deep 135 feet at mouth) channel was mined out. Remnants of high channels have been mined. Gold in upper part of gulch is angular and rough, becoming more rounded and heavier downstream. Gold assayed at 968 Au. Value of ground was 0.015 oz/ft^2 of bedrock. Was a steady producer from 1903 to 1910. Production, 1903-09, was 10,063 fine ounces (Cobb, 1976, OF 76-340. p. 155). **Geologic description:** According to Maddren (1913), the placer deposits in the Nolan Creek basin -- including Smith Creek -- occur in the shallow gravels of the modern channel, as bench deposits, and in deep, frozen gravels. Mining of the shallow gravels began in 1903, two years after their discovery. Maddren (1913) described the gold from upper Smith Creek as mostly rough, angular grains; the gold becomes progressively more rounded and coarser downstream. Some of the gold near the head of the Smith Creek is coated by a white substance that Maddren did not identify although he suspected that it might be lime. The gravel along the lower 2 miles of the modern channel of Smith Creek was selectively worked in richer spots during the earliest mining activity, but there is little information about this early mining. Gravel in the modern channel that was mined in the 1920's and 1930's were 6 to 20 feet thick. The gravel was frozen, coarse and subangular, and contained a few boulders. The gold was both coarse, rough and pitted, and smooth, rounded and waterworn. The value of the ground was about $0.30 per square foot of bedrock (gold at $20.67 per ounce) and the average fineness of the gold was about 950. The deep channel on Smith Creek rose from a depth of 135 feet near the mouth to eventually coincide upstream with the modern channel. By 1937, this deep channel was considered to be mined out by the methods available at the time. Reed (1938) reported that remnants of the bench channel on Smith Creek had been mined at several locations along the creek. The gravel was frozen and generally coarse and subangular. The gold from one of these locations was mostly coarse, unworn and pitted, but some was smooth, rounded and worn. This gold occurred throughout the gravel but most was in the lower few feet; there were no large nuggets. The fineness was about 950. Based on analysis of six gold samples, Eden (2000) concluded that the gold occurs in two different populations defined by either high or low silver content. The fineness of these six samples ranged from 925 to 986. Eden also suggests that the gold in the Nolan-Hammond area has been derived from two gold-bearing vein systems in the area. The rocks in the valley of Smith Creek are micaceous schist, phyllite, and slate (Eden, 2000). (The Alaska Resource Data File, Wiseman Quadrangle).

Spring Creek

(Gold) **Quadrangle:** Wiseman (C3, C4); **Coordinates:** 67.5113°N, 151.5287°W; **Map:** 5-17; **Worked:** 1903-Present; **Purity:** 920-940 Au; **Yield:** 0.0087-0.0435 oz/ft^2; **Production:** Through 1937=2,875 ozs; **Access:** Trail from Wild Lake; **Summary:** Tributary of Wild Lake. Bedrock is schist. Gold was discovered in 1903. Production through 1937 was about 2,875 fine ounces from open cuts and drift mines. Mining in 1938 was reported and there probably has been some since World War II (Cobb, 1976, OF 76-340. p. 157). **Geology:** Placer gold was discovered on Spring Creek in 1903, and the creek was mined extensively from 1904 through at least 1937 (Reed, 1938). Several mining operations described by Reed indicate that the depth to bedrock along Spring Creek varied from 5 to 25 feet and occasionally as much as 50 feet. The gravels were frozen and coarse and waterworn with numerous boulders on top of the gravel. The gold was coarse and unworn. It occurred in the lower 3 feet of the gravel and in the upper foot of the bedrock. The gold values determined from mining operations in the 1920's and 1930's were from $0.18 to $0.90 per square foot of bedrock; the fineness ranged from about 920 to 940. Maddren noted that $7,000 in gold (about 350 ounces) was produced from 1907 to 1909. Reed (1938) estimated that the total production through 1937 had been $59,500 (about 2975 ounces). There are no figures available for any later production. Kurtak and others (1999) collected six panned-concentrate samples along a 1.2-mile stretch of Spring Creek. The samples contained no visible gold but assays of the six samples averaged 1,260 ppb Au. Reed (1938) described the bedrock in Spring Creek as principally quartzite, and graphitic and greenstone schists. Dillon and others (1986) map the Spring Creek drainage area as Middle Devonian(?) siliceous clastic rocks. (The Alaska Resource Data File, Wiseman Quadrangle).

Steep Gulch

(Gold, Lead) **Quadrangle:** Wiseman (B1); **Coordinates:** 67.4814°N, 150.0665°W; **Map:** 5-4; **Worked:** 1900s-; **Nugget Size:** 50 ozs; **Access:** Light duty road from Wiseman, Alaska; **Summary:** Tributary of the Hammond River. Gold from Steep Gulch was probably from bench gravels of Hammond River. Much of gold was very coarse with nuggets over 50 ounces reported. Concentrates contain pyrite, some exhibiting free gold. Placers work continuously from discovery to World War II (Cobb, 1976, OF 76-340. p. 114-115).

Suckik Creek

(Gold) **Quadrangle:** Wiseman (B1); **Coordinates:** 67.19444° N 151.75976° W Map: 5-21; **Worked:** 1948-; **Access:** Trail from Koyukuk River at 66.9410°N, 151.5078°W, near Evansville/Bettles, Alaska; **Summary:** Suckik Creek (Sickik Creek on the modern (1970) A-4 topographic map and Suckik Creek on the 1:250,000 map) is south-flowing tributary to Timber Creek. The mouth of Suckik Creek is directly across Timber Creek from Button Mountain. There is no information as to the exact location of prior mining on the creek. The site is arbitrarily placed about a mile upstream from its mouth. **Geology:** The only information about placer mining on Suckik Creek is a reference to a 2-man mining operation in 1948 that included shoveling-in and prospecting (Stewart, 1949). The area of Suckik Creek is predominantly underlain by Proterozoic or lower Paleozoic(?) schist and local marble, and by smaller areas of Devonian, Ambler metavolcanic rocks (Dillon and others, 1986). (The Alaska Resource Data File, Wiseman Quadrangle).

Surprise Creek

(Gold) **Quadrangle:** Wiseman (C3, C4); **Coordinates:** 67.5179°N, 151.5249°W; **Map:** 5-17; **Worked:** 1904-; **Purity:** 920 Au; **Access:** Trail from Wild Lake; **Summary:** Tributary of Wild Lake. Gold discovered and first mined in 1904. Stream heads on Mathews Dome, where bedrock is largely greenstone schist with many quartz veins. Stream flows through various kinds of schist from Mathews Dome to Wild Lake. Some of gold is very coarse; some is flour gold. All gold in lower

3 feet of gravel and on bedrock. Small boulder of float quartz "liberally spangled with gold" found in late summer of 1929. Sporadic small-scale mining from 1904 until as recently as 1937 from about 1,350 feet - 6,900 feet from Wild Lake (Cobb, 1976, OF 76-340. p. 158). **Geology:** Surprise Creek was sporadically mined from 1904 through about 1937. Mining has been sporadic and in many cases the workings failed to reach bedrock. Where it was reached, the depth to bedrock is 12 to 14 feet but the bedrock surface apparently dips to the north under talus and its depth becomes progressively greater. The gravel in the creek is not frozen. It is very coarse and angular, and contains many large boulders at the top of the section. The gold is coarse, many nuggets occur with the fine gold. The gold occurs in the lower 3 feet of the gravel and on top of the bedrock; its fineness is about 920. Smith (1932) reported that a boulder of quartz float found in Summit (Surprise) Creek was 'liberally spangled with gold'. Samples collected by Kurtak and others (1999) along Surprise Creek include panned concentrates that contain up to 3,888 ppb gold, limonite-stained quartz-carbonate float that contains 163 ppb gold, and a piece of malachite-stained schist. The rocks in Surprise Creek are described by Reed (1938) as quartzite, greenstone, and graphitic and micaceous schists, cut by several quartz veins in the lower valley. Quartz also constitutes a substantial portion of the detrital material in the creek. Dillon and others (1986) map the area as Middle Devonian siliceous clastic rocks. (The Alaska Resource Data File, Wiseman Quadrangle).

Swift Creek

(Gold) **Quadrangle:** Wiseman (B1); **Coordinates:** 67.4913°N, 150.0986°W; **Map:** 5-4; **Worked:** 1901-; **Purity:** 880-924 Au; **Production:** 1902-1909=348 ozs; **Access:** Light duty from Wiseman, Alaska; **Summary:** Tributary of Hammond River that heads opposite Smith and Fay Gulches. The placer deposits extend from the mouth of Swift Creek about a mile upstream; the workings continue a short distance up a small, left-limit gully that joins Swift Creek about 0.8 mile above its mouth. Heads in mineralized area east of Nolan Creek. An old high channel merged with that of Hammond River. Mining, all by hand methods, was between 1901 and 1912, in 1927, and possibly in other years when it was reported with that on Hammond River. Gold is coarse and covered with a white mineral substance. Production, 1902-09, was 348 ounces (Cobb, 1976, OF 76-340. p. 159). **Geology:** Gold was produced from Swift Creek as early as 1901 (Schrader, 1904). The richest gold placer deposits were near the mouth, although the mines produced only wages and small amounts of coarse gold (Maddren, 1913). The modern stream bed has been mined as well as a high channel that coalesced with an eroded high channel of the Hammond River. Some of the gold had a white coating similar to that found on the gold from Smith Creek. Mosier and Lewis (1986) report that the gold was 880 fine from the high channel and 924 fine from the modern channel. Of three panned-concentrate samples from Swift Creek, only one, near the mouth, showed any visible gold (Kurtak and others, 1999). The rocks in the area are predominantly chlorite-mica schist, quartz-mica schist, and phyllite (Eden, 2000). Swift Creek drains mineralized areas on Smith Creek Dome. (The Alaska Resource Data File, Wiseman Quadrangle).

Thompson Pup

(Gold) **Quadrangle:** Wiseman (B1); **Coordinates:** 67.4902°N, 150.1862°W; **Map:** 5-1, 5-4; **Worked:** 1901-; **Purity:** 777-926 Au; **Production:** 1901-1999=3,100 ozs; **Access:** Light duty from Wiseman, Alaska; **Summary:** Tributary of Fay Creek that drains mineralized area. Thompson Pup reported to have mined out by 1937. Gold recovered from frozen ground was rough and angular with quartz attached. Early (1901-1909) production was 1,451 fine ounces which includes production from Fay Creek. Some recent (probably post-World War II) mining reported (Cobb, 1976, OF 76-340. p. 106). **Geology:** Thompson Pup is generally described together with the Nolan Creek placers, not as a separate placer deposit. There is, however, little doubt that there are placer gold deposits along most of Thompson Pup. Work on Thompson Pup was noted by Holdsworth (1957) and Williams (1950) describe a hydraulic and ground sluicing operation on Thompson Gulch. Eden (2000) reports that a total of 3,100 ounces of gold were produced from Thompson Pup between 1904 and 1999. Williams also reports that there were several channels and that the fineness of the gold being recovered was unusually high; the gold averages about 940 fine with nuggets up to 982 fine. Mosier and Lewis (1986), however, indicate that the fineness is generally lower, in the range of 777 to 926. Eden (2000) indicates that the gold placer deposits extend up Thompson Pup for approximately a half mile above Fay Creek, and that the gold from Thompson Pup is crystalline. The crystalline nature of the gold distinguishes it from other placer gold in the Nolan Creek area and suggests that the gold has not traveled far from its source. Kurtak and others (1999) analyzed arsenopyrite crystals from placer concentrates from Thompson Pup; they contain as much as 1,964 ppb gold, but they may have been contaminated. The rocks in the area consist of micaceous and chloritic schists, phyllite, and metasiltstone (Eden, 2000). (The Alaska Resource Data File, Wiseman Quadrangle).

Timber Creek

(Gold) **Quadrangle:** Wiseman (A4); **Coordinates:** 67.0680°N, 151.8329°W; **Map:** 5-25; **Access:** From John River at 67.0625°N, 151.8331°W; **Summary:** Timber Creek is a large, south-flowing stream which joins the John River west of Ninemile Hills. The locations of placer occurrences on Timber Creek are not described in detail. The site is arbitrarily placed about three-quarters of a mile north of the mouth of the creek.Probably was mining from river bars in the early days. Location on stream not given. Reported yield was 0.194 ounces per man per day using hand methods (Cobb, 1976, OF 76-340. p. 161). **Geology:** Although Reed (1938) did not visit Timber Creek, he noted local reports that the creek was auriferous. According to Reed (1938), in the early 1900's no claims were allowed on the creek by popular agreement and the creek was reserved for men who had not made enough money during the first half of the summer to get them through the winter. Reed related that the gold was recovered from the bars of the stream by rocking and that a man could recover $4 per day (gold at $35/ounce). There is no other record of mining on this creek. (The Alaska Resource Data File, Wiseman Quadrangle).

Tramway Bar

(Gold) **Quadrangle:** Wiseman (A1); **Coordinates:** 67.0880°N, 150.4981°W; **Map:** 5-6, 5-20; **Worked:** 1885-1965; **Yield:** 0.0097 oz/ft^2; **Production:** 1900-1909 =387 ozs; **Access:** Trail from landing strip on the west bank of rhe Middle Fork of the Koyukuk River at 67.0922°N, 150.5010°W, or by trail from the Dalton Highway at at 67.0798°N, 150.3506°W; **Summary:** Tramway Bar is located on the west side of the Middle Fork, Koyukuk River, about 0.4 mile south-southwest of Chapman Island; Tramway Bar is labeled on the Wiseman A-1 topographic map. The placer deposit is located on a bench about 80 to 100 feet above river level.One of the earliest gold discoveries in the district with the discovery of gold predating 1890. Gold-bearing gravel on bench cut on nonauriferous conglomerate and sandstone 80 - 100 feet above Middle Fork of Koyukuk was mined sporadically on a small scale from 1900 or earlier until 1965 or later. Mining was for a width of 50 feet for 1 mile along canyon wall. Bedrock appears to be Kenai conglomerate and sandstone. bedrock dips 20' NW. Gravel coarse. Coarse gold is scattered through gravel; fine, flaky gold is mainly on bedrock. Origin of gold is not known. Production, 1900-09, was 387 fine ounces (Cobb, 1976, OF 76-340. p. 162). **Geology:** The Tramway Bar mine is one of the first placer gold occurrences known in the area; it was discovered as early as 1885. The gravel that contains the gold is presumed to have been deposited by streams flowing in the Middle Fork Koyukuk River valley prior to downcutting of the Middle Fork canyon. The deposit at Tramway Bar consists of about 4-7 feet of coarse, waterworn, bouldery gravel. It lays on a bench cut in conglomerate and sandstone, along the west side of the Middle Fork canyon wall about 50 - 100 feet above the modern river channel. These gravels were mined across a 50 foot width for about a mile. One mining operation, active in 1937 described by Reed (1938), recovered both coarse, worn, and rounded gold, and very fine, flaky gold. The coarse gold occurs throughout a 3-5 foot section of coarse gravel, but mostly in the lower part. The fine gold occurs mostly in a layer on top of decomposed conglomerate bedrock. No large nuggets accompanied the coarse gold. The early miners were said to have recovered only the coarse gold. Maddren (1913) suggested that the gold at Tramway Bar was derived from the outwash of Chapman Creek. Reed (1938) thought that the coarse gold might have been reconcentrated from outwash material in the Middle Fork valley, and that the fine gold was derived from erosion of the conglomerate bedrock. Maddren reported production of about $8,000 in gold (approximately 400 ounces) between 1900 and 1909. Gold values in 1937 were said to be about $0.31 per square foot of bedrock (Reed, 1938). Mining continued sporadically at least until 1965 (Cobb, 1973). (The Alaska Resource Data File, Wiseman Quadrangle).

Twelvemile Creek

(Gold) **District:** Koyukuk; **Quadrangle:** Wiseman (A1); **Coordinates:** 67.1915°N, 150.4602°W; **Map:** 5-22; **Worked:** 1900-1986; **Purity:** 914 Au; **Yield:** 0.0214 oz/yd^3; **Production:** Through 1909=242 ozs; **Access:** Trail from Dalton Highway at 67.1866°N, 150.2917°W and across the Middle Fork of the Koyukuk River by boat; **Summary:** Tributary of the Middle Fork of the Koyukuk River. Placers apparently occur intermittently along the creek from a point beginning near its mouth, extend upstream along Twelvemile Creek, and then continues up the Upper Fork (probably called Middle Fork by Reed, 1938) for an undetermined distance. The coordinates are at the junction of Upper Fork and Twelvemile Creek. Bedrock is schist. Some rich ground has been mined, but work was very difficult because of water in gravel. There are both creek and bench placers. Production through 1909 was 242 fine ounces. High channels under benches 20 and 145 feet above creek have been mined. In 1937 a bench with bedrock 14 feet higher than creek bed was being hydraulicked. Attempts to mine deep channels were not successful. Mining on one of forks was reported but no evidence was visible in 1937. Includes references to forks of Twelvemile Creek and to Potato Creek (Cobb, 1976, OF 76-340. p. 163). **Geology:** Reed (1938) states that Twelvemile Creek contains placer deposits in the modern channel, in a deep channel, and in high bench channels. Reed's descriptions of mining and prospecting efforts suggest that the gold in the deposits is sporadically distributed. The depth to bedrock in the modern channel of Twelvemile Creek (in areas that are not coincident with the deep channel) is about 5 feet. The gold-bearing gravels on the lower parts of the modern channel are relatively fine with few boulders. Those on the upper parts contain more boulders. There are two high-bench channels, an upper one, called 'primary', and a lower one, called 'secondary'. The secondary bench is about 60 feet wide and 20 feet above the modern stream; it occurs only along the left limit of the Upper Fork (the Middle Fork of Reed, 1938). The primary bench is about 125 feet above the secondary. It is 600 feet wide, and lies along the left limit of lower Upper Fork and along Twelvemile Creek downstream as far as the Lower Fork (the North Fork of Reed, 1938). Below Lower Fork, high channels occur along the right limit of Twelvemile Creek at about the same level as the primary bench. An adit into the primary channel on the lower Upper Fork explored gravels containing gold values of $0.75 per cubic yard (gold at $35 per ounce). Although a deep channel is apparently present along Twelvemile Creek it isn't clear whether it contains any gold. Reed (1938) reported only one active mining operation in 1937, along lower Upper Fork (Middle Fork). The depth to bedrock at that location was 8 to 21 feet, and very fine, flat pieces of gold were distributed throughout the gravel. The values were about $0.35 per square foot of bedrock and the average fineness was 914. According to Reed (1938), several other tributaries of Twelvemile Creek, including West Fork (West Fork and South Fork join to form Twelvemile Creek), North Fork (probably the Lower Fork on modern topographic map), and Potato Creek (a small left limit tributary about a mile below Upper Fork) have been explored and/or mined with varying degrees of success. The bedrock in the Twelvemile Creek area is mainly graphitic and quartz schist (Reed, 1938) . Dillon and others (1986) mapped it as Mississippian to Triassic metagraywacke and phyllite, and upper Devonian, Hunt Fork schist. From about 1900 to 1910, the modern creek bed was mined by hand methods. Later mining included shafts, open cuts, and drifting (Reed, 1938) There was exploration drilling in 1934 and hydraulic mining in 1937 (Reed, 1938). Some exploration work was conducted on the south fork of Twelvemile Creek in 1986. (The Alaska Resource Data File, Wiseman Quadrangle).

Twocent Creek

(Gold) **Quadrangle:** Wiseman (C3); **Coordinates:** 67.53037° N 151.41102° W Map 5-2 **Worked:** 1900s-; **Summary:** Twocent Creek is not named on the Wiseman C-3 topographic map. Based on descriptions by Reed (1938), Twocent Creek, is probably the short, west tributary of Flat Creek whose mouth is about 0.2 mile south of the junction of Deadman and Flat Creeks. Another, less likely, possibility is that Twocent Creek is one of two short, west tributaries to Flat Creek about a mile farther downstream. The coordinates are the most likely locality in the NW1/4 section 8, T. 31 N., R. 17 W., of the Fairbanks Meridian. **Geology:** Reed (1938) reported that miners in the early 1900's could pan small amounts of gold from this creek -- two cents worth per pan at the old $20.67 per ounces price of gold. Dillon and others (1986) map the rocks in this area as Devonian Hunt Fork Shale. (The Alaska Resource Data File, Wiseman Quadrangle).

Union Gulch

(Gold) **Quadrangle:** Wiseman (B1); **Coordinates:** 67.4524°N, 150.0999°W; **Map:** 5-4; **Worked:** 1901-1998; **Nugget Size:** 32 ozs; **Production:** 1900-1909=1,850 ozs; **Access:** Light duty road from Wiseman, Alaska; **Summary:** Tributary of the Middle Fork of the Koyukuk River. Gold was discovered in 1901. A nugget weighing 32 ounces was found that same year. Drains mineralized area east of Nolan Creek. Coarse gold. Mined in early 1900's. Production through 1909 was 1,693 fine ounces. Mining in 1934 and 1937 was reported (Cobb, 1976, OF 76-340. p. 165). **Geology:** Placer gold deposits on Union Gulch are limited to the modern stream channel (Reed, 1938). Gold was discovered in 1901 when a 32 ounce nugget was recovered. Maddren (1913) reported $35,000 of gold (about 1850 oz.) produced from 1900 through 1909, with $30,000 of that produced in 1902. Mining was reported in 1934 and 1937, but Mulligan (1974) indicated that the property was inactive as of 1974. Bliss and others (1986) reported some mining activity from 1980 to 1982 and Szumigala and Swainbank (1999) reported production from Union Gulch in 1998. The gold is coarse and at a shallow depth (Heiner and Wolff, 1968). Three, panned-concentrate samples from Union Gulch contained abundant magnetite, some sulfides, and one very fine gold particle (Kurtak and others, 1999). All of these panned-concentrate samples were geochemically anomalous in gold (Kurtak and others, 1999). The rocks in the area consist of quartzite, quartz-mica schist and phyllite (Eden, 2000). (The Alaska Resource Data File, Wiseman Quadrangle).

Unnamed Creek

(Gold) **Quadrangle:** Wiseman (A1); **Coordinates:** 67.1932°N, 150.2812°W; **Map:** 5-22; **Yield:** 0.0025 oz/yd^3; **Access:** From Dalton High way at mile marker 168.7 at 67.1932°N, 150.2812°W; **Summary:** Tributary of The Middle Fork of the Koyukuk River. Designated as a **Recreation Mining Area** by the State of Alaska from its mouth upstream to its source. Average yield of gravel is estimated to be 0.0025 oz/yd^3 of gold.

Vermont Creek, Right Fork

(Gold) **Quadrangle:** Wiseman (C1); **Coordinates:** 67.5139°N, 150.1390°W; **Map:** 5-4; **Worked:** 1901-1998; **Yield:** 0.0063-0.242 oz/ft^2; **Nugget Size:** 10 ozs; **Production:** 1901-1909=8,321 ozs; **Access:** Light duty road from Wiseman, Alaska; **Summary:** Small tributary of Hammond River that heads against a fork of Nolan Creek in a mineralized area. Creek flows through a narrow canyon to a fairly broad valley cut in floor of Hammond River valley. Bedrock is phyllite, some mineralized with pyrite; quartz veinlets, one of which contains free gold and sulfides. Mining has been on claim at forks, on 2 claims on east fork, and on 2 claims below forks. Gold on west fork not in paying quantities. Lower part of course is in Hammond River bench deposits, but gold probably came from Vermont Creek. Shallow creek gravels and deep frozen gravels near Hammond River have been mined. Gold coarse and rounded; some nuggets weighing more than 10 ounces recovered. Gold was on and in upper 4 inches of schist bedrock. Mined area averaged about 0.0063 oz/ft^2 of bedrock except for a deep (30 - 90 feet) channel in valley of Hammond River which averaged 0.242 oz/ft^2 of bedrock. In lower part of creek frozen gravels 90 feet deep were drift mined; gold coarse. In upper part of creek, creek gravels were mined by open-cut methods. Creek mined from 1901 until as recently as 1969. Present channel considered to have been worked out before 1937. Production from 1901-09 was 8,321 fine ounces (Cobb, 1976, OF 76-340. p. 166). **Geology:** Placer gold deposits on Union Gulch are limited to the modern stream channel (Reed, 1938). Gold was discovered in 1901 when a 32 ounce nugget was recovered. Maddren (1913) reported $35,000 of gold (about 1850 oz.) produced from 1900 through 1909, with $30,000 of that produced in 1902. Mining was reported in 1934 and 1937, but Mulligan (1974) indicated that the property was inactive as of 1974. Bliss and others (1986) reported some mining activity from 1980 to 1982 and Szumigala and Swainbank (1999) reported production from Union Gulch in 1998. The gold is coarse and at a shallow depth (Heiner and Wolff, 1968). Three, panned-concentrate samples from Union Gulch contained abundant magnetite, some sulfides, and one very fine gold particle (Kurtak and others, 1999). All of these panned-concentrate samples were geochemically anomalous in gold (Kurtak and others, 1999). The rocks in the area consist of quartzite, quartz-mica schist and phyllite (Eden, 2000). (The Alaska Resource Data File, Wiseman Quadrangle).

Washington Creek

(Gold) **Quadrangle:** Wiseman (C1); **Coordinates:** 67.5225°N, 150.3210°W; **Map:** 5-1; **Worked:** 1902-1983; **Production:** Through 1905=241 ozs; **Access:** Trail from Wiseman, Alaska; **Summary:** Tributary of the Glacier River. The placers extend for about a mile upstream and a mile downstream. Coarse, smooth gold in gravels was discovered in 1902 and 241 fine ounces mined by 1905. All mining was small scale. None has been reported since 1904. High gravels on right limit were prospected, but results are not known (Cobb, 1976, OF 76-340. p. 169). **Geology:** Maddren (1913) reports that gold was discovered on Washington Creek in August, 1902, and that production between 1902 and 1909 was about $5,000 (approximately 260 ounces). Maddren reports that while some coarse, smooth gold was found in the gravels of Washington Creek, the amount of gold was insufficient to encourage serious mining. Reed (1938) reports some mining in the present stream bed from 1900 to 1910, but notes that the amount of gold recovered was not sufficient to warrant continued work. Heavy equipment for mining was being moved to Washington Creek from Mascot Creek in 1982 or 1983 (Bliss and others, 1988). Shafts were sunk by early miners to prospect for high channels along the lower creek but the results are not known. Brosge and Reiser (1972) collected a stream-sediment sample from Washington Creek that contains 700 to 3000 ppm antimony. Kurtak and others (1999) collected several panned concentrate samples from Washington Creek that contain traces of pyrite and magnetite, but no visible gold. The rocks in the area are black siltstone and phyllite (Dillon and others, 1986). (The Alaska Resource Data File, Wiseman Quadrangle).

Wild River

(Gold) **Quadrangle:** Wiseman (B4); **Coordinates:** 67.4024°N, 151.5062°W; **Map:** 5-23; **Worked:** 1904-1940; **Access:** Trail from Wild Lake; **Summary:** Tributary of the Koyukuk River. Mining or prospecting reported in many years between 1904 and 1940 probably was on tributaries rather than main stream. See also Birch Creek, Jay Creek, Lake Creek, Rye Creek, Spring Creek, and Surprise Creek (Cobb, 1976, OF 76-340. p. 171).

Wilson Creek

(Gold) **Quadrangle:** Wiseman (A1); **Coordinates:** 67.0646°N, 150.1747°W; **Map:** 5-11; **Worked:** 1900s-; **Access:** Boat from the Dalton Highway, South Fork of the Koyukuk River Bridge at 67.0186°N, 150.2885°W; **Summary:** Tributary of the South Fork of the Koyukuk River. Wilson Creek is a south-flowing tributary to South Fork, Koyukuk River. The mouth of Wilson Creek is about 3 or 4 miles west-northwest of Eagle Cliff and 6 miles south of Rosie Creek Pass. The location of placer prospects on the creek are unknown; the coordinates are for a location that is arbitrarily placed near the mouth of Wilson Creek. Prospects but no paying quantities of gold, reported (Cobb, 1976, OF 76-340. p. 172). **Geology:** Maddren (1913) reported that gold is can be found in Wilson Creek but not in paying quantities. Reed (1938) did not visit the creek. He repeated Maddren's (1913) information and indicated that no one was working on the creek in 1937. (The Alaska Resource Data File, Wiseman Quadrangle).

Wiseman Creek

(Gold) **Quadrangle:** Wiseman (B1); **Coordinates:** 67.4259°N, 150.1420°W; **Map:** 5-1, 5-4, 5-10; **Worked:** 1900s-; **Yield:** 0.0029-0.0048oz/yd^3; **Access:** Light duty road from Wiseman, Alaska; **Summary:** Tributary of the Middle Fork of the Koyukuk River. Small creek in a wide valley filled with 200 - 300 or more feet of gravel, sand, and silt. Prospect shafts and long cuts near valley wall found only small amounts of flour gold. Prospect shaft sunk in 1908-09 was 260 feet deep. It bottomed in mixed gravel and fragments of black, slaty country rock. A little gold was found, but not enough to encourage drift mining. Another shaft 335 feet deep and a pipe driven another 30 feet did not hit bedrock. From 1924 to 1929 a company built a 3-mile-long ditch and made 2 cuts trying to find a deep channel under a moraine next to the west side of the valley. Only a little flour gold was found and the project was abandoned in 1929. Reports of mining on Wiseman Creek, in 1904, are probably to unsuccessful prospecting or to mining on Nolan Creek, the major, and very productive tributary, of Wiseman Creek. See also Nolan Creek (Cobb, 1976, OF 76-340. p. 174). **Geology:** Gold prospecting has been concentrated in the area between the mouth of Nolan Creek and the Middle Fork Koyukuk River. Prospecting is severely hampered by thick glacial deposits in the valley. Only one shaft, located about a mile below Nolan Creek, is known to have reached bedrock. This shaft was 260 feet deep and encountered only colors of gold in 30 feet of washed gravel in the lower 40 feet of the shaft. A second shaft was sunk to a depth of 335 feet and a pipe driven another 30 feet without reaching bedrock. In the 1920's an attempt was made to find a deep channel under the glacial deposits by making two open cuts along the edge of the valley about 1.5 mile northwest of the town of Wiseman. These cuts found only fine, flour gold with values estimated to be $0.06 to $0.10 per cubic yard of gravel (gold at $20.67 per ounce) (Reed, 1938). (The Alaska

Resource Data File, Wiseman Quadrangle).

Useful Conversions

Conversions

	DWT	GM	GR	T OZ	K
1 Pennyweight (dwt)	1	1.555	24	0.05	7.8119
1 Grain (gr)	0.04167	0.0648	1	0.002083	0.325521
1 Gram (gm)	0.643	1	15.4324	0.032151	1.2515
1 Carat (K)	0.12801	0.79907	3.072	0.0064	1
1 Troy Ounce (T oz)	20	31.10348	480	1	156.25
1 Troy Pound (T lb)	240	373.24	5760	12	18.75
1 Milligram (mg)	0.000643	0.001	0.0154324	0.3258 E-4	0.0012515
1 Parts/Million (ppm)	0.58 /ton	0.902/ton	13.92/ton	0.029/ton	4.531/ton

The price of gold, international market, is based on a Troy ounce (T oz). One T oz weights approximately 10% more than 1 avoirdupois (English) ounce commonly refereed to as ounce (oz). There are 12 T ozs in a Troy pound (T lb) as compared to 14.583 T oz in an avoirdupois pound. In addition to T ozs, gold may be weighed in a number of other units, such as, grains (gr), grams (gm), carats, and pennyweights (dwt).

To convert:

Grains to grams	multiply by	0.0648
Grams to grains	multiply by	15.4324
Pennyweights to grams	multiply by	1.5552
Grams to pennyweights	multiply by	0.6430
Grams to troy ounces	multiply by	0.032151
Troy ounces to avoirdupois ounces	multiply by	1.0971
Avoirdupois ounces to troy ounces	multiply by	0.9115
Avoirdupois ounces to grams	multiply by	28.3500
Grams to avoirdupois ounces	multiply by	0.03527
Avoirdupois pounds to kilograms	multiply by	0.4536
Kilograms to avoirdupois pounds	multiply by	2.2046
Avoirdupois pounds to grains	multiply by	7000.00
Grains to avoirdupois pounds	multiply by	0.000143
Carats to troy ounces	multiply by	0.0064

With 1 cubic yard (yd^3) gravel	=	3000 av lbs	=	3645 T lbs
Then:				
1 ppm	=	0.875 dwt/ yd^3	=	0.04375 t ozs/ yd^3
1 T oz/yd^3	=	22.86 ppm		
1 dwt/ yd^3	=	1.14 ppm		
1 yd^3	=	27 cubic feet (ft^3)		
1 yd^3	=	400 stuck full 12" gold pans		
5 gals	=	0.067 ft^3	=	0.02 yd^3 (loose)

Table of Elements

Ag	Silver	Ga	Gallium	Pt	Platinum
Al	Aluminum	Ge	Germanium	Rb	Rubidium
Ar	Argon	Hg	Mercury	Rh	Rhodium
As	Arsenic	Ir	Iridium	Ru	Ruthenium
Au	Gold	K	Potassium	S	Sulfur
B	Boron	La	Lanthanum	Sb	Antimony
Ba	Barium	Mg	Magnesium	Sc	Scandium
Be	Beryllium	Mn	Manganese	Si	Silicon
Bi	Bismuth	Mo	Molybdenum	Sn	Tin
C	Carbon	Na	Sodium	Sr	Strontium
Cd	Cadmium	Nb	Niobium	Te	Tellurium
Ce	Cerium	Ni	Nickel	Th	Thorium
Co	Cobalt	O	Oxygen	Ti	Titanium
Cr	Chromium	Os	Osmium	Tl	Thallium
Cu	Copper	P	Phosphorous	W	Tungsten
F	Flourine	Pb	Lead	Y	Yttrium
Fe	Iron	Pd	Palladium	Zn	Zinc

The purity of gold is defined one of two ways. Jewelers define the purity of gold using a 24 point system called a Karat (K) where 24 K equals 100% pure gold, 18K equals 75% pure, 14K equals 58% pure and 10K equals 42% pure gold. Geologist, miners, and refiners work on a 1000 point system. A fineness of 1000 is 100% pure gold, 500 equals 50% pure, 100 equals 10% pure, etc.
With gold, at 1000 fine, priced at $400.00/T ounce:

1 dwt	=	$20.00
1 gr	=	$00.83
1 gm	=	$12.57

Properties of Gold

Symbol: Au
Atomic Number: 79
Atomic Weight: 196.967
Melting Point: 1063°C (1945°F)
Specific Gravity: 19.2

Suction Dredge Capacities

Whenever known, the average yield of gold, per cubic yard of gravel (oz/yd^3), will be listed in the summary section of each stream. The actual amount of gravel a suction dredge can process per hour is only about 25% of the figures given by manufacturer. This is because manufacturers base their figures on very optimistic conditions rather than compact stream bed material containing imbedded rocks of various sizes that have to be removed by hand. Hand processing of stream bed material drastically reduces the amount of material that is moved through the dredge. To calculate the average production with suction dredge use the figures below:

2"	0.50 yd^3/hr	5"	3.00 yd^3/hr
3"	1.00 yd^3/hr	6"	4.00 yd^3/hr
4"	2.00 yd^3/hr	8"	8.00 yd^3/hr (VW engine powered)

Every stream has areas along its course where the conditions are right for the gold to concentrate. Because a suction dredge moves such a small amount of gravel per hour it is important to locate these areas of concentrations and limit your work to that area.

Areas where gold is naturally concentrated are known as paystreaks. There have been many books written on the subject of paystreaks and how to "read" the stream. Stream gradient plays a large role in where gold settles out. Consequently, lower water velocities are the sites of gold deposition. As a general rule placer gold is concentrated at stream gradients less than 440 ft/mi.

References

Berg, H. C., and Cobb, E. H., 1967, Metalliferous lode deposits of Alaska: U.S. Geol. Survey Bull. 1246, 254 p.

Bottage, Robert; Availability of Land for Mineral Exploration and Development in Alaska, October 30, 1986, 7p.

Brosge, W. P., and Reiser, H. N., 1960, Progress map of the geology of the Wiseman quadrangle, Alaska: U.S. Geol. Survey open-file map.

———1962, Preliminary geologic map of the Christian quadrangle, Alaska: U.S. Geol. Survey open-file map.

———1964, Geologic map and section of the Chandalar quadrangle, Alaska: U.S. Geol. Survey Misc. Geol. Inv. Map I375, scale 1:250,000.

———1965, Preliminary geologic map of the Arctic quadrangle, Alaska: U.S. Geol. Survey open-file map. 1968, Geochemical reconnaissance maps of granitic rocks, Coleen and Table Mountain quadrangles, Alaska: U.S. Geol. Survey open-file rept., 4 sheets.

———1970a, compilers, Chemical analyses of rock and soil samples from the Chandalar and eastern Wiseman quadrangles, Alaska: U.S. Geol. Survey open-file rept., 8 p.

———1970b, compilers, Chemical analyses of stream sediment samples from the Chandalar and eastern Wiseman quadrangles, Alaska: U.S. Geol. Survey open-file rept., 5 p., 51 computer sheets.

Brosge, W. P., Brabb, E. E., and King, E. R., 1970, Geologic interpretation of reconnaissance aeromagnetic survey of northeastern Alaska: U.S. Geol. Survey Bull. 1271-F, p. F1-F14.

Brosge, W. P., Lanphere, M. A., Reiser, H. N., and Chapman, R. M., 1969, Probably Permian age of the Rampart Group, central Alaska: U.S. Geol. Survey Bull. 1294-B, p. B1-B18.

Burand, W. M., and Saunders, R. H., 1966, A geochemical investigation of Minook Creek, Rampart district, Alaska: Alaska Div. Mines and Minerals Geochem. Rept. 12, 18 p.

Cass, J. T., 1959c, Reconnaissance geologic map of the Ruby quadrangle, Alaska: U.S.. Geol. Survey Misc. Geol. Inv. Map I-289, scale 1:250,000.

———1959d, Reconnaissance geologic map of the Melozitna quadrangle, Alaska: U.S. Geol. Survey Misc. Geol. Inv. Map I-290, scale 1:250,000.

Chapman, R. M., Coats, R. R, and Payne, T, G., 1963, Placer tin deposits in central Alaska: U.S. Geol. Survey open-file rept., 53 p.

Cobb, E. H., 1969a, compiler, Metallic mineral resources maps of eleven Alaska quadrangles [Afognak, Beaver, Bering Glacier, Bettles, Karluk, Kenai, Middleton Island, Taylor Mountains, Trinity Islands, Tyonek, Yakutat]: U.S. Geol. Survey open-file rept., 20 p.

Cobb, E. H., 1976, United States Department of the Interior Geological Survey Summary of References to Mineral Occurrences (Other Than Mineral Fuels and Construction Materials) in the Chandalar and Wiseman Quadrangles, Alaska: U.S. Geol. Survey Open-File Report 76-340 p.

Cobb, Edward H., 1978, United States Department of the Interior Geological Survey, Summary of References to Mineral Occurrences (Other Than Mineral Fuels and Construction Materials) in the Beaver, Bettles, and Medfra Quadrangles, Alaska; Open-file Report 78-94, p. 55.

Cobb, Edward H., 1981, United States Department of the Interior Geological Survey Summaries of Data on and Lists of References to Metallic and Selected Nonmetallic Mineral Occurrences in the Tanana Quadrangle, Alaska Supplement To Open-File Report 77-432 Part A -- Summaries Of Data To June 1, 1981, U.S. Geological Survey Open-File Report 81-1313A, p.

Eakin, H. M., 1913a, A geologic reconnaissance of a part of the Rampart quadrangle, Alaska: U.S. Geol. Survey Bull. 535, 38 p.

———1916, The Yukon-Koyukuk region, Alaska: U.S. Geol. Survey Bull. 631, 88 p.

———1918, The Cosna-Nowitna region, Alaska: U.S. Geol. Survey Bull. 667, 54 p.

Herreid, Gordon, 1969, Geology and Geochemistry, Sithylemenkat Lake area, Bettles quadrangle, Alaska: Alaska Div. Mines and Geology Geol. Rept. 35, 22 p.

Joesting, H. R., 1943, Supplement to Pamphlet No. 1-Strategic mineral occurrences in interior Alaska: Alaska Dept. Mines Pamph. 2, 28 p.

Maddren, A. G., 1913, The Koyukuk-Chandalar region, Alaska: U.S. Geol. Survey Bull. 532, 119 p.

Mertie, J. B., Jr., 1934, Mineral deposits of the Rampart and Hot Springs districts, Alaska: U.S. Geol. Survey Bull. 844-D, p. 163-226.

———1937b, The Yukon-Tanana region, Alaska: U.S. Geol. Survey Bull. 8721 276 p.

Patton, W. W., Jr., 1966, Regional geology of the Kateel River quadrangle, Alaska: U.S. Geol. Survey Misc. Geol. Inv. Map I-437, scale 1:250,000.

Patton, W. W., Jr., and Csejtey, Bela, Jr., 1970, Analyses of stream-sediment samples from western St. Lawrence Island,

Alaska: U.S. Geol. Survey open-file rept., 50 p.
Prindle, L. M., 1908a, The Fairbanks and Rampart quadrangles, Yukon-Tanana region, Alaska, with a section on Placers of the Rampart region, by F. L. Hess, and a paper on Water supply of the Fairbanks region, by C. C. Covert: U.S. Geol. Survey Bull. 337, 102 p.
Prindle, L. M., and Hess, F. L., 1906, The Rampart gold placer region, Alaska: U.S. Geol. Survey Bull. 280, 54 p.
Reiser, H. N., Lanphere, M. A., and Brosge, W. P., 1965, Jurassic age of a mafic igneous complex, Christian quadrangle, Alaska, in Geological Survey research 1965: U.S. Geol. Survey Prof. Paper 525-C, p. C68-C71.
Smith, P. S., 1942a, Occurrences of molybdenum minerals in Alaska: U.S. Geol. Survey Bull. 926-C, p. 161-207.
Thomas, B. I., 1965, Reconnaissance sampling of the Avnet manganese prospect, Tanana quadrangle, central Alaska: U.S. Eur. Mines open-file rept., 8 p.
Thomas, B. I., and Wright, W. S., 1948a, Investigation of the Morelock Creek tin placer deposits, Fort Gibbon district, Alaska: U.S. Eur. Mines Rept. Inv. 4322, 8 p.
Waters, A. E., Jr., 1934, Placer concentrates of the Rampart and Hot Springs districts: U.S. Geol. Survey Bull. 844-D, p. 227-246.
Wayland, R. G., 1961, Tofty tin belt, Manley Hot Springs district, Alaska: U.S. Geol. Survey Bull. 105&I1 p. 363-414.
Williams, J. R., 1962, Geologic reconnaissance of the Yukon Flats district, Alaska: U.S. Geol. Survey Bull. 1111-H, p. 289-331.

Other Books In The Series

Hard Cover Books and Digital Books may be purchased directly from https://www.ironfiregold.com/

Alaska Gold Mining Books

	Paper Copy	Digital
Gold Placer Deposits Near Fairbanks Alaska	$30.00	$10.00
Gold Placer Deposits In EastCentral Alaska	$25.00	$10.00
Gold Placer Deposits Near Talkeetna Alaska	$25.00	$10.00
Gold Placer Deposits In NorthEast Alaska 2nd Edition	$25.00	$10.00
Gold Placer Deposits Near Anchorage Alaska	$25.00	$10.00
Gold Placer Deposits Near Nome Alaska	$30.00	$10.00
Gold Placer Deposits In SouthEast Alaska	$25.00	$10.00
Gold Placer Deposits In SouthWest Alaska	$25.00	$10.00
Gold Placer Deposits In WestCentral Alaska	$25.00	$10.00
Gold Placer Deposits In NorthWest Alaska	$25.00	$10.00
Gold Placer Deposits In NorthCentral Alaska	$20.00	$10.00
The Gold Miners Guide to Easy Pickens In Alaska	$30.00	$10.00

Gold Mining Books of Other States

	Paper Copy	Digital
Placer Gold Deposits In New Mexico	$25.00	$10.00
Placer Gold Deposits In Utah	$20.00	$10.00
Placer Gold Deposits In Nevada	$25.00	$10.00
Placer Gold Deposits In Arizona	$35.00	$10.00
Gold Placer Deposits In Colorado	$35.00	$10.00
Placer Gold Deposits In Wyoming	$25.00	$10.00

For more information on the contents of each book visit Iron Fire's Web site at:

https://www.ironfiregold.com/